CREDIBLE THREATS IN NEGOTIATIONS

THEORY AND DECISION LIBRARY

General Editors: W. Leinfellner (*Vienna*) and G. Eberlein (*Munich*)

Series A: Philosophy and Methodology of the Social Sciences

Series B: Mathematical and Statistical Methods

Series C: Game Theory, Mathematical Programming and Operations Research

Series D: System Theory, Knowledge Engineering an Problem Solving

---

SERIES C: GAME THEORY, MATHEMATICAL PROGRAMMING AND OPERATIONS RESEARCH

VOLUME 32

Scope: Particular attention is paid in this series to game theory and operations research, their formal aspects and their applications to economic, political and social sciences as well as to socio-biology. It will encourage high standards in the application of game-theoretical methods to individual and social decision making.

The titles published in this series are listed at the end of this volume.

CREDIBLE THREATS IN NEGOTIATIONS

A Game-theoretic Approach

by

HAROLD HOUBA

Centre for World Food Studies,
Vrije Universiteit, Amsterdam,
The Netherlands

and

WILKO BOLT

Research Department,
De Nederlandsche Bank, Amsterdam,
The Netherlands

SPRINGER-SCIENCE+BUSINESS MEDIA, B.V.

A C.I.P. Catalogue record for this book is available from the Library of Congress.

ISBN 978-1-4419-5304-9 ISBN 978-0-306-47539-9 (eBook)
DOI 10.1007/978-0-306-47539-9

Printed on acid-free paper

Contents

List of Figures xi
List of Tables xiii
Preface xv
Acknowledgments xix
On the authors xxi

1. THE ESSENCE OF NEGOTIATION 1
   1.1. Introduction 1
   1.2. Real life negotiations: motivating examples 3
   1.3. Outline of the book 7
2. A BARGAINING MODEL WITH THREATS 15
   2.1. Introduction 15
   2.2. The bargaining problem 16
      2.2.1 The contract space 16
      2.2.2 Disagreement actions 18
      2.2.3 Utility functions 19
      2.2.4 Mutual and conflicting interests 20
      2.2.5 Pareto efficiency 21
      2.2.6 Individual rationality 23
      2.2.7 Generic uniqueness 23
      2.2.8 Utility representation 26
   2.3. A bargaining game with threats 28
      2.3.1 The order of moves 29
      2.3.2 The players' information 31
      2.3.3 Information sets and strategies 32
      2.3.4 Outcomes and utilities 35
      2.3.5 Reinterpretation as expected utilities 36

2.3.6 An appropriate equilibrium concept 38
2.3.7 Limit sets of equilibria 40
2.3.8 Markov strategies 41
2.3.9 Strategies represented by tables 42
2.4. Related Literature 44

Part I Exogenous disagreement outcomes

3. THE ALTERNATING OFFERS PROCEDURE 49
3.1. Introduction 49
3.2. Alternating offers 50
3.3. Markov perfect equilibrium 51
3.3.1 An important fixed point problem 51
3.3.2 Dynamic programming 56
3.3.3 Optimal response 57
3.3.4 Optimal proposals 60
3.3.5 Characterization in utility representation 62
3.3.6 Characterization in the contract space 64
3.4. Subgame perfect equilibrium 67
3.4.1 The method of Shaked and Sutton 67
3.4.2 Characterization of the SPE 69
3.4.3 First-mover advantage 70
3.4.4 Computation of the SPE contract 71
3.5. Applications 72
3.5.1 Dividing a dollar 73
3.5.2 A barter economy 75
3.6. Related literature 77

4. THE NASH PROGRAM 81
4.1. Introduction 81
4.2. Nash's bargaining solution 82
4.2.1 Utility representation 82
4.2.2 Two geometrical properties 85
4.2.3 Bargaining in the contract space 88
4.2.4 Computation of axiomatic contracts 93
4.2.5 Two critical remarks 94
4.2.6 A reinterpretation 97
4.2.7 Alternative axioms for Nash's bargaining solution 99

4.2.8 Alternative axiomatic solutions 101
4.3. Strategic bargaining and Nash's bargaining solution 102
4.3.1 Nash's demand game 103
4.3.2 Interpretation of demands 105
4.3.3 Convergence in alternating offers 107
4.3.4 Convergence in the contract space 110
4.4. The two approaches are complementary 112
4.5. Related Literature 114

5. COMPREHENSIVE BARGAINING PROBLEMS 117
5.1. Introduction 117
5.2. Comprehensive bargaining problems 118
5.3. Markov perfect equilibrium 120
5.3.1 The fixed point problem 120
5.3.2 MPE in utility representation 122
5.4. Subgame perfect equilibrium 124
5.4.1 Bounds for SPE utilities 124
5.4.2 Equilibrium switching 126
5.4.3 SPE with equilibrium switching 127
5.4.4 SPE with delay 130
5.5. Nash program 132
5.5.1 Generalized Nash's solutions 132
5.5.2 Limit set of SPE utility pairs 134
5.5.3 Convergence or nonconvergence, that's the question 137
5.6. Contract space 138
5.6.1 SPE contracts 138
5.6.2 Sufficient conditions for uniqueness 139
5.7. Related Literature 141

6. COMPARATIVE STATICS 145
6.1. Introduction 145
6.2. Utility functions and the contract space 146
6.2.1 Mutual and conflicting interests 146
6.2.2 Imperfectly divisible goods 149
6.3. Nonstationary bargaining problems 152
6.3.1 Alternating probabilities of breakdown 152
6.3.2 Alternating disagreement points 155
6.4. Alternative bargaining procedure 160

6.4.1 Markov process 160
6.4.2 Strategic timing of proposals 163
6.5. Related Literature 167

Part II Endogenous Threats

7. COMMITMENT AND ENDOGENOUS THREATS 175
7.1. Introduction 175
7.2. Optimal threats with commitment 177
7.2.1 Nash's original variable-threat game 177
7.2.2 The variable-threat game with alternating offers 180
7.3. Credible threats without commitment 184
7.3.1 Nash's variable-threat game: no commitment 185
7.3.2 Variable threats with alternating offers: no commitment 188
7.3.3 Bounds for SPE utilities 190
7.3.4 The set of SPE utility pairs 192
7.3.5 SPE with delay 196
7.3.6 A comparison between models 197
7.4. Numerical examples 198
7.5. Related literature 200

8. BARGAINING OVER WAGES 203
8.1. Introduction 203
8.2. A model of wage negotiations 204
8.2.1 Wage bargaining: some facts and assumptions 205
8.2.2 The wage bargaining model 207
8.3. Wage bargaining with efficient holdouts 209
8.3.1 Markov perfect equilibrium 210
8.3.2 The minimum-wage and maximum-wage contract 210
8.3.3 Equilibria with lengthy strikes 213
8.3.4 Intermezzo: unequal discount factors 214
8.4. Dutch wage bargaining: an application 217
8.4.1 Work-to-rule as a substitute for strike 218
8.4.2 Equilibria with lengthy work-to-rule 220
8.4.3 Backdating 223
8.5. Related literature 225

9. THE POLICY BARGAINING MODEL 229
9.1. Introduction 229
9.2. Subgame perfect equilibria 230
9.2.1 Markov perfect equilibrium 230
9.2.2 Worst SPE strategies: Example 231
9.2.3 Worst SPE strategies: General case 233
9.2.4 Optimal disagreement actions 237
9.2.5 Conditions for uniqueness 241
9.2.6 Characterization of SPE utilities 242
9.3. Policy Bargaining 244
9.3.1 The policy bargaining model 245
9.3.2 Characterization of SPE utilities 246
9.3.3 Renegotiation of Agreements 247
9.3.4 Nonbinding agreements 248
9.4. Numerical Examples 251
9.5. Related literature 253

10. DESTRUCTIVE THREATS 257
10.1. Introduction 257
10.2. Difference games 258
10.2.1 General framework 259
10.2.2 The great fish war 261
10.2.3 Linear-quadratic difference games 263
10.2.4 Pareto efficient joint policies 266
10.3. Negotiations for quota 268
10.3.1 Optimal disagreement catches 268
10.3.2 Optimal proposals 270
10.3.3 Numerical Solutions 271
10.3.4 The set of SPE utility pairs 273
10.4. Multiple state variables 275
10.4.1 Motivating Example 276
10.4.2 Optimal disagreement actions 278
10.4.3 Optimal proposals 279
10.4.4 Breakdown quadratic value functions 280
10.4.5 To negotiate or not? 281
10.5. Concluding remarks 283
10.6. Related literature 286

Appendices 291

Proofs of Selected Theorems 291

References 311

Index 317

List of Figures

3.1	The Nash product curve and the fixed point (x^*, y^*)	53
4.1	Nash's bargaining solution $N(S, d)$	85
4.2	An illustration of the second geometrical property of Nash's bargaining solution	87
4.3	The set $D(s^*)$ for $s^* = (\alpha, \alpha)$ in example 4.11.	89
4.4	Covergence to Nash's bargaining solution $N(S, d)$	109
5.1	A strongly comprehensive bargaining set S^{IR}	119
6.1	The set S^+ of the bargaining problem (S, d)	147
6.2	An effective disagreement point outside the bargaining set	157

List of Tables

3.1 The MPE strategies of proposition 3.12 63
3.2 The generically unique MPE strategies of proposition 3.15 66
5.1 Equilibrium switching and punishment strategies 129
5.2 SPE strategies supporting delay 131
7.1 The MPE strategies of proposition 7.13 189
7.2 Player 1's worst pair of SPE strategies of proposition 7.15 191
7.3 A pair of SPE strategies that induce immediate agreement 193
7.4 SPE strategies that support delay and induce agreement on s^* at $t = t(\sigma)$. 195
8.1 The MPE strategies of proposition 8.2 209
8.2 A pair of SPE strategies that induce the maximum-wage contract 211
8.3 SPE strategies that support lengthy strikes before agreement is reached 215
8.4 A pair of modified alternating-strike strategies that restore equilibrium 217
9.1 The MPE strategies of proposition 9.1 231
9.2 The strategies that represent player 1's worst SPE 235
9.3 Perpetual disagreement SPE strategies 247
9.4 SPE strategies supporting efficient joint policies and inefficient deviations 249
10.1 The MPE strategies of theorem 10.7 271
10.2 MPE weights of theorem 10.7 for $\alpha = 0.50$ and differnet values for δ_1 and δ_2 272
10.3 Worst SPE strategies of proposition 10.8 273

Preface

The game-theoretic modelling of negotiations has been an active research area for the past five decades, that started with the seminal work by Nobel laureate John Nash in the early 1950s. This book provides a survey of some of the major developments in the field of strategic bargaining models with an emphasize on the role of threats in the negotiation process. Threats are all actions outside the negotiation room that negotiators have ate their disposal and the use of these actions affect the bargaining position of all negotiators. Of course, each negotiator aims to strengthen his own position. Examples of threats are the announcement of a strike by a union in centralized wage bargaining, or a nation's announcement of a trade war directed against other nations in negotiations for trade liberalization.

This book is organized on the basis of a simple guiding principle: The situation in which none of the parties involved in the negotiations has threats at its disposal is the natural benchmark for negotiations where the parties can make threats. Also on the technical level, negotiations with variable threats build on and extend the techniques applied in analyzing bargaining situations without threats. The first part of this book, containing chapter 3-6, presents the no-threat case, and the second part, containing chapter 7-10, extends the analysis for negotiation situations where threats are present. A consistent and unifying framework is provided first in 2.

The unifying framework in this book extends the alternating offers procedure, which became well-known through the work of Ariel Rubinstein in the early 1980s, to allow for threats. The alternating offers procedure assumes a rigid chess-game-like structure in which negotiators alternate in making proposals until either an agreement is reached or the negotiations end in perpetual disagreement. For the no-threat case, this procedure yields a unique prediction that applies to economic common sense. For instance, impatient negotiators do worse than patient ones and risk aversion has a similar effect upon the bargaining outcome. Although the alternating offers procedure is quite simple, its outcome is remarkable robust with respect to changes in the procedure. For example, the alternating offers procedure can be regarded as the unspecified dynamic negotiation process John Nash envisioned to underlie his 'static' demand game and this explains why both models yield the same outcome. As a second example, if the rigid sequence of 'who proposes when' is replaced by some general stochastic process, then the same outcome is obtained if the stochastic process approaches the alternating offers procedure. Third, the outcome of the alternating offers procedure can also be obtained as the outcome of some complicated continuous-time procedure with endogenous timing of proposals. The latter model can also mimic the rigid structure of the alternating offers process itself. With respect to changes in the negotiators' preferences the outcome of the alternating offers procedure is less robust.

The so-called Nash program is also discussed and it can be formulated as investigating the possible connection between two seemingly unrelated approaches to analyze bargaining problems. One is the strategic approach in which bargaining procedures are explicitly modelled as discussed above. The other is the axiomatic approach that simply treats the negotiation process as a black box and imposes axioms upon the bargaining outcome that should reflect 'reasonable' properties that any bargaining outcome is supposed to have. The rationale of the Nash program is that both approaches should strengthen each other, meaning that a list of reasonable axioms should correspond to the outcome of at least one strategic bargaining procedure and vice versa. Remarkably, the outcome of the alternating offers procedure corresponds to the list of four axioms proposed by John Nash in 1950.

The idea that threats can be an important aspect of negotiations was also suggested by John Nash in the early 1950s and he was the first to analyze a variable-threat game. The choice of threats depends upon whether or not the negotiators can commit themselves upon these actions prior to the negotiations, where commitment means that the parties cannot back out of the chosen actions if called upon to carry out their threats. In situations where such commitment is possible, a unique outcome for the negotiations is predicted. However, in situations where the parties fail commitment the threats have to be credible, meaning that these actions have to be equilibrium actions. Possible multiplicity of equilibria translates into a multiplicity of predicted outcomes in the negotiations. In this book it is shown that the standard results obtained by John Nash need some minor modifications if his demand game, specifying the unmodelled negotiation process, is replaced by the alternating offers procedure. The results for variable threats mentioned thus far are all based upon models in which the choice of the threats and the behaviour in the negotiations are separated from each other in time and this is certainly not the case in many situations of interest to economists.

Negotiations where threats are chosen and carried out simultaneously during the negotiations are complicated, because past behaviour at the negotiation table may influence the choice of threats and vice versa. Such situations also lack the absence of commitment and this means that threats have to be credible. This class of negotiation situations are studied in an order of increasing complexity. The simplest situation features one party that has a 'monopoly' on using threats and these situations are classified as (centralized) wage bargaining situations with the union deciding whether or not to go on strike. The threat of a strike can only be credible if the union's benefits in terms of a better agreement outweigh the cost of organizing the strike. If this condition is met, a multiplicity of outcomes is predicted, including those involving a delay in reaching an agreement and featuring strike during the negotiations. An application to Dutch wage negotiations in this book yields some new perspectives. The application focuses upon the role of alternative types of industrial action the union possibly could take, backdating of wage contracts and the rather lengthy periods of negotiations without strike.

The symmetric situation where each party has threats at its disposal, for example in trade liberalization negotiations among nations where each nation could start a trade war, adds more complexity to the analysis, especially in establishing the minimum result each party can secure itself. In this book it is made clear how these complexities can be overcome and it involves application of several of the techniques applied in the no-threat case. Special attention is given to so-called policy bargaining problems, where the subject of the negotiations concerns agreement upon the future use of threats. For example, during the negotiations for a reduction of import tariffs setting the current tariffs correspond to choosing threats. Within this context important issues are addressed such as future renegotiation of agreements and how to deal with situations where agreements are not binding.

All the situations discussed thus far did not allow for the possible effect that threats may have on the future gains of trade and such negotiation situations are of interest, especially in environmental economics and ecology. These situations can be characterized by intertemporal links between threats and several cumulative variables that pass on to the future. For example, in the negotiations for fish quota among nations caught fish cannot reproduce itself and a nation's threat to expand its fleet, i.e., to fish more, is likely to destroy some part of the future gains from fishing. Another example concerns the negotiations for the reduction or abatement of greenhouse gas emissions, where the latter add to the accumulation of pollutants in the earth's atmosphere. In this book the negotiations for fish quota are elaborated for very stylized and simple biological dynamics for one single specie and, contrary to above, more patient nations do worse than less patient ones. The last inquiry in this book deals with the type of results that can be obtained if the threats are linked to several accumulated variables, such as several species of fish or the concentrations of several greenhouse gasses. Although some results are presented for these type of problems, still, a full analysis of the latter problems is lacking and the book concludes with directions for future research.

HAROLD HOUBA AND WILKO BOLT

Acknowledgments

Writing a book is not something you do in isolation. Over the past years we have experienced the support and encouragement of many colleagues, family, friends and the people at KAP.

With respect to our academic colleagues, there are two who are indispensable: Gerard van der Laan and Aart de Zeeuw. We are indebted to you both for the opportunity given to each of us to write a Ph.D. thesis, the many pleasant discussions and well-judged advices, and support through the years. The following persons also made invaluable contributions to our research through the years: Hans Haller, Steinar Holden, Eric Maskin, Hans Peters, Sjef Plasmans, Christian Schultz, Stef Tijs, Eric van Damme and, in memoriam, Elaine Bennett. In writing the final manuscript we received many valuable comments from many friends and colleagues on several chapters. They include J.R. (René) van den Brink, Theo Beekman, Gerard van der Laan, Abhinay Muthoo, Maria Steijns and Quan Wen.

Special thanks go to our former colleagues of the Department of Econometrics and Operation Research at the Vrije Universiteit for providing a pleasant academic atmosphere to finish our manuscript. We explicitly mention Rob Luginbuhl, Xander Tieman and Koos Sneek for the many joyful moments that gave us the energy to continue. We are also grateful to the staff at KAP and to Allard Winterink and Cathelijne van Herwaarden for the pleasant contacts, support and patience during the entire project, which lasted much longer than initially scheduled. For sorting out some tedious LaTeXdifficulties in the finishing stages of the book we thank Carsten Folkertsma. Finally, we mention the invaluable support of our families, especially on those many occasions that they understood that completing the manuscript got priority. We sincerely hope that the publication of our book is 'The times they are a changing'.

HAROLD HOUBA AND WILKO BOLT

On the authors

Harold Houba is economist at the Centre of World Food Studies, Vrije Universiteit in Amsterdam. He was educated at Tilburg University, both Masters and Ph.D., and has been Assistant Professor of Economic Theory for about ten years at the department of Econometrics and Operation Research at the Vrije Universiteit. The list of topics published include Bargaining Theory and Game Theory, among other topics, in journals as *European Economic Review*, *Economic Theory*, *Journal of Mathematical Economics* and *Journal of Economic Dynamics and Control*.

Wilko Bolt is economist at the Research Department of De Nederlandsche Bank in Amsterdam. He was educated at the Vrije Universiteit in Amsterdam, both Masters and Ph.D., and joined De Nederlandsche Bank afterwards. The list of topics published include Bargaining Theory and Monetary Economics, among other topics, in journals as *American Economic Review*, *European Economic Review* and *Economic Theory*.

Chapter 1

THE ESSENCE OF NEGOTIATION

1.1. Introduction

Negotiations are an important phenomenon in most modern societies and these may range from collective wage bargaining and political debate to international trade agreements, arms limitation talks and peace keeping negotiations. Also, in everyday life the significance of negotiation cannot be overlooked. Common to any bargaining relationship are both conflicting and mutual interests between individuals or organizations. These actors or players in the bargaining process have to seek a way to resolve their possible differences in order to reach an agreement. The process of negotiation creates the opportunity, although there is no guarantee of reaching an agreement, that each actor will come nearer towards the realization of its own separate objectives, while enjoying the fruits of 'the gains from trade'. Each actor will recognize that he or she does not act in a vacuum, but instead is surrounded by active decision-makers whose choices and actions during the negotiations interact with his or hers. This interdependence of choices and actions is an essential feature of all bargaining situations.

The process of negotiation is directly concerned with the interaction among human beings and deals with patterns of social behaviour. Human behaviour is often hard to comprehend and difficult to formalize. Still, the analysis presented in this book intends to be formal. Indeed, it is beyond any doubt that, besides economic factors, psychological and sociological elements are of great importance for any behavioural analysis of bargaining. However, these elements are hard to quantify and frustrate adequate formalization. Therefore, we will throughout assume that players in the process act in their best interest in all circumstances. Pursuing this point of view allows us to abstract from such psychological and sociological elements in economic situations. In fact, all theories investigated will be a rather abstract inquiry into the consequences of strategic interaction and into the elements which bear strategic importance on the process of negotiation in economic situations.

Strategic interaction among players is a basic ingredient of any bargaining process. Similar as in the parlour games of Chess and Bridge, each actor or player involved in the negotiations has to think ahead, he has to anticipate his rivals' responses upon the current move he is considering, he has to update his convictions and beliefs based on new information learned from the past, he has to contemplate whether or not to make a threat or warning, whether or not to make a promise or assurance, etcetera. Essentially, each player has to think *strategically* and act accordingly in order to enhance his own bargaining power, but at the same time each player recognizes that, in principle, all of his opponents will try to do the same thing. To tackle this problem of strategic interdependence, our main tool for analysis will be *game theory*.

Game theory is designed to deal with strategic thinking and decision-making. It sets out from assuming that the decision-makers are rational in the sense that they pursue well-defined exogenous objectives and that they reason strategically in the sense that they take into account their knowledge or expectations of other decision-makers' behaviour. Situations of conflict in which several individuals - the players - strategically interact and jointly determine the outcome, are transformed into games by including only those factors which are relevant for the specific situation as perceived by the players and leaving out all further 'irrelevant details', e.g., see Meyerson (1991) and Rubinstein (1991). Once the rules of the game are specified, it can be solved by using mathematical tools and game-theoretic concepts. A solution of the game predicts the theoretical outcome of strategic interaction among rational players: It provides a set of instructions, i.e., a strategy for each player, that prescribes each player what to do in every possible situation that may arise. These strategies enable us to understand and study the role and influence of several elements present in bargaining processes on the efficacy of the bargaining process and how it affects each player's bargaining power. Essentially, a player's bargaining power, which may loosely described as the ability to maneuver himself in a favourable bargaining position during the negotiations, determines his final share of the surplus induced by agreement. The insights gained by a game-theoretic analysis may offer explanations for observed behaviour in real life bargaining situations.

The game-theoretic models of bargaining studied can be used to address various questions in economic theory. One of these questions for example the role of risk aversion by players in negotiations? Or, why are lengthy delays that are costly to the parties involved so often observed during the negotiations before a settlement is reached, while reaching an early settlement is most likely better for all the negotiating parties? Another question involves what economic factors strengthen a negotiator's position at the bargaining table. Or, in union-firm wage bargaining situations, what is the maximum wage that a union can extract in the wage negotiation process and what negotiation strategies lead to this wage contract? These questions all deserve an answer, at least from a theoretic point of view.

The main purpose of this book is to synthesize the vast literature on game-theoretic modelling of negotiations and to provide a formal game-theoretic analysis of the strategic role of several elements on the bargaining outcome. We do not intend to be complete, which would be a heroic task, but rather we focus on parts of the literature that are not (fully) covered in the existing books on bargaining theory, e.g., Gale (2000), Muthoo (1999) and Osborne and Rubinstein (1990). In our view this concerns the strategic role of so-called *threats*. Threats can be seen as all kind of actions that can be employed outside the negotiation room during the negotiations that might possibly either strengthen the own bargaining position or deteriorate the opponents' position. Examples of a threats in union-firm wage bargaining are all forms of industrial action, such as strike, organized by the union or the threat of trade wars in trade-liberalization negotiations.

To grasp some intuition on some of the mechanisms in bargaining, three examples of real life negotiation situations will be discussed in the next section. The third section provides an outline of this book.

1.2. Real life negotiations: motivating examples

As already mentioned, negotiations are encountered in a wide variety of situations. In this section, three situations will be discussed at some length, in which the process of negotiation plays a crucial role. These examples reflect realistic complex bargaining situations where, apart from economic motives, institutional settings and procedural requirements, psychological, sociological and emotional elements also enter the negotiations. Therefore, it must be stressed that the formal models to be discussed in the further chapters do not pretend to fully capture these situations, or even partially explain the observed consequences of human behaviour. Still, each of these situations embodies the common feature of strategic interaction among people or institutions and shows particular elements which are of strategic importance to the bargaining process. As mentioned, the formal strategic analysis in this book may provide an increased understanding of which features of the process are relevant and how these may affect the bargaining process.

Among one of the most natural examples in economics, one can think of wage contract negotiations, where trade unions and employers' associations are engaged in collective bargaining. Following Windmuller (1987), the central task of collective bargaining is to reach a formal collective agreement on a set of rules which govern the terms and the conditions of the employment relation, as well as to define the relation of the parties to the process. Although collective bargaining exists in many different societies and occurs in many different forms, negotiation is always the essence of collective bargaining. Common to virtually all negotiations, collective bargaining over wages and employment takes time and time means money. Hence, trade unions and employers' associations share a mutual interest in concluding the terms and conditions of a new labour contract as soon as possible in order to avoid lost wages and production. They have, however, opposing interests regarding the wage. The union's power in the wage negotiations depends to a great extent on the ability to mobilize its members for action. The union may threaten to strike during the negotiations trying to put the pressure on. But a union which always threatens to strike but never actually does, never showing its teeth, may lose its ability to organize a massive strike, so its threat to the employers becomes less effective and loses credibility. As John Hicks already stated:

> 'Weapons grow rusty if unused, ...'
> *Hicks (1963)*

Still, it is difficult to see why rational parties should turn to wasteful mechanisms, such as strikes, as a means to distribute the gains from trade. If strikes end with a new contract, as they usually do, why could not both parties agree to this outcome in advance, avoiding the costs of a strike and sharing the benefits from increased production? Just to give an example, in the Netherlands in March 1995, there was a massive strike concerning construction workers. This strike lasted about four weeks and the employers' losses were at that time estimated at EUR 160 million (around 350 million Dutch guilders). It seems hard to imagine that these costs of delaying the negotiations were outweighed by a more favourable settlement wage for the unions, which illustrates the overall inefficiency of the final outcome. The question arises whether we can explain these apparently faulty negotiations.

A very illustrative case of business negotiations, which will serve as our second example, concerns the takeover deal between the Dutch aircraft manufacturer Fokker, the Dutch government which owned one-third of Fokker and DASA, the aerospace division of the German industrial heavyweight Daimler-Benz. In 1991, Fokker announced that it would open negotiations with DASA to discuss far-reaching cooperation, which included the possibility of a takeover by DASA. The airline industry had been struck by hard times: a prolonged recession in the airline industry caused sales to go down in a fiercely competitive market. Moreover, Fokker faced a high cost base and the strength of the guilder against a sliding dollar.[1] To guarantee continuity in the long run, Fokker, one of the world's oldest plane builders, had to join forces and DASA just appeared to be the perfect partner with its huge cash-flows to provide Fokker with the necessary financial injections. At first, when the negotiations started Fokker seemed to be in a strong bargaining position. Fokker with its vast amount of technological know-how, marketing skills and one of the leaders in the market for middle-sized jet planes could choose from various possible candidates to form a partnership with. DASA, on the other hand, had only money and a bunch of ambitious plans, but was determined to become a key role player in the European airline industry. By taking over Fokker, DASA could neutralize its otherwise biggest competitor in the market and it did not need to develop a jet plane in the same market segment on its own.[2] Hence, DASA had very strong incentives to take over Fokker, a fact which should have been exploited by Fokker in the negotiations. But, surprisingly, in the end, when a mutual agreement was signed in April, 1993, DASA took over Fokker on very favourable terms and conditions. All the Dutch experts agreed that it was a sell-out: Fokker had gone to DASA for 'peanuts'. Is it true that the Germans had played the game by the book while the Dutch had shown none or little negotiation skills. What went wrong for the Dutch?

In order to address this question, one has to realize that the world airline industry did not recover from its collapse during the negotiations and Fokker's financial end economic position further deteriorated due to its sharply increased stocks of new unsold planes. Layoffs by Fokker did not bring much relief. The higher capital costs made Fokker less attractive for takeover. So, DASA was in a position where it could strategically delay the negotiations, delay which would harm Fokker the most. A factor that contributed to the deterioration of the position of the Dutch government was the aftermath of the RSV-parliamentary investigation in the late eighties. RSV was a large industrial conglomerate, that could not survive after several decades of huge financial injections from the Dutch government. The media spoke of mismanagement by the government, the Minister of Economic Affairs had to resign and parliament was heavily criticized for its easy approval of state support to RSV time after time. As a consequence, government support to business enterprises became simply taboo and during the Fokker-DASA negotiations leading Dutch politicians more than once openly declared that they would not tolerate another RSV nor the bankruptcy of Fokker. So, the Dutch government was forced at all costs by its own parliament to avoid its fall back position of providing financial support to Fokker or, even worse, to let Fokker go bankrupt. This made any tough negotiation strategy impossible, which favoured DASA. The themes of harming the opponent and strong aversion for the fall back position are recurrent themes throughout this book.

Another area where negotiations have played and will continue to play a key role is the area of international trade policy. One of the most important trends in the world economy today is globalization. Not only have technical discoveries and innovation led to greatly reduced costs of transport and communication, but those reduced costs have in turn led to an increasingly global market place. However, the potential gains will to a large extent depend on an open multilateral trading system. In order to achieve this open multilateral free trading system, the General Agreement on Tariffs and Trade (GATT) was agreed upon in the 1940's by the participating countries and the World Trade Organization (WTO) was established. Although the GATT negotiation agenda has expanded over time, the focus has always been on market access through a reduction of trade barriers and the abolition of discrimination among products. This resulted in gradually reduced tariffs among all industrial countries. However, things changed considerably in the beginning of the 1980's when the world economy was hit by a severe recession after the second oil price increase. In response, many countries, including the US, took protectionist measures. In order to stand up to these protectionist threats delegates of all parts of the world met in Montevideo in 1986, and agreed upon an ambitious agenda for a new round of trade negotiations — the Uruguay round. This new round of negotiations covered more issues than previous rounds: it also included a protocol on telecommunications, all services and, indeed, even agriculture. Despite the diversity and complexity of the talks, it is still hard to understand why the negotiations took so long: the Uruguay round of negotiations took more than seven years and were not concluded until late in 1993.

A major difficulty in reaching international agreements is enforceability and credibility of dispute settlement mechanisms. All contracting countries must be committed to the obligations of the negotiated outcome and these commitments must 'bind' under international law. However, countries can always opt out of international treaties, so the trade agreement must in a certain sense be self-enforcing. The GATT provides dispute settlement procedures, but they are often criticized as being relatively ineffectual. If a country cheats on the trade agreement, GATT initiates an administrative process that drags on for months or years. Such enforcement procedures are likely to be ineffective. Although the Uruguay agreement has strengthened and streamlined dispute settlement procedures, the fact remains that trade sanctions are the ultimate 'stick' to punish a cheating country. The question arises whether this punishment is credible to carry out. Such retaliation may hurt the imposing countries more than the offending, and this seems true particularly for small countries. Hence, further improvement in the procedures to enforce GATT obligations is an important task facing the negotiators. The question whether punishments in order to prevent players to deviate from an intended outcome path of play are credible, is an important topic in strategic bargaining theory. It will be of concern in all the next chapters, when we want to characterize the equilibrium outcomes of various bargaining situations.

Another issue which might have delayed the trade negotiations appeared to be located in the confrontation between the US and the European Union (EU) on their respective agricultural policies. Much of the difficulty arose from the fact that, while the US came to the bargaining table as a unified party, the EU first had to agree on a negotiating platform among its members. This caused delay and by the time the European negotiators entered the bargaining room not much room for maneuver was left. Moreover, the internal EU process of choosing an appropriate platform was not always a cooperative one. It can be argued that the apparent difficulty in reaching an agreement on a common proposal on agricultural policies is due to the internal EU negotiation procedures and the responsiveness of the French government to its national farm lobby during the GATT negotiations, which seemed to be a blocking factor in the talks for a while. After some time the US reacted by threatening the EU to start a trade war directed against its member states and actually did start it by sharply increasing its import tariffs for EU agriculture products, especially those of French origin. Eventually, the French government started to give in and this made a compromise between the US and the EU possible and the GATT negotiations were concluded. An interesting element in the GATT negotiations is the fact that, while negotiating a joint policy on the reduction of tariffs and the elimination of export subsidies, countries have to compete on world markets at the same time, where they have to take strategic decisions with respect to the same tariffs and subsidies until an agreement is reached. These tariffs impose external effects on other countries and, in turn, may ultimately influence the outcome of the international trade negotiations. These trade negotiations are related to wage bargaining problems but differ in the sense that each country can go on a strike, that is, start a trade war. This theme returns in the second part of this book.

Already in the light of a few of these examples, one may realize that the potential welfare gains from improving the efficiency of the negotiation process and negotiated outcomes are enormous, perhaps even greater, as Crawford (1982) argues, than those that would result from a better understanding of the effects of macroeconomic policy. If this is true, this would render an immediate justification for our study to explore relevant determinants and variables which affect the negotiation process and corresponding outcome. In other words, a formal theory of negotiation may be regarded as a necessary inquiry into the *essence of negotiation*.

1.3. Outline of the book

The game-theoretical modelling of negotiations has been an active research area for the past five decades, where many fine contributions have been made since the seminal work of Nobel laureate John Nash, e.g., Nash (1950) and (1953). This has led to a vast game-theoretic literature on bargaining including the excellent surveys in, for example, Binmore and Dasgupta (1987), Gale (2000), Muthoo (1999) and Osborne and Rubinstein (1990). So, if this book aims to make another contribution, then the material covered should be complementary with respect to the existing surveys and, at the same time, have some overlaps in order to ensure that the material is self contained. The emphasize on variable threats enables us to realize this goal and, as illustrated by the real-life situations described in the previous section, such negotiation problems certainly deserve considerable attention.

Before spelling out the outline in further detail, we make one remark. The literature distinguishes two main game-theoretic approaches to bargaining: the *axiomatic* and the *strategic* approach. In the axiomatic approach the outcome of bargaining is defined by a set of properties, or axioms, that it is required to satisfy, e.g., Nash (1950). Most information concerning the bargaining procedure and the environment within which bargaining operates is abstracted away in the axiomatic approach. This poses some fundamental problems, since it is not at all obvious how to modify the analysis when the bargaining structure or environment changes. In contrast, the strategic approach attempts to treat concretely these missing elements by explicitly modelling the bargaining process; see, for instance, Nash (1953) or Rubinstein (1982). This typically results in dynamic models in which proposals and counter proposals are made. In this book we mainly focus on dynamic strategic bargaining models, although the link between axiomatic and strategic bargaining models will also be a topic of investigation. As such this book can be regarded as a contribution to the so-called 'Nash program', which is concerned with the study of the relations between the outcomes of strategic bargaining models and bargaining solutions corresponding to axiomatic models. The rationale of the Nash program is that both approaches should strengthen each other, meaning that a list of reasonable axioms should correspond to the outcome of at least one strategic bargaining procedure and vice versa.

This book on the role of threats in negotiations is organized on the basis of a simple guiding principle: The situation in which none of the parties involved in the negotiations has threats at its disposal is the natural benchmark for negotiations where the parties can make threats. This principle also works out nicely at the technical level, because the analysis of negotiations with variable threats builds on and extends the techniques that are applied in the no-threat analysis. This guiding principle imposed a division of the analysis into two parts. Part I, containing chapters 3-6, presents the no-threat case, whereas part II, containing chapters 7-10, focuses upon the variable threats in negotiations. but first, a consistent and unifying framework is presented and many preliminaries are discussed in chapter 2. The motivation for doing so is given in the next two paragraphs.

An important modelling practice is that the primitives of the situation should be made transparent. The main reason is that, by spelling out the underlying assumptions, the modeler gains insight into the assumptions driving the analysis and that it implicitly nominates those assumptions to be relaxed in future research. So, we start with discussing the primitives of game-theoretic bargaining problems in chapter 2 at some detail. These primitives should, roughly speaking, specify the subject of the negotiations, the preferences of the negotiators, the strategic opportunities of each party and the rules of the negotiation process. Within the game-theoretic literature on bargaining there are several contributions that start with introducing these primitives, however, an even larger part of the literature takes the so-called utility representation as its origin and imposes its assumptions directly within the latter framework, without reference to the underlying primitives. In chapter 2 we take the stance that for economic applications only the primitives as first mentioned matter, because in the end (applied) economists are interested in the physical details of the final agreement, provided it is reached. The utility representation as such cannot accomplish this because it states the agreement in terms of utility levels only. Nevertheless, the utility representation is very helpful in constructing shorter and more elegant formal proofs. Our review of the literature only yielded a handful of references in which the negotiators' preferences where taken as the starting point and the mapping from preferences into the utility representation got some attention. Moreover, many all of these references concerned the one-issue case of dividing a dollar. We regard this as a serious omission. In order to close this essential gap, we discuss the mapping of going from primitives to the utility representation in quite some detail in chapter 2. Also in later chapters we return to this crucial theme.

Throughout this book we are mainly concerned with extensions of the alternating offers procedure, as proposed in Ståhl (1972) and Rubinstein (1982). These extensions allow us to incorporate threats. A consistent and unifying framework is introduced in chapter 2. This chapter also contains a discussion of the game-theoretic concepts such as strategies, so-called equilibrium concepts and strategy representation by means of simple tables. Readers that are already familiar with game theory may prefer to skip some sections of this chapter.

The analysis in Rubinstein (1982) of the alternating offers procedure with no threats and discounting caused a boom in the literature on bargaining and it forms the subject matter treated in chapter 3. The alternating offers procedure assumes a rigid chess-game-like structure in which negotiators alternate in making proposals until either an agreement is reached or the negotiations end in perpetual disagreement.[3] This model admits a unique (subgame perfect) equilibrium in which the two parties immediately reach an efficient agreement. Moreover, the proposing party has a strategic advantage. The model is applied to the divide a dollar problem and a barter economy with two individuals and two goods. In order to facilitate economic applications, the system of equations that characterizes the equilibrium agreement is also presented.

The Nash program is treated in chapter 4 and it is concerned with the (possible) connection between the seemingly unrelated strategic and the axiomatic approach. Since the axiomatic approach has not been introduced before, it is discussed at the beginning of this chapter. Attention is restricted to the axiomatic solution as first proposed in Nash (1950) and this solution is now called Nash's bargaining solution. This solution is easy to introduce in the utility representation, as is common to do in the literature. We also start off in this fashion, but then return to characterizing this solution in terms of the primitives, discussing the contributions made in Roemer (1988) and Rubinstein, Safra, and Thomson (1992). Setting out from the primitives is the way John Nash advocated when proposing his axiomatic solution. The discussion of these contributions is at times very technical and some readers may prefer to skip some parts on first reading.

Having introduced Nash's bargaining solution enables us to establish the close resemblance of this axiomatic solution with the outcome of the alternating offers procedure. This result is similar to a result obtained in Nash (1953) for a 'static' strategic bargaining procedure, nowadays called Nash's demand game, where equilibrium selection singles out a single equilibrium that corresponds to Nash's (axiomatic) bargaining solution. Although the alternating offers procedure and Nash's demand game are seemingly different, the alternating offers procedure can be regarded as the unspecified dynamic negotiation process John Nash envisioned to underlie his 'static' demand game. This explains why both strategic models yield the same (strategic) outcome and, thus, if one of these strategic models corresponds to Nash 'bargaining solution the other should automatically do also.

The next two chapters can be regarded as performing comparative statics on the equilibrium outcome of the (strategic) alternating offers procedure with respect to different modelling assumptions. Since the material did not fit in a single chapter, it is divided over two. In the first of these two chapters changes in the players preferences are investigated. These lead to so-called nonconvex bargaining problems and the uniqueness result mentioned earlier is not guaranteed to hold in this class of problems. The analysis requires the introduction of the technique of equilibrium switching in order to establish the set of equilibria. In case of multiplicity of equilibria inefficient delay can be sustained as an equilibrium outcome, but the maximum delay is bounded from above.

The alternating offers procedure assumes a rigid structure in which negotiators alternate in making proposals. Although this procedure is quite simple, its outcome is remarkable robust with respect to changes in the procedure. As already mentioned, the alternating offers procedure can be regarded as the unspecified dynamic negotiation process underlying Nash's demand game. Also, if the rigid sequence of 'who proposes when' is replaced by some general stochastic process as in e.g., Merlow and Wilson (1995), then the same outcome is obtained if the stochastic process approaches the alternating offers procedure. Third, the outcome of the alternating offers procedure can also be obtained as the outcome of some complicated continuous-time procedure with endogenous timing of proposals as analyzed in e.g., Perry and Reny (1993) and Sákovics (1993). The latter model can also mimic the rigid structure of the alternating offers process itself. The robustness of the equilibrium outcome of the alternating offers procedure with respect to alternative procedures means that assuming extensions of the relative simple model of alternating offers instead of the more (technically) demanding negotiation procedures in the second part of this book can be justified. This book is the first to treat these alternative bargaining procedures combined in a single survey.

In part II threats are considered to be an essential part of the parties strategic possibilities. In chapter 7 it is argued that the choice of threats depends upon whether or not the negotiators can commit themselves upon these actions prior to the negotiations, where commitment means that the parties cannot back out of the chosen actions if called upon to carry out their threats.[4] For example, the US is committed to carry out publicly announced threats of a trade war, because this nation would lose its 'face' if it did otherwise with the serious consequence that it will not be believed in future negotiations with other nations. In situations where such commitment is possible, a unique outcome for the negotiations is predicted. However, in situations where the parties fail commitment the threats have to be credible, meaning that these actions have to be equilibrium actions. Possible multiplicity of equilibria translates into a multiplicity of predicted outcomes in the negotiations. In this chapter it is shown that the standard results obtained by John Nash need some minor modifications if his demand game, specifying the unmodelled negotiation process, is replaced by the alternating offers procedure.

The models treated in chapter 7 all have the drawback that the choice of threats and the actual negotiations are separated in time. Negotiations where threats are chosen and carried out simultaneously during the negotiations are complicated, because past behaviour at the negotiation table may influence the choice of threats and vice versa. Such situations also lack the absence of commitment and this means that threats have to be credible. In chapter 8 we focus on the simplest situation, where only one party has a 'monopoly' on using threats. This situation is framed as a wage bargaining game between a single union and a single firm in which only the union is able to choose threats in the form of some type of industrial action. Any threat can only be credible if the union's benefits in terms of a better agreement outweigh the cost of implementing it. If this condition is met, a multiplicity of outcomes is predicted, including those involving a delay in reaching an agreement and featuring strike during the negotiations. An application to Dutch wage negotiations in this book yields some new perspectives on backdating of wage contracts and the rather lengthy periods of negotiations without strike.

In chapter 9 the symmetric situation where each party has threats at its disposal is considered. This corresponds more to the case of the GATT negotiations as described in the previous section: Not only the US has the threat of a trade war against the EU at its disposal but also the EU can threaten the US to start such war, although the internal EU procedures make this less plausible to happen. The presence of mutual threats adds more complexity to the analysis, especially in establishing the minimum result each party can secure itself. In this chapter it is made clear how these complexities can be overcome and it involves application of several of the techniques already introduced in the first part of the book. Special attention is given to so-called *policy bargaining* problems, where the subject of the negotiations concerns agreement upon the future use of variables that can be used as threats during the negotiations. For example, consider the negotiations for a reduction of import tariffs once more, where a nation that sets its current tariffs at extremely high levels correspond to a nation that starts a trade war, i.e., carries out a threat. Recall from section 1.2 that in the GATT negotiations enforceability of agreements is an important issue due to the nonbinding character of international agreements. A related issue concerns renegotiation of agreements. These important issues are also addressed in this chapter, but the theoretical results yield bleak results for the predictive power of the model and the enforceability of such international agreements.

All the situations discussed thus far did not allow for the possible effect that threats may have on the future gains of trade and such negotiation situations are of interest, especially in environmental economics and ecology. These situations are characterized by intertemporal links between threats and several cumulative variables that pass on to the future. For example, in the negotiations for fish quota among nations caught fish cannot reproduce itself and a nation's threat to expand its fleet, i.e., to fish more, is likely to destroy some part of the future gains from fishing. Another example concerns the negotiations for the reduction or abatement of greenhouse gas emissions, where the latter add to the accumulation of pollutants in the earth's atmosphere. In the final chapter of this book the negotiations for fish quota are elaborated for very stylized and simple biological dynamics for a single specie. It tuns out that more patient nations do worse than less patient ones, which is a result that is opposite with the 'patience is beneficial' result derived in chapter 6. The last inquiry in this book deals with the type of results that can be obtained if the threats are linked to several accumulated variables, such as several species of fish or the concentrations of several greenhouse gasses. Although some results are presented for these type of problems, still, a full analysis of the latter problems is lacking and chapter 10 concludes with directions for future research. One final remark concerns the important topic of incomplete information, a topic that is not treated in this book for the simple reason that excellent surveys of this research area already exist and can be found in the books mentioned at the beginning of this outline.

Notes

1 Fokker was still at the high-cost part of the learning curve in producing the Fokker 50 and 100. Furthermore, Fokker had to pay back the development costs financed by the Dutch government through some development fund. Part of the deal was that Fokker would pay back a fixed amount per plane sold and thereby increasing its marginal costs.

2 DASA was already stakeholder in Fokker, because it supplied the wings for Fokker's aircrafts. The terms of the underlying agreement never became public, but were debated in the media because these were also thought to have played a major role.

3 The collective memory seems very reluctant in forgetting the case with fixed costs per bargaining round that is also studied in Rubinstein (1982). This version yields either the extreme 'the lowest-cost party takes it all' outcome or a multiplicity of equilibrium if both parties suffer the same costs per round. As in the literature, we will not treat this version but the fixed costs case indicates that the class of preferences for which 'nice' results can be obtained is restricted to discounting. Most economists think the latter class suffices.

4 For an informal but entertaining treatment of the concept of commitment in situations of strategic interaction we refer to Dixit and Nalebuff (1991).

Chapter 2

A BARGAINING MODEL WITH THREATS

2.1. Introduction

What is a bargaining problem? What do we mean by threats? How can the dynamic process of bargaining be formalized? These are the central questions addressed in this chapter, which also sets the stage for later chapters.

In this chapter we first define the content of the negotiations, the so-called *bargaining problem.* By content we mean a mathematical description that is general enough to take into account the following elements: The parties involved in the negotiations, the subject of the negotiations and the agreements available to the parties involved, the actions beside 'verbal' negotiating available to each party during the process -called the disagreement actions or (variable) threats- and the objective each party strives for. These four ingredients constitute a *bargaining problem.*

The central theme in this book concerns negotiations in ongoing relationships between negotiating parties where each party can employ several threats during the negotiations. For example, a union and the employers' representatives in a unionized industry who each can take several forms of industrial action, such as a strike or a lockout. Or neighbouring countries that negotiate trade liberalization where each party may threaten to start a trade war during the negotiations in order to persuade the others to give in. These examples have in common that the 'physical' threats become part of the verbal process of negotiations and that these threats become an essential part of the negotiation strategies.

The main tool applied in analyzing the above bargaining situations is game theory. The formalization of these 'real-life' situations into one well defined game with strict rules involves a compromise between assumptions that capture the essence of the situation under consideration and restrictive assumptions that facilitate a smooth analysis of the game-theoretic model. This compromise is found in the model introduced in this chapter, which also allows us to analyze closely related variants that are known in the literature.

This chapter is organized as follows. First we define the bargaining problem in the next section. Then section 2.3 presents a formal description of the bargaining model and the appropriate solution concept.

2.2. The bargaining problem

The standard approach in economics is to separate the content of the bargaining from the bargaining procedure. In essence, the content is called the bargaining problem and it specifies what the negotiations are about, what actions the players may take during the process of negotiations or in case of permanent disagreement, and which payoffs result when an agreement is struck. Broadly speaking, the following elements constitute the *bargaining problem*:

- A set N of players representing the parties involved in the negotiations,
- A set of contracts C representing the menu of possible agreements,
- A set of actions A_i for each player i, $i \in N$, to choose from either during disagreement or in case of a breakdown,
- A preference relation or utility function over contracts and actions for each player i, $i \in N$.

Throughout this book the focus is on two-player bargaining games. Formally, we assume

Assumption 2.1. *The set of players is given by* $N = \{1, 2\}$.

As in real life negotiations, most of the bargaining problems analyzed in this book are dynamic and time should be incorporated in the model, as well as time-dependent disagreement actions and contracts. Instead of introducing a lot of notation to appropriately deal with the dynamics of the bargaining problem at this stage, we mention that in most of the applications the problem on hand features stationary elements for which additional notation is not needed. If however additional notation is necessary then this is introduced right on the spot.

2.2.1 The contract space

An important assumption that is often taken for granted in the literature on bargaining theory is that the final agreement is a *binding contract* in the juridical sense of the word, meaning that there exists an institutional setting, for instance a court of law, that either enforces the implementation of the agreement or that enforces a costly compensation and penalty scheme to be paid by the offender of the contract. So, it is assumed that all parties involved comply to the contract and that the agreed upon contract will be implemented without hesitation.

Of course, in many interesting economic applications it is often too costly or just infeasible to write all-encompassing contingent contracts. In this sense, contracts are usually imperfect, incomplete or simply not available at all. For example, an employer and a salesman who negotiate the wage contract characterized by a fixed wage, several bonuses and corresponding target levels that specify payment of these bonuses can only include a payment scheme in the contract that depends upon observable and verifiable variables, such as the revenue generated by the salesman, whereas there is no reason to include a target for effort, which is typically not observable nor verifiable. Or, firms that negotiate the conditions for cartel behaviour under anti-cartel legislation should be aware that any agreement fails juridical support, because the authorities will impose heavy penalties upon them if some of them would ever try to sue one of the cartel members for breaking the illegal cartel agreement. Throughout this book we rule out that nonverifiable or noncontractable conditions can become part of any agreement. Instead, we carefully specify for each economic application the content of the available contracts that can be considered to be binding in the juridical sense.

In principle, the contract space C could be a variety of things. For example, the content of negotiations between an employer and a salesman is a payment scheme that depends upon realized sales. Such a payment scheme is mathematically speaking a function, namely one that specifies the payment as a function of sales. The contract space C is the set of all (possibly discontinuous) functions $c : \mathbb{R}_+ \to \mathbb{R}_+$. In most applications in this book we restrict attention to much simpler contract spaces C, namely those that can be represented as a set $C \subseteq \mathbb{R}^n$ for some $n \geq 1$. For example, in most models of wage negotiations in this book we assume $C = [0, 1]$ and each contract $c \in [0, 1]$ represents the wage w, with $w = c$.

The following assumption is made with respect to the set C.

Assumption 2.2. *The set C is a nonempty, compact and convex subset of $\mathbb{R}^n$ for some $n \geq 1$.*

In the current literature on bargaining, which can be traced back to Nash (1950) and (1953), contracts in the set C involve some kind of risk due to uncertainty. This risk is modelled in the form of *lotteries*; see e.g., chapter 6 in Mas-Colell, Whinston, and Green (1995). A discrete or finite lottery is essentially a discrete probability distribution over a finite set of outcomes (or the support), where an outcome is referred to as a *prize*. Formally, for some n, $n \geq 1$, a discrete lottery consists of a finite set of prizes $X = \{x^1, \ldots, x^n\}$, with $x^j \in \mathbb{R}^m$ for some $m \geq 1$, and a probability distribution over X. So, if we identify the set of contracts C as the set of all probability distributions over X, then a vector $c \in \mathbb{R}^n_+$ such that $\sum_{i=1}^n c_i = 1$ represents a lottery (over X). Hence, C represents lotteries if it is a $(n-1)$-dimensional *simplex*, i.e., $C = \left\{c \in \mathbb{R}^n_+ \mid \sum_{i=1}^n c_i = 1\right\}$.[1] In most of the applications in this book the contracts do not involve any risk.

Finally, in many economic applications of bargaining problems the bargaining problem is often solved by applying optimization methods from calculus. In these applications the contract space is usually represented by a finite number of smooth quasiconvex functions, e.g., for a definition see the mathematical appendix M.C in Mas-Colell, Whinston, and Green (1995). For example, if negotiations concern lotteries the contract space C is a simplex of lotteries, or in a two-person exchange economy with two goods the contract space C is the Edgeworth-box and in both cases all the constraints describing the set C are even linear and, thus, quasiconvex. In order to facilitate techniques from calculus later on we find it sometimes convenient to represent the contract space C by several quasiconvex functions. Formally, there exists a number $k \in \mathbb{N}$ and k quasiconvex and differentiable functions $g_j : \mathbb{R}^n \to \mathbb{R}$, $j = 1, \ldots, k$, such that

$$C = \{c \in \mathbb{R}^n \mid g_j(c) \leq 0, \quad j = 1, \ldots, k\}. \tag{2.1}$$

The quasiconvex functions g_j, $j = 1, \ldots, k$, ensure that the set C is a convex set, e.g., see the mathematical appendix M.K in Mas-Colell, Whinston, and Green (1995).

2.2.2 Disagreement actions

Negotiations can be concluded with a swift agreement or may involve a lengthy delay before an agreement is struck. It is even possible that the negotiations break down permanently without having reached an agreement at all. A key feature of the bargaining problems analyzed in this book is that the players not only, let us say, verbally negotiate with each other, but that in between bargaining sessions or in case of a final breakdown the players take some 'other' actions. These actions may be harmful to the opponent and may either influence the continuation of the negotiations or the outcome in case the negotiations break down. For example, during unionized wage negotiations the union has a choice between several forms of industrial action, such as strike, which negatively affects the employer's profit. Or, in negotiations between nations over trade liberalization each nation can threaten to start a trade war if the negotiations linger on without reaching an agreement.

Actions to be taken during negotiations or in case of a permanent breakdown are simply called *disagreement actions* or dubbed *variable threats* in Nash (1953). The word 'threat' already gives away that each player may threaten to act in such a manner that is harmful to the opponent in order to persuade the opponent to give in.

In most part of this book we assume that the disagreement actions are represented by actions corresponding to some game in *normal form,* called the disagreement game. There are several reasons for doing so. Most important, games are considered as the mathematical vehicles that enable economists to model cases in which the action chosen by one party influences the other parties' well being (and vice versa) in terms of utilities or payoffs. The presence of such interdependencies is what makes endogenous disagreement actions interesting and worthwhile analyzing. For example, in unionized wage bargaining, whether or not the union organizes a strike matters for the firm's profits. Note that normal form games also capture the special case where such interdependencies are absent, because then this can be regarded as a degenerated case in which every player has only one action to follow. Second, finite games in extensive form can be translated into their normal-form representation and, therefore, there is no loss in generality by taking games in normal form as a starting point.[2]

Since we regard disagreement actions as (possibly) mixed actions in a game in normal form, we restrict attention to action spaces that can be represented as a subset of the Euclidean space, i.e., $A_i \subseteq \mathbb{R}^{n_i}$, $i = 1, 2$, for some $n_i \geq 1$. If the game is a bimatrix game, then A_i, $i = 1, 2$, represents player i's set of mixed strategies. Or, if the players are two firms engaged in price competition, each set represents an interval of a firm's own prices. We make the following assumption.

Assumption 2.3. *The set A_i, $i = 1, 2$, is a nonempty, compact and convex subset of $\mathbb{R}^{n_i}$ for some $n_i \geq 1$.*

In terms of game theory, a pair of actions (a_1, a_2) is called an outcome of the game in normal form and it represents the 'physical' outcome of the game. For example, an outcome in the price-setting Bertrand duopoly are the prices set by the firms and such outcome differs from the associated profits. The set $A = A_1 \times A_2$ is the set of outcomes and it possesses the same properties as the sets A_i, $i = 1, 2$. In bargaining theory the chosen pair of actions $a \in A$ is called the (endogenous) *disagreement outcome* and this pair will be determined by the equilibrium concept imposed later on. A bargaining problem is said to have *exogenous* -or *fixed*- disagreement outcomes if the set A consists of a single element, a topic to which we return in chapters 3, 4, 5 and 6.

2.2.3 Utility functions

Since each player should be able to compare contracts with disagreement outcomes it would be natural to take $A \cup C$ as the domain of each player's utility functions. However, we want to express the distinction between agreement and disagreement in our notation. Therefore, we introduce two utility functions for each player i, $i = 1, 2$, namely $u_i : C \to \mathbb{R}$ and $d_i : A \to \mathbb{R}$. Nevertheless, we do think of preferences on $A \cup C$ when we think about bargaining problems.

With respect to the two utility functions, consider the following assumption.

Assumption 2.4. *The utility functions $u_i : C \to \mathbb{R}$, $i = 1, 2$, and $d_i : A \to \mathbb{R}$, are* continuous. *The utility function u_i is also* concave, *whereas d_i is* quasiconcave *in i's own action a_i.*

These utility functions correspond to a subset of preferences that are complete, transitive, continuous and convex, e.g., Mas-Colell, Whinston, and Green (1995). Note that a convex preference relation corresponds to a utility function that is quasiconcave, which contains the subclass of concave utility functions on C.

As mentioned earlier, part of the leading literature treats the contract space C as a simplex of lotteries and, therefore, requires preference relations defined over lotteries.[3] *Expected utility* functions correspond to utility functions that are linear in c, i.e., $\sum_{j=1}^{n} c_j \mu_i(x_j)$ for some function $\mu_i : X \to \mathbb{R}$ defined on the finite set of prizes X, see e.g., Mas-Colell, Whinston, and Green (1995). Then, by definition, $u_i(e_j) = \mu_i(x_j)$ for the j-th unit vector $e_j \in C$, $j = 1, \ldots, n$ and the utility $\mu_i(x_j)$ of the prize x_j can simply be regarded as the weight of probability c_j in the linear utility function. In the remainder of this book we do not explicitly deal with expected utility functions on a simplex, but whenever linear utility functions on a simplex C are considered we implicitly mean expected utility functions over a contract space C of lotteries. Since we share the popular opinion that expected utility functions over lotteries are very important we assume concave utility functions instead of strictly concave utility functions. The latter would rule out expected utility functions. Doing so, comes at a cost, which is further explained in section 2.2.7 on (non)generic uniqueness.

2.2.4 Mutual and conflicting interests

In order to make the bargaining problem interesting we introduce the notions of *mutual* and *conflicting interests* over contracts. Loosely speaking, mutual interests mean that potentially both players have something to gain from reaching an agreement, whereas conflicting interests refers to the players conflicting goals, which require a compromise in order to resolve the negotiations with an agreement.

Mutual interests formalizes the idea that the bargaining process creates the opportunity to share gains from an agreement among the players. That is, the following definition states that each possible disagreement outcome can be improved upon by some contract and that this contract may depend upon the particular disagreement outcome under consideration. Note that this does not mean that there is one contract that dominates all the disagreement outcomes. So, mutual interests imply that the players have an incentive to reach an agreement.

Definition 2.5. *A bargaining problem has mutual interests if for every $a \in A$ there exists some contract $c(a) \in C$ such that $u_i(c(a)) \geq d_i(a)$ for both $i = 1, 2$.*

A bargaining problem with mutual interests is also referred to as *essential.*

In order to define conflicting interests, the set of player i's most preferred contracts is first defined as

$$C_i^* = \{c^* \in C \mid u_i(c^*) \geq u_i(c), \forall c \in C\}, i = 1, 2.$$

The assumptions made for the set C and the utility functions u_i, $i = 1, 2$, ensure that C_i^* is nonempty. Consider the following definition.

Definition 2.6. *The players have conflicting interests if $C_1^* \cap C_2^* = \emptyset$. Common interests occur if $C_1^* \cap C_2^* \neq \emptyset$.*

In case of conflicting interests none of player 1's most-preferred alternatives belong to the set of player 2's most-preferred alternatives and vice versa. Negotiations have a role to play here in bridging the players' separate objectives. For example, the seller's notion of a favourable price is a high price whereas the buyer's notion of favourable price is a 'bargain' and negotiations have to result in an acceptable price for both parties. *Common interests* are considered less interesting, because agreements are available in which both players do not have to make any concessions and rational players would agree upon one of these agreements. Loosely speaking, there is no need to negotiate when players have common interests.

One in our view essential part of bargaining is captured by making the following assumption.

Assumption 2.7. *The players have conflicting and mutual interests.*

2.2.5 Pareto efficiency

An important concept in bargaining theory is *Pareto efficiency*. A contract is Pareto efficient if both players cannot improve upon this contract.

Definition 2.8. *The contract $c \in C$ is Pareto efficient if there does not exists a contract $\hat{c} \in C$ such that $u_i(\hat{c}) > u_i(c)$ for both $i = 1, 2$.*

The set $C^{PE} \subset C$ consists of all Pareto efficient contracts, that is

$$C^{PE} = \{c \in C \mid \nexists\, \hat{c} \in C : \; u_i(\hat{c}) > u_i(c), \quad i = 1, 2\}.$$

Most standard microeconomic textbooks provide the well known graphical illustration of the *contract curve* in the Edgeworth box, e.g., Mas-Colell, Whinston, and Green (1995) or Varian (1984). This box represents all feasible trades in a two-person exchange economy with two goods. The contract curve consists of the feasible trades that are Pareto efficient. Within the bargaining framework of this chapter one can easily see the set C as the generalization of the Edgeworth box and the set C^{PE} as the generalization of the contract curve for the arbitrary contract spaces C. This point of view enables us to apply several standard results from microeconomic theory (and convex programming) in order to derive several properties of the set C^{PE}.

Microeconomic theory implies that a typical element of the contract curve can be found by solving a maximization problem. Translated to the bargaining problem this yields the following. Let $\overline{u}_2 \in (\min_{c \in C} u_2(c), u_2(c_2^*)]$, where $c_2^* \in C_2^*$, and consider the maximization problem given by

$$\max_c u_1(c) \ , \text{s.t. } u_2(c) \geq \overline{u}_2 \text{ and } c \in C. \tag{2.2}$$

Then every solution of this maximization problem is Pareto efficient.

In many economic applications the utility functions are differentiable and the contract space C can be represented by a finite number of differentiable and quasiconvex functions as in (2.1). In that case standard optimization methods can be applied. Then maximization problem (2.2) becomes the nonlinear programming problem given by

$$\max_c u_1(c) \ , \text{s.t. } -u_2(c) + \overline{u}_2 \leq 0 \text{ and } g_j(c) \leq 0, \ j = 1, \ldots, k.$$

Since the function $-u_2$ is convex (and then also quasiconvex) this implies that the subset $\{c \in C | -u_2(c) + \overline{u}_2 \leq 0\}$ is also a convex set and the Kuhn-Tucker first-order conditions are a sufficient condition for a global maximum, see e.g., the mathematical appendix M.K in Mas-Colell, Whinston, and Green (1995). Application of the Kuhn-Tucker first-order conditions, where λ_0 and λ_j denote the Lagrange multipliers associated with the first constraint respectively the constraint involving function g_j, yields the system of equations given by

$$\begin{aligned} \frac{\partial u_1(c)}{\partial c_\ell} + \lambda_0 \frac{\partial u_2(c)}{\partial c_\ell} - \sum_{j=1}^{k} \lambda_j \frac{\partial g_j(c)}{\partial c_\ell} &= 0, \quad \ell = 1, \ldots, n, \\ \lambda_0(-u_2(c) + \overline{u}_2) &= 0, \\ \lambda_j g_j(c) &= 0, \quad j = 1, \ldots, k. \end{aligned} \tag{2.3}$$

In case it is known in advance that C^{PE} lies in the interior of C, then this prior information can be used to simplify the first-order conditions, because interior means that $g_j(c) < 0$ for all $j = 1, \ldots, k$ and, therefore, $\lambda_j = 0$ for all $j = 1, \ldots, k$. For this special case the well-known result for the (interior) contract curve in the Edgeworth box of equal marginal rates of substitution is easily generalized to the bargaining model. Formally, substitution of $\lambda_j = 0$

for all $j = 1, \ldots, k$ into (2.3) and combining the h-th and ℓ-th equation, $h, \ell = 1, \ldots, n$, to get rid of λ_0 yields

$$\frac{\partial u_1(c)/\partial c_h}{\partial u_1(c)/\partial c_\ell} = \frac{\partial u_2(c)/\partial c_h}{\partial u_2(c)/\partial c_\ell}, \qquad h, \ell = 1, \ldots, n. \tag{2.4}$$

2.2.6 Individual rationality

In two-player games in normal form each player can secure his *minmax* payoff (or utility), i.e., the utility a player can guarantee by applying a best response to the opponent's action that minimizes his highest utility. Formally,

$$m_i = \min_{a_j \in A_j} \max_{a_i \in A_i} d_i(a), \quad i, j = 1, 2 \quad \text{and} \quad j \neq i.$$

In order to avoid additional notation we simply assume that there is a unique pair of actions $a^i \in A$, $i = 1, 2$, that correspond to player i's minmax utility m_i. Formally, the *pair* of (possibly mixed) actions $a^i \in A$, $i = 1, 2$, that minmax player i are defined as

$$a^i = \arg \min_{a_j \in A_j} \max_{a_i \in A_i} d_i(a), \quad i, j = 1, 2 \quad \text{and} \quad j \neq i.$$

The set of *individually rational* contracts $C^{IR} \subset C$ denotes all contracts $c \in C$ that are weakly preferred by *both* players to their minmax utility, i.e.

$$C^{IR} = \{c \in C \mid u_i(c) \geq m_i, i = 1, 2\}.$$

Assumption 2.3 and 2.4 imply that C^{IR} is a nonempty, compact and convex set. Note that the last property follows for similar arguments as in the previous section, i.e., C is a convex set and the function $-u_i$, $i = 1, 2$, is a (quasi)convex function in the reformulated restriction $-u_i(c) + m_i \leq 0$.

In order to avoid more notation we assume that each player's set of most preferred contracts does not belong to the set of strict individual-rational payoffs. This means that every contract in let us say player 1's set of most preferred contracts C_1^* yields his opponent a utility that is at most m_2. Alternatively formulated, it means that the 'endpoints' of the set C^{PE} do not belong to the set of strict individual-rational contracts. However, the 'endpoints' C_i^*, $i = 1, 2$, may belong to the boundary of C^{IR}. That is,

Assumption 2.9. *If $c \in C_i^*$, $i = 1, 2$, then $u_j(c) \leq m_j$ for $j = 1, 2$ and $j \neq i$.*

In section 6.2.1 we argue that this assumption can be removed if some additional notation is introduced.

2.2.7 Generic uniqueness

As argued in section 2.2.3 we want to allow for lotteries and expected utility functions, i.e., linear utility functions on a simplex C of lotteries. This is accomplished by assuming concave utility functions u_i, $i = 1, 2$. However, assuming concavity instead of strict concavity comes at a cost, namely maximization problem (2.2) may yield multiple solutions. In this section we investigate this problem and conclude that it is present but that it is also very rare. In mathematical terms, uniqueness is generic, whereas multiplicity of solutions is nongeneric. First it is remarked that if u_1 and u_2 are both strictly concave functions, then maximization problem (2.2) automatically yields a unique solution for every $\overline{u}_2 \in \left(\max_{c \in C_1^*} u_2(c), u_2(c^*)\right]$, where $c^* \in C_2^*$. The reason is simple, maximization problem (2.2) has a unique solution, because u_1 is also strictly quasi-concave and (as shown before) the set $\{c \in C| -u_2(c) + \overline{u}_2 \leq 0\}$ is a convex set implying that the conditions of theorem M.K.4 in Mas-Colell, Whinston, and Green (1995) are satisfied. Moreover, the set of most preferred contracts C_i^*, $i = 1, 2$, consists of a unique contract. These results can be slightly extended to allow for just one player with a strictly concave utility function.[4] So, within the set of all concave utility functions, multiplicity of solutions can only occur for the subclass of bargaining problems in which both utility functions are concave but not strict. The latter subclass consist of utility functions that are linear, piecewise linear or possess these properties on a (small) subset of their domain.

For linear and piecewise linear utility functions it is easy to come up with examples such that the maximization problem (2.2) has more than one solution, as example 2.10 illustrates. However, these cases are rare in the sense that if we would slightly alter the utility functions by changing the parameters, or alternatively slightly alter the set C, then for 'almost all' cases uniqueness would be restored. In this sense, it is said that nonuniqueness is not robust or *nongeneric*. Or, to put it the other way round, uniqueness of the solution of maximization problem (2.2) is *generic*.

Genericity and nongenericity can be formalized by considering some measure on the space of parameters that characterize the utility functions and the set C. A measure generalizes the idea of a continuous probability distribution. *Genericity* here means that the set of parameters for which a certain property holds, has positive measure. *Nongenericity* is associated with measure zero. The following example illustrates these notions.

Example 2.10. *Consider the contract space C given by the simplex*

$$C = \left\{c \in \mathbb{R}_+^3 \mid c_1 + c_2 + c_3 = 1\right\}$$

and, for some $\alpha \in (0, 1)$, the players' utility functions are given by

$$u_1(c) = c_1 + \alpha c_2 \quad \textit{and} \quad u_2(c) = \alpha c_2 + c_3.$$

Then $0 \leq u_i(c) \leq 1$, $i = 1, 2$, for all $c \in C$. Maximization problem (2.2) is given by the linear programming problem

$$\max_{c_1, c_2, c_3} c_1 + \alpha c_2 \quad \textit{s.t.} \quad \alpha c_2 + c_3 \geq \overline{u}_2, \quad c \in C.$$

The solution depends upon α and there are three different cases.
First, $\alpha < \frac{1}{2}$. *Then* $u_i(0,1,0) = \alpha < \frac{1}{2} = u_i\left(\frac{1}{2},0,\frac{1}{2}\right)$, $i = 1,2$. *For* $c' = \left(c_1 + \frac{1}{2}c_2, 0, c_3 + \frac{1}{2}c_2\right)$ *we then have that* $u_i(c') > u_i(c)$ *for all* $c \in C$ *with* $c_2 > 0$. *So, application of definition (2.8) implies that every* c *with* $c_2 > 0$ *cannot be Pareto efficient and the unique solution of the linear programming problem is given by* $c = (1 - \overline{u}_2, 0, \overline{u}_2)$.
Second, $\alpha > \frac{1}{2}$. *Then* $u_i(0,1,0) = \alpha > \frac{1}{2} = u_i\left(\frac{1}{2},0,\frac{1}{2}\right)$, $i = 1,2$. *Consider an arbitrary* $c \in C$ *with* $c_1 > 0$ *and* $c_3 > 0$. *Choose* $c' = \left(c_1 - \frac{1}{2}\varepsilon, c_2 + \varepsilon, c_3 - \frac{1}{2}\varepsilon\right) \in C$ *for sufficiently small* $\varepsilon > 0$. *Then* $u_i(c') > u_i(c)$ *and, therefore, application of definition (2.8) rules out every* c *with* $c_1 > 0$ *and* $c_3 > 0$ *as a Pareto efficient contract. The unique solution of the linear programming problem is given by*

$$c = \begin{cases} \left(1 - \frac{\overline{u}_2}{\alpha}, \frac{\overline{u}_2}{\alpha}, 0\right), & \text{if } 0 < \overline{u}_2 \le \alpha, \\ \left(0, \frac{1-\overline{u}_2}{1-\alpha}, \frac{\overline{u}_2-\alpha}{1-\alpha}\right), & \text{if } \alpha \le \overline{u}_2 < 1. \end{cases}$$

Third, for $\alpha = \frac{1}{2}$ *we have that* $C^{PE}\left(\frac{1}{2}\right) = C$.
Summarizing, the set of Pareto efficient contracts as a correspondence of α, *denoted by* $C^{PE}(\alpha)$, *is given by*

$$C^{PE}(\alpha) = \begin{cases} \{(1-c_3, 0, c_3) \mid \ c_3 \in [0,1]\}, & \alpha < \frac{1}{2}, \\ C, & \alpha = \frac{1}{2}, \\ \{(1-c_2, c_2, 0) \text{ or } (0, c_2, 1-c_2) \mid \ c_2 \in [0,1]\}, & \alpha > \frac{1}{2}. \end{cases}$$

The linear programming problem has a unique solution provided $\alpha \ne \frac{1}{2}$. *The nonuniqueness for* $\alpha = \frac{1}{2}$ *is said to be nongeneric. Nongenericity in this example means that if we would take the measure over all the areas in the interval* $[0,1]$ *where nonuniqueness occurs (the measure is equivalent to integration of the constant unit function), then this measure is equal to* 0. *Formally, the measure* $\int_{\alpha=\frac{1}{2}} 1 d\alpha = 0$. *Alternatively, because the measure* $\int_{\alpha\in[0,\frac{1}{2})\cup(\frac{1}{2},1]} 1 d\alpha = 1$ *is positive, uniqueness holds as a generic property in this example.*

Finally, note that the set $C^{PE}(\alpha)$, $\alpha \ne \frac{1}{2}$, in example 2.10 coincides with either one or two of the three edges of the set C. So, the 'contract curve' lies on the edges of the 'Edgeworth box'. This insight is the generic case for (piecewise) linear utility functions, because generically (piecewise) linear utility functions violate the equal marginal rates of substitution relation in (2.4) for interior solutions.

In general, if the utility functions are concave but none is strictly concave, then multiplicity of Pareto efficient contracts, given fixed $\overline{u}_2$, can only occur in case the intersection of the two indifference curves and the other binding constraints at an optimum locally form a plane. Then this requires some kind of local piecewise linearity in the utility functions or the constraints that specify C, which is nongeneric.

The important message of this section that returns in later chapters is that we cannot prove uniqueness in case of (piecewise) linear utility functions, but that we are able to show the slightly weaker result of generic uniqueness. So, multiplicity can occur (and we certainly discuss it when it may happen) but it is also nongeneric. The following proposition summarizes the discussion above and is given without a formal proof.

Proposition 2.11. *Maximization problem (2.2) yields a generically unique solution.*

2.2.8 Utility representation

In the pioneering work of Nash (1950) and (1953) the bargaining problem is analyzed in the space of utility pairs. According to Nash, two bargaining problems that coincide in the utility space should be considered as being equal and, being equal, its prediction in terms of utilities should also be the same. The rational behind this view is that the utility function is one of the primitives of the bargaining problem and that the physical representation of the bargaining problem in terms of contracts is of minor importance. However, in applications economists are usually interested in the outcome of the bargaining problem in terms of physical contracts $c \in C$. So, why then study the bargaining problem in the space of utility pairs? There are two reasons for doing so. First, a significant part of the literature on bargaining follows Nash's approach, whereas another significant part analyzes the bargaining problem directly in terms of possible agreements. In order to capture both strands of literature in one book and to make the results known in the space of utility pairs more accessible we discuss both representations of the bargaining problem. Second, many of the results in this book are somewhat easier to derive in the space of utility pairs. So, in this section the bargaining problem in the space of utility pairs is introduced.

The two-player bargaining problem in *utility representation* corresponding to the bargaining problem in the previous section is denoted as (S, A, d), where $S \in \mathbb{R}^2$ is the set of vectors consisting of the players' utilities associated to the set of contracts C, the set A is the set of disagreement outcomes as defined before and $d : A \to \mathbb{R}^2$ is a vector function representing the players' utilities with $d(a) = (d_1(a), d_2(a))$. Formally,

$$S = \{\ s \in \mathbb{R}^2 |\ \exists\ c \in C \text{ s.t. } s = (u_1(c), u_2(c))\ \} \tag{2.5}$$

and

$$d(a) = (\ d_1(a),\ d_2(a)\ ),\ a \in A. \tag{2.6}$$

The vector $d(a)$ is called a *disagreement point* and $s \in S$ is called a pair of *feasible payoffs* or utilities.

For convenience, and if it leads to no confusion, the bargaining problem in utility representation (S, A, d) is simply also called the bargaining problem. Furthermore, the set C_i^*, $i = 1, 2$, of player i's most preferred contracts translates into

$$S_i^* = \{\ s \in S |\ s = (u_1(c), u_2(c)),\ c \in C_i^*\ \},\ i = 1, 2.$$

Proposition 2.11 implies that, generically, each set S_i^*, $i = 1, 2$, consists of a unique element, i.e., generically it is a singleton.

Assumption 2.2 and 2.4 ensure that the set S satisfies the standard assumptions made with respect to this set in the bargaining literature. The following proposition states these standard properties of S and we refer to e.g., Billera and Bixby (1973) and Roemer (1988) for a proof.

Proposition 2.12. *If assumptions 2.2, 2.4 and 2.7 hold, then for the associated bargaining problem (S, A, d) it holds that S is nonempty, compact and convex, and for every $a \in A$ there exists an $s \in S$ such that $s \geq d(a)$.*

The interpretation of this proposition is that S is nonempty simply means that the players have at least some alternatives over which they can bargain, corresponding to the assumption of conflicting and mutual interests.

The set of *individually rational* payoffs $S^{IR} \subset S$ denotes all payoffs $s \in S$ that yield each player i, $i = 1, 2$, a utility of at least his minmax value m_i. Formally,

$$S^{IR} = \{\ s \in S |\ s \geq (m_1, m_2)\ \}.$$

Proposition 2.12 implies that S^{IR} is nonempty, compact and convex. Then, definition 2.8 translates to: The utilities $s \in S$ are Pareto efficient if there does not exist an $\hat{s} \in S$ such that $\hat{s} > s$.

Similar as the set of Pareto efficient contracts we may now define the set of Pareto efficient payoffs, denoted as S^{PE}, as follows

$$S^{PE} = \left\{ s \in S \;\middle|\; s = (u_1(c), u_2(c)), c \in C^{PE} \right\}.$$

Assumption 2.9 rules out that $s \in S_i^*$ belong to S^{IR}. Denote $s_i^{\max}$, $i = 1, 2$, as the maximum attainable utility for player i in the set S^{IR}. That is,

$$s_i^{\max} = \max_{s \in S^{IR}} s_i, \quad i = 1, 2.$$

Player i's payoff $s_i^{\max}$ is also known as his *utopia* payoff, see e.g., Kalai and Smorodinsky (1975).

The assumptions 2.2, 2.4 and 2.9 ensure that the set S^{IR} is a convex set. This means that the subset of Pareto efficient payoffs of S^{IR}, called the *Pareto frontier*, can be described by a *strictly decreasing*, *concave* and *continuous* function. So, the Pareto frontier of S^{IR} does not contain horizontal or vertical line pieces. This function is denoted by $f_i(s_j)$, $i, j = 1, 2$, $i \neq j$, and specifies player i's maximum utility in S^{IR} given the utility s_j for player j. Formally, the Pareto frontier of S^{IR} is given by

$$\{s \in S^{IR} \mid s_2 = f_2(s_1),\, s_1 \in [m_1, s_1^{\max}]\}$$

or, since f_2 is the inverse function of f_1 (and visa versa),

$$\{s \in S^{IR} \mid s_1 = f_1(s_2),\, s_2 \in [m_2, s_2^{\max}]\}.$$

Since $(m_1, m_2) \in S$ is not Pareto efficient we have that $m_1 < f_1(m_2)$ and $m_2 < f_2(m_1)$.

2.3. A bargaining game with threats

The bargaining problem defined in section 2.2 did not include any reference to the dynamic process of negotiations nor to how each party employs its threats. In this book bargaining processes are modelled as games in extensive form, which explicitly deal with the sequential structure facing the players during the process of strategic interaction. One then asks what outcomes would result from rational and self-interested actions of the players, i.e., apply one of the standard game theoretic equilibrium concepts.

Many models of bargaining processes have been proposed in the literature. This book chooses one of those that serves as the basic model in later chapters. It extends the *alternating offers* model as introduced in Rubinstein (1982) and Ståhl (1972) to allow for endogenous threats in a manner similar to e.g., Busch and Wen (1995), Haller and Holden (1990), Houba and De Zeeuw (1995) and Okada (1991a). To formulate the two-player bargaining model with threats as a game in extensive form, we need the following information:

- A bargaining procedure that specifies the order of moves.
- The information each party knows when it makes a move,
- A set Σ_i of strategies for each player i, $i \in \{1, 2\}$,
- A set of outcomes,
- The payoffs for each player i, $i \in \{1, 2\}$, as a function of the outcomes,

The payoffs of each player in the bargaining game are related to the bargaining problem defined in the previous section. We assume that each player has *complete information* with respect to the set of contracts C, the sets of disagreement actions A_i, $i = 1, 2$, and the players' utility functions over C and $A = A_1 \times A_2$. Complete information goes beyond assuming that the players share the same information about the bargaining problem and that there are no asymmetries in information, because it also implies that the players are fully informed about the bargaining problem. The latter includes full knowledge about the opponent's preference relation, which cannot be observed. So, complete information is a strong assumption. Furthermore, it rules out ignorance, secret agendas, framing effects or other secret tricks.

Assumption 2.13. *Each player has complete information about C, A_i, u_i, d_i, $i = 1, 2$.*

2.3.1 The order of moves

Negotiations in ongoing relationships feature that none of the players can invoke a permanent termination of the negotiations. For example, in many large unionized industries neither the union nor all the firms can leave the scene if the industry is profitable and the union is stable in its membership. Two neighbouring countries remain neighbouring countries (ruling out a conquest by war). This means that the parties always face the possibility of future negotiations even if they fail to negotiate an agreement today. The alternating offer model in Rubinstein (1982) and Ståhl (1972) captures this particular aspect by assuming that each player in turn proposes a contract that his opponent may accept or reject and this process either ends in case one party accepts or the parties fail to come to an agreement and the process lasts forever. According to this model, each time a party rejects a proposal it makes a counter proposal. To quote Osborne and Rubinstein (1990):

> This bargaining procedure conforms to negotiation situations in which the players perceive that, after any rejection of an offer, there is room for a counterproposal.

In order to properly model negotiations in ongoing relationships we have to extend the alternating offer model, because it does not take into account the possibility of threats during the negotiations. These threats are easily incorporated by assuming that during the time in between the rejection of a proposal and its subsequent counter proposal, each player chooses a threat from his set of disagreement actions once. This threat or disagreement action could represent a form of industrial action in unionized wage bargaining or starting a trade war in negotiations on trade liberalization between countries. As negotiations linger on these *endogenously* determined disagreement actions may vary over time. This captures the idea that each player cannot commit himself prior to the negotiations on some particular disagreement action and that each player takes his decision with respect to the disagreement actions contingent upon the actual history of the negotiations so far. So, disagreement actions become part of the negotiation strategies of the players.

Formally, bargaining rounds, which indicate the constant time interval elapsing between two consecutive proposals, are indexed by t, $t \in \mathbb{N}$,[5] and are of length $\Delta > 0$ measured in continuous or real time. Negotiations start at round $t = 0$ and the start of round t, $t \in \mathbb{N}$, corresponds to (real) time $t\Delta$. Since each player in turn proposes a contract that his opponent may accept or reject we impose the convention that, as long as the negotiations still continue, player 1 makes a proposal at even numbered rounds, whereas player 2 proposes at the odd numbered rounds.

Consider bargaining round t with t is even. Then player 1 proposes the contract $c^t \in C$. Player 2 hears this proposal and decides either to accept it ('yes') or to reject it ('no'). If he accepts, then the negotiations are concluded and the players will implement c^t. However, if player 1's proposal is rejected, then before a counterproposal can be formulated at the next round, both players choose disagreement actions once to determine their conflict utilities in the current round. That is, player i, $i = 1, 2$, selects an action $a_i^t \in A_i$ in case of 'disagreement' at round t and obtains the per-period utility $d_i\left(a_1^t, a_2^t\right)$. A pair of actions played at round t, $t \in \mathbb{N}$, is denoted by a^t, $a^t \in A$. The negotiations then move on to round $t + 1$, which is odd.

Similar, at bargaining round t, t is odd, player 2 proposes the contract $c^t \in C$ and, then, player 1 decides whether or not to accept it, i.e., 'yes' or 'no'. Acceptance means that the negotiations are concluded with c^t, whereas if player 2's proposal is rejected, then, before a counterproposal can be formulated at $t + 1$, player i, $i = 1, 2$, chooses his disagreement action $a_i^t \in A_i$ once and obtains the per-period utility $d_i\left(a_1^t, a_2^t\right)$. The negotiations then move on to the next even round $t + 1$.

In terms of game theory, the players' disagreement outcome a^t, $a^t \in A$ in round t are the actions in a normal-form game, which is called the *disagreement game* and denoted by Γ. Formally, Γ is the normal-form game $\langle \{1,2\}, \{A_i\}_{i=1,2}, \{d_i\}_{i=1,2} \rangle$, where A_i satisfies assumption 2.3 and d_i satisfies assumption 2.4. The assumptions made are sufficient for the existence of a Nash equilibrium in the game Γ. Then the set $A^N \subseteq A$ denotes the nonempty set of Nash equilibrium actions.

The bargaining game ends if an agreement is reached in a certain round t, $t \in \mathbb{N}$, at (real) time $t\Delta$. Then the contract accepted c^t is *binding* in the juridical sense of the word and the players will implement it. The following procedure summarizes the bargaining game with threats.

Bargaining Procedure 2.14.
Round t (t even, $t \geq 0$):
First player 1 proposes the contract $c^t \in C$ and, then, player 2 either accepts ('yes') or rejects ('no') the proposed c^t. If player 2 accepts, then the negoti-

ations end and the players implement the contract c^t yielding a utility $u_i(c)$, $i = 1, 2$, in round t. However, if player 2 rejects, then the players play the disagreement game Γ that determines the disagreement outcome $d(a^t)$, $a^t \in A$, in the current round t. Then the bargaining process continues to round $t + 1$.

Round t (t odd, $t \geq 1$):
The procedure is similar except that now player 2 has the initiative to propose a partition $c^t \in C$ and, afterwards, player 1 responds.

Consider the following remark.

Remark 2.15. *Other applications start with the bargaining problem in the utility representation of section 2.2.8. Then $s^t \in S$, $t \in \mathbb{N}$, denotes the proposal made at round t of bargaining procedure 2.14.*

2.3.2 The players' information

Bargaining procedure 2.14 is not exhaustive in formalizing all the informational aspects of the negotiations and, therefore, it is at this stage not clear which part of the past moves is observed by each player and which part is not. Assumption 2.13 implies that the players have complete information and that there is no asymmetry in information at the start of the negotiations. During the negotiations asymmetries in information might arise in either of two ways: Players are forgetful about the past and players do not perfectly observe the past. With respect to forgetful players we can be brief. A common assumption in game theory is that the players have *perfect recall*. Loosely stated, perfect recall means that no player ever forgets what he once knew, and each player correctly remembers all his moves made in the past. We also impose perfect recall. Then player i, $i = 1, 2$, can distinguish between any two histories that differ with respect to player i's own disagreement actions independent of the opponent's moves.

What does each player observe from past moves made by his opponent? First, it is very natural to assume that the communication process does not distort proposals nor the response to it. Combined with perfect recall this already means that each player can distinguish between any two histories involving differences in one of the proposals made, differences in the response to proposals and differences in one own's choice in the disagreement games Γ played thus far. Then the remaining issue to be dealt with is what does each player infer at round $t+1$ or later from his opponent's disagreement action in Γ at round t. We make the following simplifying assumption, which says that each player also perfectly observes the past actions of his opponent in the disagreement game Γ.

Assumption 2.16. *Each player has perfect recall and perfectly observes all past moves, with the understanding that the actions $a^t \in A$, $t \geq 0$, in the* *disagreement game Γ at round t are observed after this normal-form game has been played and before the start of round $t+1$.*

Remark 2.17. *For a good understanding it should be noted that assumption 2.16 rules out that at any round t, $t \geq 0$, player i, $i = 1, 2$, already knows his opponent's choice $a_j^t \in A_j$, $j = 1, 2$ and $j \neq i$, in the normal-form game Γ when player i is called upon to move in Γ at round t.*

How restrictive this assumption is heavily depends upon the disagreement game in question. For a large class of games player i, $i = 1, 2$, can *reconstruct* a_j^t, $j = 1, 2$ and $j \neq i$, from the disagreement utility $d_i(a^t)$ and his own action a_i^t. For example, this is the case for every pure action in the prisoners' dilemma and, more general, the pure actions in a bimatrix game in which each number of player i's payoff matrix appears only once in the entire payoff matrix for both $i = 1, 2$. For this class of disagreement games it is as if each player perfectly observes the opponent's sequence of disagreement actions chosen in the past and we regard assumption 2.16 to be superfluous in this case.

Nevertheless, the class of disagreement games in which a player cannot reconstruct the opponent's past choices is also quite large and contains many interesting cases. For example, consider a price-setting duopoly with heterogenous products and a uncertain market size (or demand) such that the realized market size cannot be observed by the firms. Then a low demand can mean one of two things: The opponent did lower its price or a small market size occurred. Nevertheless, to our defence, assumption 2.16 is common in the early literature on repeated games, see e.g., Benoît and Krishna (1993) and Fudenberg and Maskin (1986) for more details, and it is a natural starting point for analyzing bargaining procedure 2.14. The reader should consult Busch and Wen (1995) for a discussion if assumption 2.16 is dropped.

2.3.3 Information sets and strategies

Strategies in any game in extensive form are defined as a complete plan for each player how to play the game. This plan specifies for each player a feasible or admitted move at each of the *information sets* he is supposed to make a move. In order to introduce strategies we first define the notion of a history and, then, relate each history to one of the information sets and vice versa. Finally, we introduce the players' strategies and these strategies are contingent upon the history of the bargaining game.

Following the notation in Busch and Wen (1995), we can specify three types of histories in the game. The first type is the history at the beginning of each bargaining round, say t, which consists of all the rejected proposals $c^0, \ldots, c^{t-1}$ made in all of the previous rounds and the disagreement actions $a^0, \ldots, a^{t-1}$ chosen in the previous rounds. (For $t = 0$ there simply are no past proposals.) The second type is the history at each bargaining round after a new partition has been proposed in this round. The third type refers to the history at each bargaining round after a proposed contract has been rejected. So, define

$$\begin{aligned} H_1(t) &= C^{t-1} \times A^{t-1}, && \text{(for } t \geq 1\text{)}, \\ H_2(t) &= H_1(t) \times C, && \text{(for } t \geq 0\text{)}, \\ H_3(t) &= H_2(t) \times \{no\}, && \text{(for } t \geq 0\text{)}. \end{aligned}$$

Note that $H_1(0) = \emptyset$ is the set of 'null histories' preceding round 0 with the understanding that nothing has happened before round 0. The set H of all possible histories can be represented by taking all appropriate infinite unions over bargaining rounds t of the sets of all possible histories up to round t. That is, for $j = 1, 2, 3$, and $t \in \mathbb{N}$, define

$$H(t) = H_1(t) \cup H_2(t) \cup H_3(t), \quad \text{and} \quad H = \bigcup_{t=0}^{\infty} H(t).$$

So, a history $h \in H$ is a sequence of moves in the bargaining game and belongs to one of the three types of histories. Note that we do not specify histories that have acceptances, since the bargaining game ends as soon as one of the players has accepted.

Each $h(t)$ can also be regarded as a unique path through the extensive form to the current decision node in some of the information sets. Furthermore, assumption 2.16 means that each player can tell apart any two histories $h(t), h'(t) \in H(t)$, $t \geq 0$. Consider the start of round $t \geq 0$ where one of the players is called upon to propose a contract. Then each player can tell apart the path $h_1(t)$ from the path $h'_1(t)$ meaning that each player can tell apart the decision node reached by the path $h_1(t)$ from the decision node reached by the path $h'_1(t)$. So, the latter two decision nodes belong to different information sets. In fact, it is impossible to have an information set at the start of round $t \geq 0$ with more than one decision node, called a *trivial* information set. Thus, each trivial information set corresponds to one history $h_1(t)$ and vice versa. This means that we can label each trivial information set (or its corresponding decision node) by its preceding history (or path) $h_1(t)$. Similar, each history $h_2(t)$ corresponds to one trivial information set in which the responding player is called upon to move and vice versa.

With respect to the disagreement game Γ things are slightly different. Consider the extensive-form representation of Γ in which player 1 moves before player 2, but player 2 is unaware of player 1's move when called upon to move.[6] Then player 1's information set at round t in the bargaining game when called upon to play Γ is also trivial and corresponds to the unique history $h_3(t)$ leading to it. However, when player 2 is called upon to move this player also knows $h_3(t)$ but does not yet know player 1's action $a_1^t \in A_1$. In order to be precise, we define the histories $h_4(t) = (h_3(t), a_1^t)$ and $h'_4(t) = (h'_3(t), a_1'^t)$. In case $a_1^t = a_1'^t$ we still have that player 2 can tell apart $h_4(t)$ from $h_4'(t)$ and the two decision nodes have to belong to separate information sets. Now, if $h_3(t) = h_3'(t)$, then player 2 cannot tell the difference between $h_4(t)$ and $h_4'(t)$ and the two decision nodes have to be part of the same information set, which is then not trivial. This information set contains the same number of elements as player 1's set of actions A_1. Moreover, $h_3(t)$ can be used to label all of player 2's information sets. Since the actions (a_1^t, a_2^t) are revealed to the players after the disagreement game Γ we then again have trivial information sets at the start of round $t+1$.

Strategies for the players are rules telling the players how to move at each information set and we have seen that each element in $H(t)$ corresponds to one information set at round t (and vice versa). So, by now, we have established that a strategy is contingent or conditional upon the observed history $h(t)$ of the game. Formally, a strategy σ_i for player i in the bargaining game is a function that assigns to each history an appropriate action from the relevant set. So, for $i = 1, 2$,

$$\sigma_i : H \to A_i \cup C \cup \{yes, no, wait\}$$

where 'wait' indicates that player i cannot take an action at player j's information set but just has to wait. In particular, for player 1 and t even, $t \in \mathbb{N}$, we have

$$\sigma_1(h_1(t)) \in C, \quad \sigma_1(h_2(t)) \in \{wait\}, \quad \sigma_1(h_3(t)) \in A_1,$$

and for t odd, $t \in \mathbb{N}$,

$$\sigma_1(h_1(t)) \in \{wait\}, \quad \sigma_1(h_2(t)) \in \{yes, no\}, \quad \sigma_1(h_3(t)) \in A_1,$$

for any $h_k(t) \in H_k(t)$, $k = 1, 2, 3$. Analogously, this is specified for player 2.

These strategies are defined in terms of strategies in a *multi-stage game* with *observable actions*, see e.g., Fudenberg and Tirole (1991) or Osborne and Rubinstein (1994). A multi-stage game is a game consisting of (possibly infinite) stages and at each stage all the players move simultaneously and independently of each other similar as in a normal-form game. The artificial construct of the action 'wait' trivially transforms the bargaining game into a multi-stage game and assumption 2.16 ensures that the actions are indeed observable. The reader should note that each round t, $t \geq 0$, in the bargaining game consists of three separate stages in the multi-stage game, namely one stage in which a proposal is made, one stage in with the response to it and one stage where the disagreement game is played.

For completeness we mention that multi-stage games with observable actions are said to have *almost complete information* even though these games have imperfect information meaning that not all the information sets are trivial.

Almost complete information refers to the fact that at the start of each stage all players have complete knowledge about the past play and this is the maximum of information each player can obtain in such games if all moves are taken simultaneously.

Remark 2.18. *In what follows we do no longer specify whenever a player has to choose the trivial action 'wait'.*

2.3.4 Outcomes and utilities

Any strategy profile $\sigma = (\sigma_1, \sigma_2)$ uniquely determines an outcome of bargaining procedure 2.14 and each outcome yields the players a certain payoff or utility. The payoffs in an ongoing relation are streams of per-period utilities. For example, in wage bargaining the daily wage and daily profits are streams of per-period payoffs. We assume that the players maximize the *normalized discounted sum of utilities*.

First, an outcome of bargaining procedure 2.14 is simply a (possibly infinite) path through the extensive form induced by the strategy profile $\sigma = (\sigma_1, \sigma_2)$. The length of this path is measured by the bargaining round $t(\sigma) \in \mathbb{N}$ in which the negotiations are concluded, with the convention that $t(\sigma)$ is infinite if the strategy profile σ rules out that an agreement is reached meaning that the negotiations continue forever. Furthermore, the contract $c^t(\sigma) \in C, t \leq t(\sigma)$, denotes the proposed contract at the t-th round of bargaining if both players follow σ and, similar, the disagreement actions $a^t(\sigma) \in A, t \leq t(\sigma) - 1$, denote the disagreement actions in Γ before the agreement is concluded in round $t(\sigma)$. Then the outcome or path through the extensive form induced by σ is given by

$$\left\langle c^0(\sigma), no, a^0(\sigma), \ldots, c^{t(\sigma)-1}(\sigma), no, a^{t(\sigma)-1}(\sigma), c^{t(\sigma)}(\sigma), yes \right\rangle .$$

Note that the path corresponds to one history $h_2(\sigma) \in H_2(t(\sigma))$ followed by the first and final 'yes'. So, it is without loss of generality to use $h_2(\sigma) \in H_2(t(\sigma))$ to represent the outcome of the bargaining.

We assume that the players' overall utilities only depend upon the stream of disagreement actions and the final contract reached but not depend upon the sequence of proposed and rejected contracts.[7] Then the *(payoff-relevant) outcome path* induced by strategy profile σ, denoted as $p(\sigma)$, consists of the sequence of all disagreement actions $a^t(\sigma)$ before round $t(\sigma)$ and the agreement $c^{t(\sigma)}(\sigma)$ reached in round $t(\sigma)$. We write

$$p(\sigma) = \left\langle a^0(\sigma), a^1(\sigma), \ldots, a^{t(\sigma)-1}(\sigma), c^{t(\sigma)}(\sigma) \right\rangle .$$

As mentioned, the players maximize the *normalized discounted sum of utilities*. Let δ represents player i's per-period constant discount factor with $\delta \in [0, 1)$. So, we assume a common discount factor. Then, given σ and the corresponding outcome path $p(\sigma)$, player i's overall utility corresponds to the normalized discounted sum of utilities given by

$$U_i(p(\sigma);\delta) = \delta^{t(\sigma)} u_i\left(c^{t(\sigma)}(\sigma)\right) + (1-\delta)\sum_{t=0}^{t(\sigma)-1} \delta^t d_i\left(a^t(\sigma)\right), \qquad (2.7)$$

for $i = 1, 2$. The term $1-\delta$ in the first term represents the normalization factor, which enables us to compare streams of payoffs with one-shot payoffs.

There are two special cases. First, immediate agreement. Then the players agree at round $t(\sigma) = 0$ upon the contract $c^0 \in C$ and the overall utility is simply $U_i(p(\sigma);\delta) = u_i(c^0)$. Second, perpetual disagreement meaning the players fail to reach an agreement. Then $t(\sigma) = \infty$ and the corresponding outcome path $p(\sigma)$ yields utilities

$$U_i(p(\sigma);\delta) = (1-\delta)\sum_{t=0}^{\infty} \delta^t d_i\left(a^t(\sigma)\right), \quad i = 1, 2.$$

Similar as for the bargaining problem (S, d) we can define Pareto efficiency of paths. The path $p(\sigma)$ is *Pareto efficient* if there does not exists a path $p'(\sigma)$ such that $U_i(p'(\sigma);\delta) > U_i(p(\sigma);\delta)$, $i = 1, 2$. There are two possibilities. Either $t(\sigma) = 0$ and then the path $p(\sigma)$ can only be Pareto efficient if $c^0(\sigma)$ belongs to C^{PE}. Or $t(\sigma) > 0$ and not only $c^{t(\sigma)}(\sigma) \in C^{PE}$ but also there must exist a pair of actions $a \in A$ such that $a^t(\sigma) = a$ for all t and $d_i(a) = u_i\left(c^{t(\sigma)}(\sigma)\right)$, $i = 1, 2$, because of the convexity of the utility space. If the game Γ is such that no such pair of actions a exists, then only the first case can arise.

Finally, it is often necessary to evaluate the present value of a stream of payoffs (in terms if utilities) associated with the outcome path $p(\sigma)$ having arrived at round $k \leq t(\sigma)$. Then, likewise, given outcome path $p(\sigma)$, to denote each participant's discounted payoff evaluated at round k, $0 \leq k \leq t(\sigma)$, we write

$$U_i^k(p(\sigma);\delta) = (1-\delta)\sum_{t=k}^{t(\sigma)-1} \delta^{t-k} d_i(a^t(\sigma)) + \delta^{t(\sigma)-k} u_i\left(c^{t(\sigma)}(\sigma)\right) \qquad (2.8)$$

with $i = 1, 2$. We also use the notation $U_i(\sigma;\delta)$ instead of $U_i^k(p(\sigma);\delta)$, $i = 1, 2$, if k is even and if this does not lead to confusion. Sometimes, we will leave out the discount factor δ in order to economize on the notation. That is, we will use $U_i(\sigma)$ instead of $U_i^k(\sigma;\delta)$.

2.3.5 Reinterpretation as expected utilities

The normalized discounted payoffs in (2.7) can also be reinterpreted in terms of expected utility functions. This can be seen as follows. For every $c^{t(\sigma)}(\sigma) \in$

C we may write

$$u_i(c^{t(\sigma)}(\sigma)) = (1-\delta)\sum_{t=0}^{\infty}\delta^t u_i(c^{t(\sigma)}(\sigma)),$$

which simply means that $u_i(c^{t(\sigma)}(\sigma))$ can be regarded as an infinite stream of payoffs corresponding to an everlasting contract that specifies a constant per-period utility of $u_i(c^{t(\sigma)}(\sigma))$. Then each path $p(\sigma)$ can be written as the infinite stream of payoffs given by

$$\left\langle a^0(\sigma), a^1(\sigma), \ldots, a^{t(\sigma)-1}(\sigma), c^{t(\sigma)}(\sigma), c^{t(\sigma)}(\sigma), \ldots \right\rangle$$

and (2.7) is equivalent to

$$U_i(p(\sigma);\delta) = (1-\delta)\left[\sum_{t=0}^{t(\sigma)-1}\delta^t d_i\left(a^t(\sigma)\right) + \sum_{t=t(\sigma)}^{\infty}\delta^t u_i\left(c^{t(\sigma)}(\sigma)\right)\right]$$

The interpretation of expected utility kicks in if we assume that at the end of each bargaining round t, $t \in \mathbb{N}$, there is uncertainty whether or not the bargaining game ends or continues to the next round $t+1$, $t \in \mathbb{N}$. This uncertainty is modeled by a geometric stochastic process that is history independent. At bargaining round t, $t \in \mathbb{N}$, the probability of a next bargaining round $t+1$ is δ, whereas $1-\delta$ is the probability that the negotiations end at round t once and for all. Then the negotiations terminate at round t, $t \in \mathbb{N}$, with probability $(1-\delta)\delta^t > 0$ and summing up over all rounds $t \in \mathbb{N}$ yields a total probability mass of 1. Then $U_i(p(\sigma);\delta)$ simply consists of an infinite sum of probabilities times per-period utilities, which is an expected utility function associated with the infinite path $p(\sigma)$. So, the probability distribution over the infinite sequences of per-period payoffs is fixed and differences in expected utilities are due to differences between the 'prizes' $p(\sigma)$ and $p(\sigma')$.[8] The interpretation of δ in terms of uncertainty has the additional advantage that the expected number of bargaining rounds is finite, i.e., $(1-\delta)\sum_{t=0}^{\infty}\delta^t \cdot t < \infty$. So, the players know that the negotiations will not last forever but they are uncertain about the exact date the negotiations terminate.

The interpretation of (2.7) in terms of an expected utility function is regarded as a methodological superior interpretation, see e.g., Binmore, Rubinstein, and Wolinsky (1986) and Osborne and Rubinstein (1990). The reason is that the leading theoretical literature regards C as a simplex of lotteries and u_i, $i = 1, 2$, representing expected utility preferences over lotteries. Within this framework, adding uncertainty in the negotiation process means that the overall utility functions U_i, $i = 1, 2$, still belong to the realm of expected utility functions, whereas the interpretation of δ being a discount factor assumes a mixture of uncertainty in C and time preferences over outcomes in the negotiation process. Nevertheless, both interpretations are mathematically equivalent when streams of payoffs are considered.

Finally, for the bargaining problem in utility representation (S, d, A) we write the outcome path $p(\sigma)$ as

$$p(\sigma) = \left\langle a^0(\sigma), a^1(\sigma), \ldots, a^{t(\sigma)-1}(\sigma), s^{t(\sigma)}(\sigma) \right\rangle$$

and write (2.7) as

$$U_i(p(\sigma); \delta) = (1-\delta) \sum_{t=0}^{t(\sigma)-1} \delta^t d_i\left(a^t(\sigma)\right) + \delta^{t(\sigma)} s_i^{t(\sigma)}(\sigma).$$

2.3.6 An appropriate equilibrium concept

Game theory has established that the concept of Nash equilibrium has not much bite in games in extensive form. The notion of a Nash equilibrium only checks whether strategies are robust against profitable deviations by each of the players along the equilibrium path and it does not consider the profitability of deviations off the equilibrium path. This may cause the players to rely on so called 'empty threats' in a Nash equilibrium, threats that will not be carried out when actually called upon to do so. Nash equilibria that are supported by empty threats are ruled out by applying the concept of *subgame perfect equilibrium* (SPE). Only threats that are credible, in the sense that these are carried out if required, survive. It is standard in the literature on games in extensive form with (almost) perfect information to refine the Nash equilibrium solution concept and to apply the notion of subgame perfect equilibrium.

Basically, a subgame is a game on itself. The origin of a subgame always is a trivial information set and the subgame contains all moves and decision nodes that can be reached from its origin provided that all the decision nodes in each of these information sets can only be reached starting from its origin. Any game in extensive form consists of a nonempty collection of subgames, since the game itself trivially is a subgame.

The subgames in bargaining procedure 2.14 are easily identified. Every bargaining round t, $t \in \mathbb{N}$, consists of three sorts of subgames: the subgame that starts when one of the players proposes a partition; the subgame that starts when one of the players responds to a proposal and the subgame that starts after a rejection inducing the players to play the disagreement game Γ. Since the origin of every subgame is a trivial information set and a subgame starts at every trivial subgame we label subgames with the history $h \in H$ preceding it. Let $G(\Gamma|h)$, $h \in H$, denote the subgame that follows history h. The whole bargaining game, which can be regarded as the subgame following the null-history $h_1(0) = \emptyset$, is denoted as $G(\Gamma)$.

Subgame perfectness requires that the prescribed strategy profile σ constitutes a Nash equilibrium in every subgame, even the subgames that will not be reached by the strategies. Given the i-th player's strategy σ_i, denote by $\sigma_i|_h$ the strategy that σ_i induces in the subgame $G(\Gamma|h;\delta)$ and let $U_i^h(\sigma|_h)$ denote player i's (expected) payoff conditional on the history h being reached. The following definition formalizes the idea of a subgame perfect equilibrium.

Definition 2.19. *A subgame perfect equilibrium of the bargaining game in extensive form $G(\Gamma)$ is a strategy profile σ^* such that for every player $i \in N$ and every history $h \in H$ after which player i moves, we have*

$$U_i^h(\sigma_i^*|_h, \sigma_{-i}^*|_h) \geq U_i^h(\sigma_i|_h, \sigma_{-i}^*|_h)$$

for every strategy σ_i of player i in the subgame $G(\Gamma|h;\delta)$.

To verify whether a strategy profile is a subgame perfect equilibrium we apply a strong result that is known as the *one-stage-deviation principle*, see e.g., Fudenberg and Tirole (1991) for infinite-horizon multi-stage game with observed actions and discounting and Hendon, Jacobsen, and Sloth (1996) for the general case with imperfect information. The strategy σ_i', $i = 1, 2$, is said to be a one-stage deviation from the strategy σ_i if the strategy σ_i' coincides with σ for all $h \in H$ except one particular history $h' \in H$ after which σ_i' prescribes a different move than σ_i.

Definition 2.20. *Consider the strategy σ_i, $i = 1, 2$. Strategy σ_i', $i = 1, 2$, is a one-stage deviation from strategy σ_i if there exists a $h' \in H$ such that $\sigma_i'(h) = \sigma_i(h)$ for all $h \in H \backslash \{h'\}$ and $\sigma_i'(h') \neq \sigma_i(h')$.*

One-stage deviations in bargaining procedure 2.14 are easily identified, because $h \in H$ determines $t \in \mathbb{N}$ and one of three sorts of histories. Fix $t \in \mathbb{N}$. The first type has σ_i and σ_i' only differ if some $h_1'(t) \in H_1(t)$ is reached and, then, the strategies prescribe a different contract to propose. Note that both strategies prescribe the same moves elsewhere in the extensive form. Similar, the second type corresponds to one $h_2'(t) \in H_2(t)$ and σ_i' prescribes the opposite of σ_i at $h_2'(t)$, because of the binary decision '*yes*' or '*no*'. The third type specifies how to play Γ and the strategies σ_i and σ_i' only differ if $h_3'(t) \in H_3(t)$ was the path leading to playing the game Γ.

In particular, in order to verify that σ^* is a subgame perfect equilibrium of our game $G(\Gamma)$, it suffices to check whether there are any histories $h \in H$ where some player $i \in N$ can gain by deviating from the moves prescribed by σ_i^* at the subgame reached by history h and conforming to σ_i^* thereafter. The proof of this result is essentially the principle of dynamic (discounted) programming, based on backward induction. For a proof of this result, the reader is referred to Fudenberg and Tirole (1991), Osborne and Rubinstein (1994) or Hendon, Jacobsen, and Sloth (1996).

Theorem 2.21. *(one-stage-deviation principle)*
The strategy profile σ^* is a subgame perfect equilibrium of the game $G(\Gamma)$ if and only if for every player i, $i = 1, 2$, and every history $h \in H$ after which player i moves, we have

$$U_i^h\left(\sigma_i^*|_h, \sigma_{-i}^*|_h\right) \geq U_i^h\left(\sigma_i|_h, \sigma_{-i}^*|_h\right)$$

for every one-stage deviation σ_i from σ_i^* such that $\sigma_i(h) \neq \sigma_i^*(h)$ in the subgame $G(\Gamma|h)$.

Finally, in many applications the set of subgame perfect equilibrium strategies is quite large and many of the equilibrium strategies induce the same pair of equilibrium utilities. For that reason we are often interested in the set of subgame perfect equilibrium utility pairs given a particular value of δ. Since definition 2.19 is formulated in terms of a weak inequality it automatically holds that the set of equilibrium utilities is a closed set. Given δ, this set is defined by

$$E(\delta) = \{s \in S \mid (s_1, s_2) = (U_1(p(\sigma^*); \delta), U_2(p(\sigma^*); \delta)), \sigma^* \text{ is SPE}\}. \tag{2.9}$$

2.3.7 Limit sets of equilibria

Bargaining procedure 2.14 is defined as a sequence of bargaining rounds where each round lasts $\Delta > 0$ measured in real time. The presence of Δ causes a small friction in the bargaining process, because whenever the responding player rejects the last proposal this player has to wait Δ of time before he is allowed to make his counter proposal and delay is costly since players discount payoffs. Alternatively, the responding player faces a risk of breakdown in rejecting the current proposal. Taking the limit Δ goes to 0 then means that we let this friction vanish and, in what follows, we frequently do so.

In order to investigate limit properties, it is convenient to write $\delta = e^{-\theta\Delta}$, $i = 1, 2$, where $\theta > 0$ represents player i's (subjective) discount rate. Then Δ goes to 0 seems equivalent with δ goes to 1. However, sometimes we are interested in equilibria that feature delay and then $k \geq 1$ rounds of delay with a length of $k \cdot \delta$ in real time vanish in the limit. More interesting is whether or not a delay of length $\ell > 0$ in real time can be sustained as an equilibrium outcome and this is accomplished by having the k rounds of delay depend upon Δ as well. That is, we take some $\ell > 0$ and define the number of rounds of delay associated with ℓ as $k(\Delta) = \ell/\Delta$. Obviously, $k(\Delta)$ goes to infinity as Δ goes to 0. This means that taking the limit Δ goes to 0 involves something more than just taking the limit δ goes to 1, but in some cases both are equivalent and then we simply take δ goes to 1. Since we are often interested in sets of equilibrium payoffs as defined in (2.9) we investigate somewhat loosely stated

$$E = \lim_{\delta \to 1} E(\delta) = \lim_{\Delta \to 0} E(e^{-\theta\Delta}),$$

where $E \subseteq S$ denotes the limit set of SPE utility pairs.

The limit sets E derived in this book belong to one of the following two classes of sets: Either E consists of a single element or it is a connected subset in S^{IR} with positive measure. In deriving the latter type of limit sets it is often relatively easy to show that every utility pair in the interior of this limit indeed belongs to the limit set, whereas for utility pairs on the boundary a lot of additional arguments and several distinctive cases are necessary to complete the formal proof. Following the literature we will not do the latter. Instead, we adopt the following point of view. Whether payoffs on the boundary of the limit set E can actually be supported by subgame perfect equilibrium strategy profiles will generally depend on the parameters of the bargaining problem. In order to avoid too much technicalities we concentrate on the *closure* of the set of SPE payoffs, i.e., the smallest closed set that contains all SPE payoffs. Formally, $E \subseteq Cl\{E\}$, where $Cl\{\cdot\}$ stands for the closure of the set $\{\cdot\}$. Note that the measure of the set $Cl\{E\}\backslash E$ is equal to 0 and not much is lost. Then it suffices to show that every interior utility pair of E belongs to this set.

Finally, whenever the limit Δ goes to 0 is taken, we do not only consider specific sequences of lotteries, i.e., the probability of a breakdown at round t, $t \in \mathbb{N}$, is $(1-\delta)\delta^t$, but also in the set of prizes the path $p(\sigma(e^{-\theta\Delta}))$ depends upon Δ through the strategy profile σ.

2.3.8 Markov strategies

Strategies can be of a complexity that is beyond one's imagination but equilibrium analysis often yields rather simple strategies. One such popular class of strategies are called *stationary* or *Markov*. These strategies are loosely speaking *history-independent*, although we should take into account that the responding player has to either accept or reject the current proposal c^t associated with the history $h_2(t) = (h_1(t), c^t) \in H_2(t)$ and in that case history independent naturally means that the 'yes' or 'no' decision following $h_2(t)$ is independent of $h_1(t)$ but obviously not of c^t. Then a history-independent strategy prescribes the somewhat naive behaviour in which a player does not condition his behaviour upon past play but, instead, whenever this player faces a situation that is 'similar to' a situation faced earlier this player will act in exactly the same manner as before. Or, loosely speaking, each player chooses the same move for all t and for all $h_k(t) \in H_k(t)$, $k = 1, 2, 3$. For example, whenever player 1 has the opportunity to propose this player simply repeats his proposal made at round 0 forever after irrespective of the number of times this proposals was rejected.

Since each player's 'yes' or 'no' decision is binary this allows us to partition the set of proposals C into two subsets for each player: one subset of contracts considered as being acceptable and one subset of contracts that are definitely not. Formally, for player i, $i = 1, 2$, the (possibly empty) set $C_i \subseteq C$ is the set of (proposed) contracts player i's response is 'yes' with the understanding that all contracts proposed from the complementary set $C\backslash C_i$ are followed by 'no'. In terms of strategies, whenever player i, $i = 1,2$, has to respond to an offer $c^t \in C$ of its opponent, its strategy can be rewritten as

$$\sigma_i\left(h_2(t)\right) = \begin{cases} yes, & \text{if } c^t \in C_i, \text{ for all } h_2(t) = (h_1(t), c^t) \in H_2(t), \\ no, & \text{if } c^t \in C\backslash C_i, \text{ for all } h_2(t) = (h_1(t), c^t) \in H_2(t). \end{cases}$$

In what follows we economize on notation and we always drop the second line from the description of a strategy. Furthermore, we also drop lines such as 'for all $h_k(t) \in H_k(t)$', for $k = 1, 2, 3$.

A stationary or Markov strategy is defined as follows.

Definition 2.22. *Let $c^i \in C$, $i = 1,2$, $C_i \subseteq C$ and $a_i \in A_i$. At round $t \in \mathbb{N}$ player i's Markov strategy in bargaining procedure 2.14 is given by*

$$\sigma_i\left(h_t\right) = \begin{cases} \textit{proposes} & c^i, \\ \textit{responds yes} & \text{if } c^t \in C_i, \\ \textit{plays} & a_i, \end{cases}$$

with the understanding that i) player i proposes the contract c^i for all $t \in \mathbb{N}$ and $h_1(t) \in H_1(t)$ where i is the proposing player, ii) player i responds 'yes' to c^t for all $t \in \mathbb{N}$ and $h_2(t) = (h_1(t), c^t) \in H_2(t)$ where i is the responding player if and only if $c^t \in C_i$ and iii) player i chooses disagreement action a_i in Γ for all $t \in \mathbb{N}$ and $h_3(t) \in H_3(t)$.

From this definition it is easy to see that player i's Markov strategy, $i = 1,2$, is fully specified by the proposal c^i, the set of acceptable proposals C_i^* and the disagreement action a_i.

Finally, we define a *Markov perfect equilibrium* as a profile of Markov strategies that induce a Nash equilibrium in every subgame of play. Hence, an Markov perfect equilibrium is also a subgame perfect equilibrium.

Definition 2.23. *A Markov perfect equilibrium (MPE) is a subgame perfect equilibrium in Markov strategies.*

2.3.9 Strategies represented by tables

Strategies are plans of actions or rules that prescribe a move for every history $h \in H$. The strategies needed in the equilibrium analysis have a relatively simple structure in which the players' moves prescribed by such strategies are solely conditional upon some finite set of publicly known variables, called *states*, that adequately summarize each history in the set H. Throughout this book it suffices to consider at most four states and because of the limited number of states it is convenient to represent strategies by tables. In this section we explain how to read such tables.

The key idea is that the strategies under consideration allow us to partition the set of histories H into a finite collection of subclasses, that within each subclass the prescribed moves are the same and that each subclass can be identified with a state. Let us first consider the trivial example of the Markov strategies stated in definition 2.22. For Markov strategies the set of histories H can be trivially partitioned into a collection that contains only one subclass, namely H itself. So, we only need one state for the 'subclass' H, which we will call M. Since $h_1(0) = \emptyset \in H$ it is clear that the state at round 0, called the *initial* state, is the state M. Since every possible history $h(t)$ also belongs to H it is clear that the state remains the state M for ever. Then it is said that the state M is *absorbing*. We now prescribe the Markov strategy for player i, $i = 1, 2$, as follows: In the state M player i always proposes the contract c^i, player i always accepts the proposal c^t at round t if $c^t \in C_i$ and player i always chooses the disagreement action $a_i \in A_i$. The following table summarizes these Markov strategies and the moves in state M can be found in the column named M.

State		M	
Player 1	proposes	c^1	always
	responds yes		if $c(t) \in C_1$
	plays	a_1	always
Player 2	proposes	c^2	always
	responds yes		if $c(t) \in C_2$
	plays	a_2	always
Transitions		absorbing	

In general, the set of histories is partitioned into more than one subclass and, therefore, there are two or more states. To illustrate such strategies, consider the following strategies. As the negotiations start, player i, $i = 1, 2$, proposes his (generically unique) most preferred contract $c^{i*} \in C_i^*$, is not willing to settle for less than his most preferred contract and alternates between $a_i^* \in A_i$ and $a_i^{**} \in A_i$ in, respectively, even and odd numbered rounds. If one of the players only once proposes a contract that differs from his most preferred contract, then immediately different behaviour is triggered. Then both players propose the same contract $c \in C$ and are only willing to accept this particular contract and no other contract. As long as they do not agree, player i's disagreement action is $a_i \in A_i$, $i = 1, 2$. (Note that the associated table is printed on the next page.)

This is made formal as follows. Suppose that H is partitioned into the two nonempty subclasses of histories H^A and H^B, where $H_k^A(t)$, $t \in \mathbb{N}$, $k = 1, 2, 3$, consists of all histories in which both players proposed their most preferred contract up to round t independent of the actions in Γ. So, $H^B = H \backslash \Pi_{t=0}^{\infty} H_1^A(t) \times H_2^A(t) \times H_3^A(t)$. We label the two states as A and B and we impose that $h_1(0) \in H^A$. So, the initial state is A. As the negotiations proceed, the state evolves between the two states and its progression is governed by a rule called the *state transition*. Since $H^B \neq \emptyset$, there must be some sequence of events that makes the state move from A to B. In fact, such event occurs if one of the players does not propose his most preferred contract. If the state becomes B and remains B forever after then the state B is absorbing.

State		A		B	
Player 1	proposes	c^{1*}		c	
	responds *yes*		if $c(t) \in C_1^*$		if $c(t) = c$
	plays	a_1^*	t odd	a_1	t odd
	plays	a_1^{**}	t even	a_1	t even
Player 2	proposes	c^{2*}		c	
	responds *yes*		if $c(t) \in C_2^*$		if $c(t) = c$
	plays	a_2^*	t odd	a_2	t odd
	plays	a_2^{**}	t even	a_2	t even
Transitions		go to B: If $c(t) \notin C_1^*, t$ even If $c(t) \notin C_2^*, t$ odd		absorbing	

2.4. Related Literature

Bargaining theory as it is known today was first proposed in Nash (1950) and Nash (1953). In these seminal articles the bargaining problem is introduced in the utility representation of section 2.2.8 although it is clearly stated that utilities are regarded to represent (von Neumann-Morgenstern) expected utilities over risky agreements. The definition of a bargaining problem in terms of contracts is a straightforward extension of the definition in Osborne and Rubinstein (1990), where the latter assumes a fixed disagreement outcome. In Roemer (1988) the bargaining problem is stated in terms of an economic environment, where the latter is defined as a two-person exchange or barter economy with production and an arbitrary number of goods that naturally has a fixed disagreement point.

For the translation of the bargaining problem in terms of contracts to its utility representation in section 2.2.8 we refer to Roemer (1988). The latter reference is the notable exception in the literature where sufficient conditions on utility functions and the contract space are introduced such that the set of utility pairs is guaranteed to be a convex set. Next, our interpretation of expected utility functions as a special case of linear utility functions on a simplex is also unusual, but in our view it nicely illustrates that the mathematical framework of the theoretical literature and the more applied economics literature are basically the same. The problem of generic uniqueness of Pareto efficient contracts then naturally arises, which has never been addressed in the literature as far as we know.

Bargaining procedure 2.14 originated in a series of articles. Within the context of wage bargaining it was first introduced in Haller (1991) and further analyzed in e.g., Bolt (1995), Fernandez and Glazer (1991), Haller and Holden (1990) and Houba and Bolt (2000). The disagreement game Γ in these wage bargaining models is degenerated in the sense that the union can choose between several forms of industrial action whereas the firm is condemned to 'continue'. The generalization in which both the union and the firm have nontrivial sets of disagreement actions in Γ can be found in Busch and Wen (1995). All the mentioned models impose a linear Pareto efficient frontier for the bargaining problem in utility representation. The extension to a nonlinear Pareto frontier was first partially studied in Okada (1991a) and Okada (1991b), which appeared together with Haller (1991) in the same book, and the full analysis was conducted in Houba (1997). In Stefanski and Cichocki (1986), Houba and De Zeeuw (1995) and Houba, Sneek, and Várdy (2000) a further generalization can found in which the contract space C depends upon the past history of play in the sequence of disagreement games Γ. All these models return in later chapters.

For a thorough and fundamental treatment of game theory we may refer to, for instance, Fudenberg and Tirole (1991), Kreps (1990), Mas-Colell, Whinston, and Green (1995), Meyerson (1991) and Osborne and Rubinstein (1994). It is standard in game theory to assume expected utility, discounting of streams of payoffs and perfect recall. For an axiomatic underpinning of expected utility functions we refer to e.g., Mas-Colell, Whinston, and Green (1995). Discounted utility functions are axiomatized in e.g., Fishburn and Rubinstein (1982), Osborne and Rubinstein (1990) and Osborne and Rubinstein (1994). A recent contribution on nonexpected utilities over lotteries in the context of bargaining can be found in e.g., Hanany and Safra (2000). The familiar game theoretic equilibrium concepts need to be redefined in the context of nonexpected utilities e.g., see Gintis (2000) and Shalev (2000). For games without perfect recall we refer to the special issue on this topic in Games and Behavior (1997). For an analysis of imperfect recall in bargaining procedure 2.14 under the additional assumption of a fixed disagreement point we refer to Andrelini and Felli (2001).

The equivalence of discounting and expected utilities of section 2.3.5 only holds in the context of streams of per-period payoffs. This equivalence is lost if, for example, the players do not consume per-period payoffs during the negotiations but instead they consume only once either the final agreement at the date it is concluded or the disagreement point at the round the negotiations break down (but not before). The reader is referred to e.g., Binmore, Rubinstein, and Wolinsky (1986) and Osborne and Rubinstein (1990).

Finally, we mention that the communication process in the context of negotiations is not perfect in Ma and Manove (1993), where it is assumed that the time needed to transmit a proposal between the players is random including the risk that a proposal will never reach the other player.

Notes

1 In Nash (1950) the assumption of a convex set of lotteries is defended by assuming that there exists some randomization device $\lambda \in [0,1]$ that allows the players to agree upon so-called compound lotteries. The compound lottery $c'' = \lambda c + (1-\lambda)c'$, $c, c' \in C$, puts probability λ on the lottery $c \in C$ and probability $1-\lambda$ on the lottery $c' \in C$. Since C is a convex set, it automatically follows that $c'' \in C$. So, the presence of the randomization device λ ensures that the set C of probability vectors is a convex set.

2 Of course, in terms of equilibrium concepts one has to restrict attention to a refinement of the Nash equilibrium concept that is considered to be appropriate for the game in extensive form.

3 Similar remarks apply to the functions d_i, $i = 1,2$, if the disagreement actions $a_i \in A_i$ correspond to mixed strategies in some normal form game.

4 If, for example, u_1 is strictly concave while u_2 is not then the weighted average of the utility functions is still strictly concave for all $\omega \in (0,1]$. Only for $\omega = 0$ multiple solutions to maximization problem (2.2) may occur. Since all the latter solutions form the set C_2^* and $C_2^* \cap C^{IR} = \emptyset$ by assumption 2.9 we do not regard this particular multiplicity as a serious problem.

5 Not all books on mathematics treat 0 as a natural number. We impose the convention that $0 \in \mathbb{N}$ and that 0 is an even number.

6 The other extensive-form representation of Γ has player 2 move before player 1 and player 1 cannot observe player 2's action chosen when called upon to move. By reversing the roles of both players the arguments of the main text still apply to this case.

7 More general preferences over outcomes are not yet explored in the literature. At this point economist may learn something from psychologists. For example, more general preferences might incorporate players with aspirations or regret. One interesting situation we think of is a situation in which one of the players's initial aspiration made him reject a certain proposed contract, whereas later on he regrets not having accepted it and has to settle on a contract that is worse.

8 For a good understanding, before we had that the set of prizes X is fixed in case C represents a simplex of probability distributions over this set, whereas in the bargaining game there is one probability distribution for each path in the large set of prizes given by $\{p(\sigma)|$ for all $\sigma\}$.

I

EXOGENOUS DISAGREEMENT OUTCOMES

Chapter 3

THE ALTERNATING OFFERS PROCEDURE

3.1. Introduction

Although variable threats are an essential part of bargaining procedure 2.14 it is also legitimate to ask what the bargaining outcomes are for the simpler case where the disagreement actions cannot be influenced by the players. The latter case is referred to as exogenous, or fixed, disagreement actions. In terms of the previous chapter this case corresponds to the degenerate case in which the set of disagreement actions consists of only one element, leaving no freedom of choice to the players than to implement the one action available to them. Since the players alternate in making proposals to each other until an agreement is reached this bargaining procedure is known in the literature as alternating offers, a name that nicely covers this procedure.

There are several reasons for studying the special case of alternating offers in this chapter and also in chapters 4, 5 and 6. A more or less historical motivation would be that in the game theoretic study of bargaining John Nash started out with exogenous disagreement actions, being the 'fall-back levels' for the players in case bargaining would break down; see Nash (1950). A more compelling argument is that the results derived for the model with exogenous disagreement actions will lead to a better understanding of the results for the endogenous model introduced in later chapters, and so serves as a benchmark and allows comparison between the different models. Moreover, as will become apparent in later chapters, several of the techniques introduced in this chapter can be easily adapted to derive results in bargaining problems with endogenous disagreement actions. For the latter reason we provide a detailed derivation in this chapter in order to fully explain the technical details that are often taken for granted in later chapters.

The first step of the full analysis in this chapter consists of characterizing equilibria in which players adopt the relative simple Markov strategies. Such Markov perfect equilibria are found by employing the concept of dynamic programming, which is introduced in section 3.3.2. We show that alternating offers admits only one of such equilibria and that the proposals of this unique equilibrium correspond to a fixed point problem, which is introduced and investigated in section 3.3.1 before the characterization of Markov perfect equilibria starts. In characterizing this unique equilibrium we find it convenient to derive results for the bargaining problem in utility representation first and, then, translate these results in equilibrium contracts.

The second and final step in the analysis is to investigate the possibility of equilibria in non-Markov strategies, i.e., strategies that are not Markov strategies. The method employed was first proposed in Shaked and Sutton (1984) and later named after the authors. Application of this method yields that every subgame perfect equilibrium must be an equilibrium in Markov strategies and, given uniqueness of the Markov perfect equilibrium (MPE), the model admits a unique subgame perfect equilibrium (SPE).

Finally, this chapter is concluded with three applications, which include the well-known divide a dollar problem and a simple barter economy. But before a formal analysis can start we must first define alternating offers in the next section.

3.2. Alternating offers

An exogenous or fixed disagreement outcome reflects negotiation situations in which the players lack any control over disagreement actions. In terms of the previous chapter we assume that the set of disagreement actions in Γ simply consists of only one element, leaving no freedom of choice to the players than to implement the one action available to them.

Definition 3.1. *A bargaining problem is said to have an exogenous or fixed disagreement outcome if there exists one pair of actions* (a_1, a_2) *such that the set* $A_i = \{a_i\}$, $i = 1, 2$.

Obviously, an exogenous disagreement outcome implies that the minmax utility $m_i = d_i(a_1, a_2)$, $i = 1, 2$.

An implicit notational convention in the literature is that the point d represents the exogenous *disagreement point*. By disagreement point we mean the pair of utilities corresponding to the pair of actions (a_1, a_2). For the bargaining problem in utility representation we simply write (S, d), where $d := (m_1, m_2) = (d_1(a), d_2(a))$. The Pareto frontier of S is then described by the concave and strictly decreasing function $f_i : [d_j, s_j^{\max}] \to [d_i, s_i^{\max}]$, $i, j = 1, 2$ and $j \neq i$.

The players' strategies are rules telling each player how to move at each information set and we have seen that information sets are one-to-one related to histories. Since the players's disagreement actions in the game Γ are exogenously given there is no loss of generality by discarding these actions in our notation of strategies. Formally, for player 1 we write

$$\sigma_1(h_1(t)) \in C, \quad t \text{ even}, \quad \text{and} \quad \sigma_1(h_2(t)) \in \{yes, no\}, \quad t \text{ odd},$$

where $h_k(t) \in H_k(t)$, $k = 1, 2$. In particular, player 1's strategy is a Markov strategy of definition 2.22 if there exists a $c^1 \in C$ and a set $C_1 \subseteq C$ such that for all $h_k(t) \in H_k(t)$, $k = 1, 2$:

$$\sigma_1(h_1(t)) = c^1, \quad t \text{ even}, \quad \text{and} \quad \sigma_1(h_2(t)) = yes, \text{ if } c(t) \in C_1, \quad t \text{ odd}.$$

Player 2's (Markov) strategy σ_2 is defined analogously.

As notation for Markov strategies for bargaining problems in utility representation (S, d) we write $x \in S$ respectively $y \in S$ as player 1's and player 2's history-independent proposal and the set $T_i \subset S$, $i = 1, 2$ as player i's (possibly empty) set of proposals that he considers acceptable. Formally,

$$\sigma_1(h_1(t)) = x, \quad \text{and} \quad \sigma_1(h_2(t)) = yes, \text{ if } s(t) \in T_1$$

and

$$\sigma_2(h_1(t)) = y, \quad \text{and} \quad \sigma_2(h_2(t)) = yes, \text{ if } s(t) \in T_2.$$

Finally, we slightly strengthen assumption 2.7 by imposing $d \notin S^{PE}$. The latter simply means that (one of) the inequalities in definition 2.5 is strict. This is without loss of generality, because $d \in S^{PE}$ would make the bargaining problem trivial in the sense that $S^{IR} = \{d\}$ and rational players never agree upon a utility pair $s \in S$ outside S^{IR}. Furthermore, in trivial bargaining problems the infinite stream of disagreement outcomes d associated with perpetual disagreement is payoff equivalent to agreement upon $s = d$ and both can be sustained as subgame perfect equilibrium paths.

3.3. Markov perfect equilibrium

In this section we show that there exists a generically unique MPE and discuss the properties of such equilibrium. In order to derive our results we first analyze the bargaining problem in utility representation. For this problem we introduce a fixed point problem and, then, relate it to the MPE. Then the results for the bargaining problem in utility representation are translated into the contract space which yields the generic uniqueness of the MPE contract.

3.3.1 An important fixed point problem

In this section we introduce a fixed point problem formulated in terms of the bargaining problem in utility representation and show that this problem admits a unique fixed point. We also derive some properties of this unique fixed point. The results obtained in this section are not only essential to the analysis in this chapter, but rather lie at the heart of characterizing the set $E(\delta)$ of equilibrium utilities of bargaining procedure 2.14.

Consider the bargaining problem in utility representation (S, d). In order to define the fixed point problem it is necessary to introduce two vector functions named p and q. The vector function $p : S^{IR} \to S^{IR}$ is defined as

$$p_2(y) = (1-\delta)d_2 + \delta y_2, \quad p_1(y) = f_1(p_2(y))$$

and the vector function $q : S^{IR} \to S^{IR}$ is defined as

$$q_1(x) = (1-\delta)d_1 + \delta x_1, \quad q_2(x) = f_2(q_1(x)).$$

Note that $d, x, y \in S^{IR}$ indeed imply that $p(y)$ and $q(x)$ belong to S^{IR}. Moreover, by definition the images $p(y)$ and $q(x)$ lie on the Pareto frontier of S^{IR}, i.e., $p(y), q(x) \in S^{IR} \cap S^{PE}$.

The four-dimensional vector function $p \times q : S^{IR} \times S^{IR} \to S^{IR} \times S^{IR}$ is defined as the function $(p_1(y), p_2(y), q_1(x), q_2(x))$, where the latter is written as $(p(y), q(x))$. Similar, (x, y) is short hand notation for (x_1, x_2, y_1, y_2). The fixed points of the function $p \times q$ play a fundamental role in the analysis. The point $(x^*, y^*) \in S^{IR}$ is a *fixed point* of the function $p \times q$ if

$$(x^*, y^*) = (p(y^*), q(x^*)). \tag{3.1}$$

The following lemma states that the function $p \times q$ admits a unique fixed point. The proof of this lemma is rather simple because it fully exploits the geometry of the bargaining problem. It is first established that the proposals x^* and y^* have the same Nash product, where the *Nash product* of the vector $s \in \mathbb{R}^2$ is defined as the product $(s_1 - d_1)(s_2 - d_2)$. The Nash product can be regarded as the symmetric Cobb-Douglas 'utility' function $z_1 z_2$, where $z_i = s_i - d_i$, $i = 1, 2$. The associated 'indifference' curves of this Cobb-Douglas function are simply called Nash-product curves. Now, if x^* and y^* have the same Nash product, then both of these two points are intersections of two curves, namely the Pareto frontier of S^{IR} and some Nash product curve. See figure 3.1 for an illustration. Since the set S is convex and the Nash product is both strictly increasing and strictly quasi-concave in $s \in \mathbb{R}^2$, it follows that each Nash product curve intersects the Pareto frontier of S^{IR} at most twice. In the proof this fact is exploited to show that it is impossible that two pairs (x^*, y^*) of fixed points exist that lie on two different Nash product curves. The proof of this lemma, as all the proofs of further results, is deferred to appendix A.

Lemma 3.2. *The function* $p \times q$ *admits a unique fixed point* (x^*, y^*).

We continue by deriving some of the mathematical properties of the fixed point (x^*, y^*).

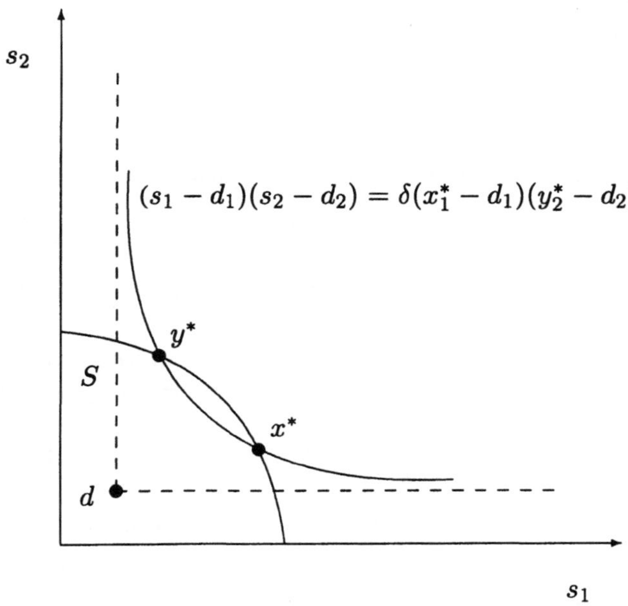

Figure 3.1. For every fixed point (x^*, y^*) of $p \times q$ the vectors x^* and y^* both lie on the Pareto frontier and on the same Nash product curve.

Lemma 3.3. *Let (x^*, y^*) be the fixed point of the function $p \times q$. Then*

i) $x^* \neq y^*$

ii) *y^* lies north-west of x^* on the Pareto frontier, i.e., $y_1^* < x_1^*$ and $y_2^* > x_2^*$.*

iii) *Player 1's components x_1^* and y_1^* are increasing in d_1 and decreasing in d_2 whereas player 2's components x_2^* and y_2^* are decreasing in d_1 and increasing in d_2.*

The first property states that two vectors are distinct and the second means that player 1 strictly prefers x^* to y^* while player 2 strictly prefers y^* to x^*. The third property of lemma 3.3 regards the fixed point as a function of the disagreement point d, which is an important parameter of the bargaining problem (S, d). If the disagreement point would shift horizontally to the right, then both vectors x^* and y^* accrue a higher utility to player 1 at the expense of player 2's utility. A vertical downward shift of the disagreement point has a similar effect for player 1 and 2. So, if player 1 would be able to shift the exogenous disagreement point this player would move it in the south-east direction.

The following examples illustrate how to compute the fixed point.

Example 3.4. *Consider* (S,d) *with* $S = \{s \in \mathbb{R}^2_+ \mid s_1 + s_2 \le 1\}$, $d \ge 0$ *and* $d_1 + d_2 < 1$. *Then (3.1) with* $f_1(s_2) = 1 - s_2$ *and* $f_2(s_1) = 1 - s_1$ *imply that the fixed point solves*

$$x_2 = p_2(y) = (1-\delta)d_2 + \delta y_2, \quad x_1 = p_1(y) = 1 - (1-\delta)d_2 - \delta y_2,$$
$$y_1 = q_1(x) = (1-\delta)d_1 + \delta x_1, \quad y_2 = q_2(x) = 1 - (1-\delta)d_1 - \delta x_1.$$

Solving for the four unknowns x_1, x_2, y_1 *and* y_2 *yields*

$$x^* = \left(d_1 + \tfrac{1}{1+\delta}(1-d_1-d_2),\ d_2 + \tfrac{\delta}{1+\delta}(1-d_1-d_2)\right),$$
$$y^* = \left(d_1 + \tfrac{\delta}{1+\delta}(1-d_1-d_2),\ d_2 + \tfrac{1}{1+\delta}(1-d_1-d_2)\right).$$

If we define the fraction $\gamma = \frac{1}{1+\delta}$ *and the net gains from trade, or the* effective surplus, *as* $s(d) = 1 - d_1 - d_2 > 0$, *then we can rewrite the fixed point as*

$$x^* = d + \gamma \cdot s(d) \begin{bmatrix} 1 \\ \delta \end{bmatrix}, \quad y^* = d + \gamma \cdot s(d) \begin{bmatrix} \delta \\ 1 \end{bmatrix}. \tag{3.2}$$

It is left to the reader to verify the properties stated in lemma 3.3.

Example 3.5. *Consider* (S,d) *with* $d = 0$ *and the set* S *equal to the convex hull of the four points* $(1,0)$, $(0,1)$, $(0,0)$ *and* (α_1,α_2), $\alpha_1,\alpha_2 \in (0,1)$ *and* $\alpha_1 + \alpha_2 \ge 1$. *The piecewise linear functions* f_1 *and* f_2 *are given by*

$$f_1(s_2) = \begin{cases} 1 - \frac{1-\alpha_1}{\alpha_2}s_2, & \text{if } s_2 \in [0,\alpha_2], \\ \frac{\alpha_1}{1-\alpha_2} - \frac{\alpha_1}{1-\alpha_2}s_2, & \text{if } s_2 \in (\alpha_2, 1], \end{cases}$$

and

$$f_2(s_1) = \begin{cases} 1 - \frac{1-\alpha_2}{\alpha_1}s_1, & \text{if } s_1 \in [0,\alpha_1], \\ \frac{\alpha_2}{1-\alpha_1} - \frac{\alpha_2}{1-\alpha_1}s_1, & \text{if } s_1 \in (\alpha_1, 1]. \end{cases}$$

Lemma 3.3 implies $y_1^* < x_1^*$, *which reduces the number of cases to the following three:* $y_1^* < \alpha_1 < x_1^*$, $y_1^* < x_1^* < \alpha_1$ *and* $\alpha_1 < y_1^* < x_1^*$.
First, $y_1^* < \alpha_1 < x_1^*$ *(or* $x_2^* < \alpha_2 < y_2^*$*). Then (3.1) becomes*

$$x_2 = \delta y_2, \quad x_1 = 1 - \tfrac{1-\alpha_1}{\alpha_2}x_2,$$
$$y_1 = \delta x_1, \quad y_2 = 1 - \tfrac{1-\alpha_2}{\alpha_1}y_1,$$

and solving this system of equations yields the unique solution

$$x^* = \left(\frac{\alpha_1(\alpha_2 - \delta(1-\alpha_1))}{\alpha_1\alpha_2 - \delta^2(1-\alpha_1)(1-\alpha_2)}, \frac{\delta\alpha_2(\alpha_1 - \delta(1-\alpha_2))}{\alpha_1\alpha_2 - \delta^2(1-\alpha_1)(1-\alpha_2)}\right),$$
$$y^* = \left(\frac{\delta\alpha_1(\alpha_2 - \delta(1-\alpha_1))}{\alpha_1\alpha_2 - \delta^2(1-\alpha_1)(1-\alpha_2)}, \frac{\alpha_2(\alpha_1 - \delta(1-\alpha_2))}{\alpha_1\alpha_2 - \delta^2(1-\alpha_1)(1-\alpha_2)}\right).$$

Note that $\alpha_1+\alpha_2 \geq 1$ suffices to guarantee that the numerator and denominator of x_1^* and y_1^* are positive.[1] *Consistency with $y_1^* < \alpha_1 < x_1^*$, as imposed at the beginning of this case, requires $\alpha_1, \alpha_2 \geq \frac{\delta}{1+\delta}$.*
First, $y_1^* < x_1^* < \alpha_1$ (or $\alpha_2 < x_2^* < y_2^*$). Then (3.1) becomes

$$\begin{aligned} x_2 &= \delta y_2, \quad x_1 = \tfrac{\alpha_1}{1-\alpha_2} - \tfrac{\alpha_1}{1-\alpha_2} x_2, \\ y_1 &= \delta x_1, \quad y_2 = 1 - \tfrac{1-\alpha_2}{\alpha_1} y_1, \end{aligned}$$

and solving this system of equations yields the unique solution

$$x^* = \left(\tfrac{\alpha_1}{(1+\delta)(1-\alpha_2)}, \tfrac{\delta}{1+\delta} \right), y^* = \left(\tfrac{\delta\alpha_1}{(1+\delta)(1-\alpha_2)}, \tfrac{1}{1+\delta} \right).$$

This solution is consistent with $y_1^* < x_1^* \leq \alpha_1$ if and only if $\alpha_2 \leq \frac{\delta}{1+\delta}$.
Third, $\alpha_1 < y_1^* < x_1^*$ (or $x_2^* < y_2^* < \alpha_2$). Then (3.1) becomes

$$\begin{aligned} x_2 &= \delta y_2, \quad x_1 = 1 - \tfrac{1-\alpha_1}{\alpha_2} x_2, \\ y_1 &= \delta x_1, \quad y_2 = \tfrac{\alpha_2}{1-\alpha_1} - \tfrac{\alpha_2}{1-\alpha_1} y_1, \end{aligned}$$

and solving this system of equations yields the unique solution

$$x^* = \left(\tfrac{1}{1+\delta}, \tfrac{\delta\alpha_2}{(1+\delta)(1-\alpha_1)} \right), y^* = \left(\tfrac{\delta}{1+\delta}, \tfrac{\alpha_2}{(1+\delta)(1-\alpha_1)} \right).$$

which is consistent with $\alpha_1 < y_1^* < x_1^*$ if and only if $\alpha_1 \leq \frac{\delta}{1+\delta}$.
To summarize, the area in the (α_1, α_2)-space on and above the line $\alpha_1+\alpha_2 = 1$ is partitioned into three areas and the expression for the unique pair (x^*, y^*) of fixed points is different in each of these areas. These three areas are: $\alpha_1 \leq \frac{\delta}{1+\delta}$ (third case), $\alpha_2 \leq \frac{\delta}{1+\delta}$ (second case) and $\alpha_1, \alpha_2 \geq \frac{\delta}{1+\delta}$ (first case). The unique fixed point is a continuous *function of α_1 and α_2 and x^* (α_1, α_2) is given by*

$$\begin{cases} \left(\tfrac{1}{1+\delta}, \tfrac{\delta\alpha_2}{(1+\delta)(1-\alpha_1)} \right), & \alpha_1 < \tfrac{\delta}{1+\delta}, \\ \left(\tfrac{\alpha_1(\alpha_2-\delta(1-\alpha_1))}{\alpha_1\alpha_2-\delta^2(1-\alpha_1)(1-\alpha_2)}, \tfrac{\delta\alpha_2(\alpha_1-\delta(1-\alpha_2))}{\alpha_1\alpha_2-\delta^2(1-\alpha_1)(1-\alpha_2)} \right), & \alpha_1, \alpha_2 \geq \tfrac{\delta}{1+\delta}, \\ \left(\tfrac{\alpha_1}{(1+\delta)(1-\alpha_2)}, \tfrac{\delta}{1+\delta} \right), & \alpha_2 < \tfrac{\delta}{1+\delta}, \end{cases}$$

and y^* (α_1, α_2) is given by

$$\begin{cases} \left(\tfrac{\delta}{1+\delta}, \tfrac{\alpha_2}{(1+\delta)(1-\alpha_1)} \right), & \alpha_1 < \tfrac{\delta}{1+\delta}, \\ \left(\tfrac{\delta\alpha_1(\alpha_2-\delta(1-\alpha_1))}{\alpha_1\alpha_2-\delta^2(1-\alpha_1)(1-\alpha_2)}, \tfrac{\alpha_2(\alpha_1-\delta(1-\alpha_2))}{\alpha_1\alpha_2-\delta^2(1-\alpha_1)(1-\alpha_2)} \right), & \alpha_1, \alpha_2 \geq \tfrac{\delta}{1+\delta}, \\ \left(\tfrac{\delta\alpha_1}{(1+\delta)(1-\alpha_2)}, \tfrac{1}{1+\delta} \right), & \alpha_2 < \tfrac{\delta}{1+\delta}. \end{cases}$$

Finally, note that the expression of example 3.4 for d $= 0$ *corresponds to* $\alpha_1 = \alpha_2 = \frac{1}{2}$.

3.3.2 Dynamic programming

In sections 3.3.3-3.3.6 it is shown that each MPE for the bargaining problem in utility representation is closely related to the fixed point problem (3.1) and vice versa. This result is derived by applying the one-stage-deviation principle or, equivalent, dynamic programming, see e.g., Fudenberg and Tirole (1991), Osborne and Rubinstein (1994) and Hendon, Jacobsen, and Sloth (1996). In this section the latter technique is introduced and it is employed in the following sections and some of the later chapters.

The general idea of *dynamic programming* is to introduce the value function $V_i : H \times T \to \mathbb{R}$, $i = 1, 2$, that represents the SPE utility of player i at each subgame (or information set) $h_1(t)$, $t \in T$, and apply the principle of backward induction or subgame perfectness to relate the value functions of a particular stage to the value functions of the preceding stage as follows. Consider finite games in extensive form first. Dynamic programming starts by assigning each player's utility at each end node to this player's value function for such final history. Then, in an iterative manner, one calculates one of the Nash equilibria of some subgame that only includes smaller subgames that can be replaced by value functions that are already derived in one of the previous iterations. This iterative procedure is repeated until the value functions at the origin of the finite game in extensive form are calculated and this takes as many iterations as there are subgames in the finite game tree.

It is easy to see that dynamic programming as just explained solves for Nash equilibrium actions of each subgame by taking the behaviour in each of this subgame's subgames as being fixed and the players' utilities associated with the latter behaviour are replaced by the value functions. Loosely speaking, this means that only the behaviour at the 'beginning' of a particular subgame before any of this subgame's subgames is reached matters in the calculation. In these calculations each player's behaviour is optimized against the (equilibrium) strategy of the other player(s). This means that, for finite multi-stage games with observed actions, varying only the behaviour at the beginning of a subgame while keeping the behaviour in each of the subgame's subgames fixed coincides with the notion underlying the one-stage deviation of definition 2.20. Thus, dynamic programming yields each player's optimal one-stage deviation against the other players' strategies. Therefore, dynamic programming yields behaviour that cannot be improved upon by one-stage deviations and by proposition 2.21 these strategies are SPE strategies.

As mentioned, for finite games dynamic programming requires as many iterations as there are subgames in this game. However, for infinite game trees, as is the case for the alternating offer model, there are i) infinitely many subgames and ii) paths of infinite length without end nodes meaning that the initialization of the value functions at end nodes is incomplete. To overcome this problem, ad hoc strategies and value functions are introduced and then equilibrium conditions upon these strategies and value functions are derived. It should be mentioned that this ad hoc solution is one of trial and error and that there is no systematic way in calculating equilibria, see for an exception e.g., Tsutsui and Mino (1990) where a systematic method for linear-quadratic differential games is proposed.

We conclude this section by stating the functional form of the value functions that correspond to Markov strategies in the alternating offer model. Consider a pair of Markov strategies σ of section 3.2 and value functions $V_i : H \times T \to \mathbb{R}$, $i = 1, 2$. One advantage of going through the extra trouble of introducing value functions is that at the start of our analysis it is not known whether or not the MPE strategies induce agreement and the value function introduced below is rich enough to capture every possible outcome that is consistent with Markov strategies. Since Markov strategies are history independent it follows that each player's utility evaluated at the information set or subgame $h_k(t) \in H_k(t)$, $k = 1, 2$ and $t \in T$, only depends upon t being even or odd and in case $k = 2$ also upon the last proposal on the negotiation table. In our notation we distinguish between even and odd rounds by specifying which of the two players is the proposing player at round $t \in T$. Value functions that are consistent with Markov strategies are also history independent up to the proposing player, the last proposal if it is still on the table and the exact stage reached in each round, i.e., the stage where a player proposes and the stage where a player responds. Given this simple structure, it suffices to introduce four numbers, denoted by V_i^j, $i, j = 1, 2$, and four functions $\hat{V}_i^j : S \to \mathbb{R}$, such that

$$V_i\left(h_1(t), t\right) = \begin{cases} V_i^1, & \text{if } t \text{ is even,} \\ V_i^2, & \text{if } t \text{ is odd,} \end{cases} \tag{3.3}$$

and

$$V_i\left((h_1(t), s^t), t\right) = \begin{cases} \hat{V}_i^1(s^t), & \text{if } t \text{ is even,} \\ \hat{V}_i^2(s^t), & \text{if } t \text{ is odd.} \end{cases} \tag{3.4}$$

Note that the superscript specifies which player is the proposing player at $t \in T$ and the proposing player contains the essential information whether t is even or odd. The SPE concept guarantees that no player accepts an agreement that yields a utility lower than the infinite stream of disagreement payoffs, called individual rationality, which implies that $(V_1^j, V_2^j), (\hat{V}_1^j(s^t), \hat{V}_2^j(s^t)) \in S^{IR}$, $j = 1, 2$.

3.3.3 Optimal response

In this section the optimal behaviour of the responding player is derived with dynamic programming. Recall that we distinguish between even numbered rounds in which player 2 responds to player 1's proposal and the opposite case in odd numbered rounds. In order to avoid duplication of arguments we only derive the optimal response in the former case.

Consider the bargaining round $t \in T$ is odd. Then player 2 proposes in this round and player 1 is the proposing player in the next round $t+1$ (is even). So, for player i, $i = 1, 2$, his history-independent MPE utility from the next round onward is V_i^1. Given the latter numbers dynamic programming means deriving the value function $\hat{V}^1(s^t) = (\hat{V}_1^1(s^t), \hat{V}_2^1(s^t))$ as a function of $V^1 = (V_1^1, V_2^1)$.

Consider the information set or subgame $h_2(t) = (h_1(t), s^t) \in H_2(t)$, which means that player 1 responds to player 2's proposal $s^t \in S$. Player 1 faces a maximization problem with a binary variable in $\{yes, no\}$ and the MPE strategies should coincide with the maximum of this problem. If player 1 accepts, then his utility becomes s_1^t, whereas rejecting it yields an expected utility of $q_1\left(V^1\right) = (1-\delta)\, d_1 + \delta V_1^1$. So, player 1's dynamic programming problem is given by

$$\hat{V}_1^2\left((h_1(t), s^t), t\right) = \max_{\{yes,no\}} \begin{cases} q_1\left(V^1\right), & \text{if } no, \\ s_1^t, & \text{if } yes. \end{cases} \tag{3.5}$$

It is easy to see that the optimal strategy for player 1 prescribes yes if $s_1^t > q_1\left(V^1\right)$ and no in case $s_1^t < q_1\left(V^1\right)$. Furthermore, player 1 is indifferent between his two alternatives if s_1^t and $q_1\left(V^1\right)$ are equal and both yes and no are optimal. Formally, player 1's utility maximizing best response correspondence at information set $(h_1(t), s^t)$ is given by

$$\begin{cases} \{no\}, & \text{if } s_1^t < q_1\left(V^1\right), \\ \{no, yes\}, & \text{if } s_1^t = q_1\left(V_1^1, V_2^1\right), \\ \{yes\}, & \text{if } s_1^t > q_1\left(V_1^1, V_2^1\right). \end{cases}$$

Every strategy, i.e., function, that belongs to this correspondence is an optimal strategy for player 1 and, therefore, (only) two strategies meet this requirement, which we denote $\sigma_1\left((h_1(t), s^t)\right)$ and $\sigma_1'\left((h_1(t), s^t)\right)$. These optimal strategies are

$$\sigma_1\left((h_1(t), s^t)\right) = yes, \quad \text{if } s_1^t \geq q_1\left(V^1\right),$$

and

$$\sigma_1'\left((h_1(t), s^t)\right) = yes, \quad \text{if } s_1^t > q_1\left(V^1\right).$$

These strategies imply that the set of acceptable proposals $T_1\left(V^1\right)$ associated with σ_1 is given by

$$T_1\left(V^1\right) = \left\{s \in S \mid\ s_1 \geq q_1\left(V^1\right)\right\} \tag{3.6}$$

whereas the set $T_1'(V^1)$ corresponding to σ_1' is given by

$$T_1'(V^1) = \{s \in S | \; s_1 > q_1(V^1)\}. \tag{3.7}$$

Note that both sets are history independent, which is consistent with our initial ad hoc choice of imposing Markov strategies. Furthermore, both sets are nonempty, because $d_1 \leq q_1(V^1) < s_1^{\max}$.

Both Markov strategies imply that player 1 uses a simple threshold rule with respect to responding to the last proposal on the negotiation table. It simply prescribes '*yes*', i.e., accept, if this player's utility of the offer on the table exceeds his threshold level, which is measured in utility, and this rule describes '*no*' in case the offer falls short of it. Each player's threshold level is determined by the value function at the start of the next bargaining round and it is equal to the expected or present value this player would obtain by delaying the negotiations for one more round and both players follow their Markov strategies from the next round onward. The corresponding present value for player 1, who is the responding player at t odd, is equal to $q_1(V^1)$.

The dynamic programming problem in this section is concluded by deriving the pair of value functions $\hat{V}_i^2 : S \to \mathbb{R}$, $i = 1, 2$, as a function of V^1 that correspond to the optimal strategies σ_1 and σ_1', respectively, given by

$$\hat{V}^2(s^t) = \begin{cases} s^t, & \text{if } s^t \in T_1(V^1), \\ (1-\delta)d + \delta V^1, & \text{if } s^t \notin T_1(V^1), \end{cases} \tag{3.8}$$

and

$$\hat{V}'^2(s^t) = \begin{cases} s^t, & \text{if } s^t \in T_1'(V^1), \\ (1-\delta)d + \delta V^1, & \text{if } s^t \notin T_1'(V^1), \end{cases} \tag{3.9}$$

From these expressions it follows that the value functions are indeed independent of the history $h_1(t)$, $t \in T$ is odd. Note that both value functions are discontinuous at $s_1^t = q_1(V^1)$.[2]

Finally, for $t \in T$ is even, similar arguments imply that player 2's optimal response in each Markov strategy as a function of V^2 is characterized by the two sets of acceptable proposals, which are denoted $T_2(V^2)$ and $T_2'(V^2)$, given by

$$T_2(V^2) = \{s \in S | \; s_2 \geq p_2(V^2)\} \tag{3.10}$$

$$T_2'(V^2) = \{s \in S | \; s_2 > p_2(V^2)\}. \tag{3.11}$$

As before, player 2's threshold level between *yes* or *no* is his expected utility $p_2(V^2)$ of delaying the negotiations for one more round and continuation of the Markov strategies from the next round onward. Finally, the associated pairs of value functions $\hat{V}^1$ and $\hat{V}'^1$ are straightforward and omitted.

3.3.4 Optimal proposals

In this section we proceed by deriving the optimal proposal for player 2 as the proposing player at odd numbered rounds. We do so for each of the two sets of acceptable proposals as derived in the previous section, i.e., $T_1\left(V^1\right)$ and $T_1'\left(V^1\right)$. We first characterize the optimal proposal for the case $T_1\left(V^1\right)$ and, next, show that no optimal proposal exists in the other case.

Consider the set of acceptable proposals $T_1\left(V^1\right)$ and $T_1'\left(V^1\right)$ at the information set $h_1(t) \in H_1(t)$ for $t \in T$ is odd. The value function $\hat{V}^2$ in (3.8) specifies each player's utility if player 2 proposes $s^t \in S$. Given $\hat{V}^2(s^t)$ we derive V^2 as a function of the former value functions. The Markov strategies specify the proposal $y \in S$, i.e., $s^t = y$, for player 2 whenever this player proposes. Player 2 faces a maximization problem with $y \in S$ as its variable and the MPE strategies should maximize player 2's utility over y given his opponent's optimal (Markov) reaction and both players follow Markov strategies from the next stage onward. It is easy to see that player 2's utility of proposing $y \in S$ is given by the second component of the value function $\hat{V}^2$, because if player 2 proposes $y \in T_1\left(V^1\right)$ then player 1 accepts and player 2's utility is y_2, whereas otherwise playerer 2's utility is equal to $(1-\delta)d_2 + \delta V_2^1$. So, player 2's maximization problem is given by

$$y \in \left\{\arg\max_{s^t \in S} \hat{V}_2^2(s^t)\right\}. \tag{3.12}$$

This maximization problem looks rather simple, but the discontinuity of the value function $\hat{V}_2^2(s^t)$ is a technicality that has to be dealt with. Fortunately, this complication can be circumvented, because the next lemma rules out all $y \notin T_1\left(V^1\right)$ as an optimal proposal.

Lemma 3.6. *If $y \in S$ is a maximizer of (3.12), then $y \in T_1\left(V^1\right)$.*

This lemma means that player 2 would be wise to propose an offer his opponent can't refuse and that this player should refrain from making unacceptable proposals. It also means that the negotiations, having reached subgame $h_1(t)$ at t is odd, immediately end in this subgame.

Lemma 3.6 also simplifies maximization problem (3.12), because

$$\max_{s^t \in S} \hat{V}_2^2(s^t) = \max_{s^t \in T_1(V^1)} s_2^t. \tag{3.13}$$

The maximization problem on the right hand side is well defined and, therefore, attains a maximum on the compact and nonempty set $T_1\left(V^1\right)$. The next lemma characterizes the unique solution of maximization problem (3.12).

Lemma 3.7. *Maximization problem (3.12) has a unique solution given by $y = q\left(V^1\right) \in T_1\left(V^1\right)$.*

This lemma states that player 2 offers his opponent a utility that is just sufficient to meet his opponent's threshold level $q_1\left(V^1\right)$ for acceptance. Player 2 obtains the maximum attainable utility given his opponent's threshold, which is uniquely determined. Furthermore, player 2 is strictly better off by making his optimal proposal instead of proposing an unacceptable proposal, i.e.,

$$\max_{s^t \in S} \hat{V}_2^2(s^t) = \max_{s^t \in T_1(V^1)} s_2^t > (1-\delta)d_2 + \delta V_2^1,$$

because disagreement at round t corresponds to the Pareto inefficient pair of utilities $(1-\delta)d + \delta V^1$ (recall the additional assumption in section 3.2 that d is Pareto inefficient). Thus, player 2's optimal proposal is Pareto efficient in two ways: First, the optimal proposal is a Pareto efficient proposal in the set S^{IR} and, second, since player 2's optimal proposal will be accepted by his opponent there is no inefficiency due to costly delay in any subgame $h_1(t)$, $t \in T$ is odd.

The next proposition summarizes the results obtained in this section. This proposition also states the value V^2 at the subgame $h_1(t) \in H_1(t)$ and these are equal to the players' utilities associated with the immediate acceptance of y at t.

Proposition 3.8. *If $T_1(V^1)$ describes player 1's optimal acceptance decision in round t, $t \in T$ and t is odd, then $y = q(V^1) \in T_1(V^1)$ is player 2's optimal proposal and $V^2 = y$.*

For the case $T_2(V_1^2, V_2^2)$ at $t \in T$ is even it similarly follows that player 1's optimal proposal x belongs to the set of acceptable proposals $T_2(V_1^2, V_2^2)$ and that the negotiations end with an immediate agreement at subgame $h_1(t)$. Furthermore, the optimal proposal $x = p(V^2)$ and the associated value V^1 are given by $V^1 = x$. The following proposition is stated without proof.

Proposition 3.9. *If $T_2(V_1^2, V_2^2)$ describes player 2's optimal acceptance decision in round t, $t \in T$ and t is even, then $x = p(V_1^2, V_2^2) \in T_2(V_1^2, V_2^2)$ is player 1's optimal proposal and $V^1 = x$.*

To conclude this section, we now turn attention to the case that is associated with $T_1'\left(V^1\right)$ as derived in section 3.3.3. The analysis in this case is very similar to the previous case and we briefly go through the similarities and only discuss the differences in detail. First, maximization problem (3.12) translates into

$$y \in \left\{ \arg\max_{s^t \in S} \hat{V}_2^{\prime 2}(s^t) \right\}. \tag{3.14}$$

Next, the same arguments as in the proof of lemma 3.6 imply that each unacceptable proposal $y \notin T_1'\left(V^1\right)$ cannot be an optimal proposal for player 2. Therefore, maximization problem (3.14) is equivalent to

$$\max_{s^t \in S} \hat{V}_2^{\prime 2}(s^t) = \max_{s^t \in T_1'(V^1)} s_2^t. \tag{3.15}$$

However, the maximization problem on the right hand side is not well defined, because the set $T_1'(V^1)$ is not closed. As a consequence, the maximization problem on the left hand side of (3.15) does not have a solution.

The mathematical problem is that $f_2(q_1(V^1))$ is the supremum but not the maximum on $T_1'(V^1)$. So, Markov strategies in which player 1 at t is odd rejects below and at the threshold level cannot occur in any MPE simply because there does not exist a utility maximizing proposal for player 2 in this case.[3] Formulated in economic terminology, if player 1 at t is odd would attempt to force player 2 to give him more than the threshold level by rejecting at the threshold level, then this credible threat will not result in even the tiniest improvement in player 1's utility. Therefore, player 1's attempt to do slightly better is futile and will not occur.[4] The next proposition formally states this result.

Proposition 3.10. *There does not exist a MPE in which player 1's set of acceptable proposals at* $t \in T$ *odd is given by* $T_1'(V^1)$.

Corollary 3.11. *There does not exist a MPE in which player 2's set of acceptable proposals at* $t \in T$ *even is given by* $T_2'(V_1^2, V_2^2)$.

To summarize the results of this section, of the two cases $T_1(V^1)$ and $T_1'(V^1)$ derived in the previous section the latter cannot occur in any MPE, which reduces the number of cases one in the sequel with respect to the optimal response to just.

3.3.5 Characterization in utility representation

In this section the results obtained thus far are combined in order to show that alternating offers admits a unique SPE in Markov strategies, i.e., a unique MPE, for the bargaining problem in utility representation. The unique pair of equilibrium proposals (x, y) coincides with the unique fixed point of (3.1).

Combining the results of the previous sections yield the following result. First, proposition 3.8 and 3.9 imply that $V^1 = x$ and $V^2 = y$. Substitution of x and y for the value functions in these propositions then yields that the Markov strategies are consistent with dynamic programming provided (x, y) is a fixed point of the function $p \times q$. Then the uniqueness result of lemma 3.2 immediately yields: Alternating offers admits a unique MPE. We state the following proposition without proof.

Proposition 3.12. *The strategies of table 3.1 constitute the unique MPE, where* (x^*, y^*) *is the unique fixed point of lemma 3.2.*

Proposition 3.8 and 3.9 also imply that $x^* \in T_2(y^*)$ and $y^* \in T_1(x^*)$. This implies that MPE strategies have player 2 accept player 1's proposal x^* and that player 1 accepts player 2's proposal y^*. Thus, MPE strategies prescribe equilibrium behaviour in which the negotiations are immediately concluded with an

Player 1	proposes	x^*	always
	responds *yes*		if $s_1^t \geq y_1^*$
Player 2	proposes	y^*	always
	responds *yes*		if $s_2^t \geq x_2^*$

Table 3.1. The MPE strategies of proposition 3.12, where (x^*, y^*) denotes the unique fixed point of $p \times q$.

agreement at round 0 upon the pair of utilities x^* proposed by player 1. Moreover, if we would start the MPE strategies at the subgame labeled $h_1(t) \in H_1(t)$, $t \in T$, then these strategies induce immediate agreement in this subgame.

Next, proposition 3.12 implies that each player uses a simple threshold rule with respect to responding to the last proposal on the table. It simply prescribes '*yes*', i.e., accept, if this player's utility of the offer on the table gives him at least his threshold level, which is measured in utility, and this rule describes '*no*' in case the offer falls short of it. Each player's threshold level is determined endogenously by the model and it is equal to the expected or present value this player would obtain by delaying the negotiations for one more round and both players follow their MPE strategies from the next round onward, which means agreement upon either x^* or y^* in the next round. The corresponding present value for player 1 as the responding player is equal to $y_1^* = q_1(x^*)$ and, similar, $x_2^* = p_2(y^*)$ for player 2.

The responding player's decision to accept or reject is credible. If the proposal falls short of the threshold level, then the responding player is better off by rejecting it and delaying the negotiations for one more round. Therefore, the threat to reject those offers is credible and the proposing player has to back down if he wants to reach an agreement at the current round. The responding player fails this credible threat above the threshold, because by accepting any proposal that yields a higher utility than the threshold level this player can secure this higher utility just proposed simply by accepting whereas rejecting will yield the lower utility of delaying the agreement for one more round. The latter utility is exactly equal to the MPE threshold level. We conclude this section with an example.

Example 3.13. *Consider the bargaining problem (S, d) of example 3.4. We already established that*

$$x^* = d + \gamma \cdot s(d) \begin{bmatrix} 1 \\ \delta \end{bmatrix}, \quad y^* = d + \gamma \cdot s(d) \begin{bmatrix} \delta \\ 1 \end{bmatrix}, \tag{3.16}$$

where $\gamma = \frac{1}{1+\delta}$ *and the effective surplus* $s(d) = 1 - d_1 - d_2 > 0$. *The set* T_i, $i = 1, 2$, *is given by*

$$T_i = \{s \in S | \; s_i \geq d_i + \delta \cdot \gamma \cdot s(d)\}.$$

Then the MPE strategies are given by

$$\sigma_1(h_t) = \begin{cases} \textit{proposes} & d + \gamma \cdot s(d) \begin{bmatrix} 1 \\ \delta \end{bmatrix} \\ \textit{responds yes,} & \textit{if } s_1(t) \geq d_1 + \delta \cdot \gamma \cdot s(d), \end{cases}$$

and

$$\sigma_2(h_t) = \begin{cases} \textit{proposes} & d + \gamma \cdot s(d) \begin{bmatrix} \delta \\ 1 \end{bmatrix} \\ \textit{responds yes} & \textit{if } s(t)_2 \geq d_2 + \delta \cdot \gamma \cdot s(d). \end{cases}$$

These strategies result in an immediate agreement upon x^*.

3.3.6 Characterization in the contract space

Proposition 3.12 states a uniqueness result in terms of utilities without specifying the associated MPE contract (or possibly multiple contracts). Equilibrium contracts are of interest in many economic applications and in this section the characterization and the generic uniqueness of such contracts is derived.

Basically, the problem how to translate the equilibrium proposals x^* and y^* and the threshold levels of proposition 3.12 into the corresponding MPE contracts c^1 and c^2 is relatively easy, i.e., solve for c^1 and c^2 in

$$\left(u_1(c^1), u_2(c^1)\right) = x^* \quad \text{and} \quad \left(u_1(c^2), u_2(c^2)\right) = y^* \tag{3.17}$$

Furthermore, the sets T_1 and T_2 translate into

$$\begin{aligned} C_1 &= \left\{c \in C \left| u_1(c) \geq (1-\delta)\, d_1 + \delta u_1\left(c^1\right)\right.\right\}, \\ C_2 &= \left\{c \in C \left| u_2(c) \geq (1-\delta)\, d_2 + \delta u_2\left(c^2\right)\right.\right\}. \end{aligned}$$

Note however that (3.17) is a misleading answer from a computational point of view, because it implies the following elaborated procedure to calculate the MPE contracts: First, the bargaining problem has to be translated into its utility representation (S, d) in order to derive explicit expressions for the functions f_1 and f_2, next, the fixed point problem (3.1) has to be solved and, finally, the system of equations (3.17) has to be solved. Obviously, the conclusion must be that the route via (S, d) enables us to swiftly derive theoretical results, but it is less suited for practical purposes.

In order to avoid the above objections we combine proposition 3.12 with our discussion of the utility maximizing behaviour in (3.13) underlying the

Markov strategies in the utility representation. Then we immediately arrive at the following proposition, which is given without proof.

Proposition 3.14. *The Markov strategies are MPE if and only if the pair of contracts* (c^1, c^2) *solves*

$$c^1 = \arg\max_{c \in C_2} u_1(c), \quad C_1 = \left\{ c \in C \,\middle|\, u_1(c) \geq (1-\delta) d_1 + \delta u_1\left(c^1\right) \right\},$$

and

$$c^2 = \arg\max_{c \in C_1} u_2(c), \quad C_2 = \left\{ c \in C \,\middle|\, u_2(c) \geq (1-\delta) d_2 + \delta u_2\left(c^2\right) \right\}.$$

As before, player 1's proposal $c^1 \in C_2$ will be accepted by player 2 and, similar, player 2's proposal $c^2 \in C_1$. Therefore, these strategies induce immediate agreement at the first round. In every MPE the responding player accepts every proposed contract that yields a utility of at least the expected utility (or the present value measured in utility) of one period of delay. Furthermore, the proposing player proposes a contract that maximizes his utility over the set of acceptable proposals. The MPE contracts c^1 and c^2 are Pareto efficient. Similar as before, the proposing player has an incentive to propose an acceptable contract, because

$$u_1\left(c^1\right) = \max_{c \in C_2} u_1(c) > (1-\delta) d_1 + \delta u_1\left(c^2\right)$$

and

$$u_2\left(c^2\right) = \max_{c \in C_1} u_2(c) > (1-\delta) d_2 + \delta u_2\left(c^1\right).$$

It is easy to see that player i's proposal c^i, $i = 1, 2$, in proposition 3.14 is the solution of a maximization problem that is equivalent to maximization problem (2.2). This is another way to see that each player's proposal in proposition 3.14 is Pareto efficient. However, recall the discussion in section 2.2.7. Proposition 2.11 implies that each of the maximization problems in proposition 3.14 yields a generically unique contract as its solution and the nongeneric case multiple solutions cannot be ruled out. Example 3.16 illustrates this point. For completeness, we mention that multiplicity of pairs of MPE contracts (c^1, c^2) is explicitly ruled out if one of the utility functions u_i, $i = 1, 2$, is strictly concave. We state the next proposition without proof.

Proposition 3.15. *The pair of MPE proposals* $\left(c^1, c^2\right)$ *of proposition 3.14 is generically unique, each contract* c^i, $i = 1, 2$, *is Pareto efficient and the Markov strategies of table 3.2 constitute the generically unique MPE.*

The bargaining procedure yields uniqueness of the MPE strategies for the bargaining problem in the utility representation (S, d). Given the assumptions

Player 1	proposes	c^1	always
	responds *yes*		if $u_1(c(t)) \geq (1-\delta)\, d_1 + \delta u_1(c^1)$
Player 2	proposes	c^2	always
	responds *yes*		if $u_2(c(t)) \geq (1-\delta)\, d_2 + \delta u_2(c^2)$

Table 3.2. The generically unique MPE strategies of proposition 3.15, where (c^1, c^2) is the generically unique pair of MPE proposals of proposition 3.14 and $c(t)$ denotes the proposed contract in round t, $t \in \mathbb{N}$.

made in section 2.2 it is impossible to prove the same result for the pair of MPE contracts (c^1, c^2) formulated in the contract space C. The result that is nearest to uniqueness in the contract space is the slightly weaker result of generic uniqueness of such pairs.

The following example illustrates that multiplicity of pairs of MPE proposals is nongeneric.

Example 3.16. *Consider example 2.10 with $\alpha = \frac{1}{2}$. The contract space C given by the simplex*

$$C = \left\{ c \in \mathbb{R}^3_+ \mid c_1 + c_2 + c_3 = 1 \right\}$$

and the players' utility functions

$$u_1(c) = c_1 + \tfrac{1}{2}c_2 \quad \text{and} \quad u_2(c) = \tfrac{1}{2}c_2 + c_3.$$

Then we obtain the set $S = S^{IR} = \left\{ s \in \mathbb{R}^2_+ \mid s_1 + s_2 = 1 \right\}$. If we additionally suppose that $d = 0$, then we obtain the (S, d) of example 3.13. The unique pair of utilities (x^*, y^*) is given in equation (3.16) of the latter example. A simple procedure to calculate the MPE contract c^1 is to solve

$$c_1^1 + \tfrac{1}{2}c_2^1 = x_1^* = \tfrac{1}{1+\delta} \quad \text{and} \quad \tfrac{1}{2}c_2^1 + c_3^1 = x_2^* = \tfrac{\delta}{1+\delta},$$

which yields $c^1 = (\gamma_1, \gamma_2, \gamma_3)$ given by

$$\gamma_1 = \tfrac{1}{1+\delta} - \tfrac{1}{2}\gamma_2, \quad \gamma_3 = \tfrac{\delta}{1+\delta} - \tfrac{1}{2}\gamma_2, \quad \gamma_2 \in \left[0, \tfrac{2\delta}{1+\delta}\right].$$

By symmetry, $c^2 = (\gamma_3, \gamma_2, \gamma_1)$, $\gamma_2 \in \left[0, \frac{2\delta}{1+\delta}\right]$. Thus, infinitely many pairs of MPE proposals (c^1, c^2) exist, while the pair of MPE utilities (x^*, y^*) is unique. The reason for the nonuniqueness is that the set S is equal to the convex hull of the three utility pairs $(0,1)$, $\left(\frac{1}{2}, \frac{1}{2}\right)$ and $(1,0)$, which is degenerated to the line piece $s_1 + s_2 = 1$. This nonuniqueness is nongeneric, which can be seen *as follows. First, $\alpha < \frac{1}{2}$ in example 2.10 yields the same bargaining problem (S,d) as in this example but the unique pair of SPE contracts is given by*

$$c^1 = \left(\tfrac{1}{1+\delta}, 0, \tfrac{\delta}{1+\delta}\right) \quad \textit{and} \quad c^2 = \left(\tfrac{\delta}{1+\delta}, 0, \tfrac{1}{1+\delta}\right).$$

Second, $\alpha > \frac{1}{2}$ in example 2.10 yields the bargaining problem (S,d) of example 3.5 with the unique SPE contract

$$c^1 = \left(\tfrac{(1-\delta)\alpha}{(1-\delta)\alpha+\delta}, \tfrac{\delta}{(1-\delta)\alpha+\delta}, 0\right)$$

and

$$c^2 = \left(0, \tfrac{\delta}{(1-\delta)\alpha+\delta}, \tfrac{(1-\delta)\alpha}{(1-\delta)\alpha+\delta}\right).$$

3.4. Subgame perfect equilibrium

In the previous section the existence and uniqueness of a SPE in Markov strategies is shown. An important question is do there also exist SPE strategies that are not Markov strategies. The answer is negative. The MPE of the previous section is the only SPE of the model. In this section we will derive this result. Before we do so, we want to introduce a powerful technique that will be applied to prove this uniqueness result. This technique is first introduced in Shaked and Sutton (1984).

3.4.1 The method of Shaked and Sutton

The elegance of the method of Shaked and Sutton is that the analysis is conducted in the utility space S^{IR} instead of the very complicated strategy space. Furthermore, it exploits the fact that the set of utility pairs corresponding to a SPE at round 0 is equal to the set of SPE utility pairs at the subgame that starts at the beginning of every even round. To put it differently, the latter set is independent of the history leading to this particular subgame. A similar statement can be made with respect to the sets of SPE utility pairs at round 1 and all odd rounds.

The first lemma makes these latter statements formal. But first we define $E(h_1(t),t) \subseteq S^{IR}$, $t \in T$ and $h_1(t) \in H_1(T)$, as the set of utility pairs that correspond to a SPE that starts at the *beginning* of round t given the history $h_1(t)$, i.e., the subgame induced by $h_1(t)$, where one of the players has the initiative to propose. That is, for every $s^* \in E(h_1(t),t)$, $t \in T$ and $h_1(t) \in h_1(t)$, there exists a pair of SPE strategies $\sigma^*|_{h_1(t)}$ such that $\left(u_1^t\left(p\left(\sigma^*|_{h_1(t)}\right)\right), u_2^t\left(p\left(\sigma^*|_{h_1(t)}\right)\right)\right) = s^*$. Obviously, the existence of the MPE in proposition 3.12 implies that each set $E(h_1(t),t)$ is nonempty. In particular, $x^* \in E(h_1(t),t)$ for all t even and $y^* \in E(h_1(t),t)$ for all t odd.

The following lemma states that the only part of the history that matters for the set $E(h_1(t), t)$ is whether t is even or odd. Thus, $E(h_1(t), t)$ is independent of the rest of the information contained in $h_1(t)$. Loosely speaking, $E(h_1(t), t)$ is history independent.

Lemma 3.17. *There exist two nonempty sets $E_j \subseteq S^{IR}$, $j = 1, 2$, such that for all $t \in T$ and $h_1(t) \in H_1(t)$:*

$$E(h_1(t), t) = \begin{cases} E_1, & t \text{ even,} \\ E_2, & t \text{ odd.} \end{cases}$$

Moreover, $x^* \in E_1$ and $y^* \in E_2$, where (x^*, y^*) is the unique fixed point of $p \times q$.

At this point the shape of the set E_j, $j = 1, 2$, is not known. What can be said is that there exists a smallest rectangular in $\mathbb{R}^2$ that contains the set $E_j \subseteq S^{IR}$, $j = 1, 2$. In order to formalize the smallest rectangular we define $\underline{s}_i^j$, respectively, $\overline{s}_i^j$, $i, j = 1, 2$, as the lower and upper bound upon the i-th component of any vector s in this rectangular, i.e.,

$$\underline{s}_i^j = \inf \{s_i \mid s \in E_j\} \quad \text{and} \quad \overline{s}_i^j = \sup \{s_i \mid s \in E_j\}.$$

Then whatever the shapes of the sets E_1 and E_2 are, these are contained in the intersection of S^{IR} and the smallest rectangular. Formally,

$$E_j \subseteq \left\{ s \in \mathbb{R}^2 \;\middle|\; \underline{s}_i^j \leq s_i \leq \overline{s}_i^j, i = 1, 2 \right\} \cap S^{IR}.$$

The method of Shaked and Sutton consists of relating the bounds $\underline{s}_i^1$ and $\overline{s}_i^1$, $i = 1, 2$, to the bounds $\underline{s}_i^2$ and $\overline{s}_i^2$ and vice versa in an attempt to derive conditions that substantially shrink the smallest rectangular. Then there are two distinct cases. In the first case the lower left corner $(\underline{s}_1^j, \underline{s}_2^j)$, $j = 1, 2$, of the smallest rectangular coincides with its upper right corner $(\overline{s}_1^j, \overline{s}_2^j)$ and, thereby, the smallest rectangular coincides with the single element $(\underline{s}_1^j, \underline{s}_2^j) = (\overline{s}_1^j, \overline{s}_2^j)$. If the latter point lies in S^{IR}, then the model has a unique pair of SPE utilities in each E_j, $j = 1, 2$. The second case corresponds to multiple pairs of SPE utilities and corresponds to $(\underline{s}_1^j, \underline{s}_2^j) \neq (\overline{s}_1^j, \overline{s}_2^j)$. In the latter case some additional analysis has to be done in order to derive the exact shape of E_j, $j = 1, 2$. Nevertheless, the method of Shaked and Sutton is also useful in case of nonuniqueness, because it determines for each player a lower and a upper bound upon his set of SPE utilities meaning that equilibrium play cannot force a player below his lower bound and that there is no way in which he can do better than his upper bound. The most extreme case of nonuniqueness (measured in terms of the smallest rectangular) is $\underline{s}_i^j = d_i$, $i, j = 1, 2$, and $\overline{s}_i^j = s_i^{max}$, $i, j = 1, 2$, meaning that the entire set S^{IR} is contained in the smallest rectangular, see e.g., example 9.19.

3.4.2 Characterization of the SPE

In this section the method of Shaked and Sutton is applied to show that the set of SPE utilities in each subgame consists of a unique pair of utilities that corresponds to the pair of MPE utilities of proposition 3.12. It then immediately follows that the generically unique pair of MPE strategies of proposition 3.14, which are formulated in terms of MPE contracts, is the generically unique pair of SPE strategies.

The next proposition states that each set $E_j, j = 1, 2$, of lemma 3.17 consists of a single element and characterizes this element. This result is derived by showing that the lower-left corner of the smallest rectangular that contains E_j, $j = 1, 2$, coincides with the upper-left corner of this rectangular.

Proposition 3.18. *Let (x^*, y^*) be the unique fixed point of $p \times q$. Then $E_1 = \{x^*\}$ and $E_2 = \{y^*\}$.*

Proposition 3.18 states a uniqueness result in terms of utility pairs but, in principle, there could be many SPE strategies in the complex strategy space that support this outcome. So, we have to investigate SPE strategies as well, which is very easy to do. Since the present value or expectation of the unique pair of SPE utilities in the set E_j, $j = 1, 2$, is Pareto efficient it immediately follows that all SPE strategies σ^* correspond to Pareto efficient paths $p(\sigma^*)$. Since the disagreement point d is Pareto inefficient (recall section 3.2) the discussion in section 2.2.7 immediately implies that any Pareto efficient path corresponds to an immediate agreement upon a pair of utilities $s^* \in S^{PE}$. So, this already discards all strategies with delay in reaching an agreement as well as all strategies with immediate agreement upon an Pareto inefficient agreement $s \in S \backslash S^{PE}$. Similar, lemma 3.17 implies that in each subgame the players' pair of utilities corresponds to a Pareto efficient path in this subgame and, therefore, SPE strategies have to induce immediate agreement upon some $s \in S^{PE}$ in every subgame as well. Moreover, the SPE utilities in each subgame are either x^* in even rounds and y^* in odd rounds, which implies history independence of SPE proposals. But then all SPE strategies are necessarily Markov strategies and we already obtained uniqueness of the MPE. The following theorem is given without a formal proof.

Theorem 3.19. *The MPE strategies of proposition 3.12 constitute the unique SPE strategies and feature immediate agreement in every subgame upon an Pareto efficient pair of utilities. In particular, immediate agreement upon x^* at round* 0 *is the unique SPE outcome.*

Thus, the alternating offer model admits a unique pair of SPE strategies $\sigma^* = (\sigma_1^*, \sigma_2^*)$ and this equilibrium is given by the Markov strategies in table 3.1. The translation into the generically SPE formulated in terms of SPE proposals is straightforward.

Theorem 3.20. *The generically unique MPE strategies of proposition 3.15 constitute the generically unique SPE strategies and feature immediate agreement in every subgame upon an Pareto efficient pair of contracts. In particular, immediate agreement upon* c^1 *at round* 0 *is the generically unique SPE outcome.*

Finally, note that the (generically) unique SPE strategies imply that the players *immediately agree* at the start of the negotiations and that there is no inefficient delay. As a consequence, the alternating offer model with complete information cannot explain delay in negotiations.

3.4.3 First-mover advantage

The bargaining procedure has a fixed order of moves and player 1 is entitled to make the first move, i.e., propose first. In this section the notion of a first-mover advantage is defined and it is shown that player 1 has such an advantage at the expense of his opponent.

In general the idea of the notion of a first-mover advantage is to switch the players' roles in the game in order to let the opponent become the first-mover and, then, the sets of equilibrium payoffs of both games are compared. To be precise, consider an arbitrary game in extensive form with two players, label player 1 as the first-mover in this game and call this game the 'original' game. Then define the modified game as the game that results after the roles of the players are switched meaning player 1 (2) moves in the modified game when player 2 (1) moves in the original game. Then player 2 is the first-mover in the modified game. In order to avoid additional notation, let $E_j, j = 1, 2$, be the set of utility pairs corresponding to the equilibria in the game where player j is the first mover. Then player 1 is said to have a first-mover advantage in the original game if all his equilibrium utilities in E_1 dominate all his equilibrium utilities in E_2. In terms of earlier notation we could say that player 1 has a first-mover advantage if $\underline{s}_1^1 > \overline{s}_1^2$. Similar, player 2 has a first-mover advantage in the modified game if all his equilibrium utilities in E_2 dominate all his equilibrium utilities in E_1. Alternatively, we could also say that player 2 has a second-mover disadvantage in the original game, i.e., $\overline{s}_2^1 < \underline{s}_2^2$. Translated to the alternating offer model we introduce the following definition.[5]

Definition 3.21. *Player* 1 *has a first-mover advantage if each* $s \in E_1$ *is preferred by this player to each* $s' \in E_2$*, whereas player* 2 *has a second-mover disadvantage if every* $s' \in E_2$ *is preferred by this player to each* $s \in E_1$.

The reader should be aware that, in general, player 1's first-mover advantage does not automatically imply a second-mover disadvantage for player 2.[6] Note however that Player 1's first-mover advantage does imply that $E_1 \cap E_2 = \emptyset$. Finally, a first-mover advantage for player 1 and a second-mover disadvantage for player 2 imply that the set E_2 lies to the north-west of the set E_1.

Lemma 3.3 and proposition 3.18 imply that $E_2 = \{y^*\}$ lies north-west of $E_1 = \{x^*\}$ meaning $x_1^* > y_1^*$ and $y_2^* > x_2^*$. So, player 1 has a first-mover advantage and player 2 a second-mover disadvantage.

Proposition 3.22. *Player 1 has a first-mover advantage and player 2 a second-mover disadvantage.*

The first-mover advantage can be explained as follows. Despite the symmetric bargaining procedure in which players alternate in making proposals over time there is an asymmetry between the proposing player and the responding player in every bargaining round. For the responding player, rejecting an offer is costly because it involves either consumption of the inefficient disagreement point for one more round or the positive risk of a permanent breakdown. To see this, let player 2 be the responding player. The function $p_2(y)$ can be rewritten as $y_2 - (1 - \delta)(y_2 - d_2)$ and the second term represents the loss measured in utility due to a delay of one more round. So, given a utility of y_2 in the next bargaining round, player 2 as the responding player is willing to give up at most $(1 - \delta)(y_2 - d_2)$ in order to avoid these costs. In the unique SPE the proposing player, i.e., player 1 here, exploits this by extracting these costs from player 2's 'share' y_2. When faced by the SPE proposal x^* player 2 faces a *fait accompli*, which cannot be made undone. So, the asymmetry between the proposing and responding player in every subgame is in favour of the proposing player who offers the responding player's proposal (read: agreement) of the next round minus the responding player's costs of delay.

3.4.4 Computation of the SPE contract

In many economic applications standard optimization methods can be applied, because it is additionally assumed that the utility functions are twice differentiable and the contract space C can be represented by a finite number of twice differentiable and quasiconvex functions as in (2.1). In this section we impose the latter assumptions and translate the two maximization problems of proposition 3.14 into the associated Kuhn-Tucker first-order conditions that can be solved with standard numerical methods.

In (2.1) the contract space C is represented by a finite number k of twice differentiable and quasiconvex functions $g_j : C \to \mathbb{R}$, $j = 1, \ldots, k$, i.e.

$$C = \{c \in \mathbb{R}^n \mid g_j(c) \leq 0, \quad j = 1, \ldots, k\},$$

Then the two maximization problems of proposition 3.14 can be rewritten as the two nonlinear programming problems given by

$$\max_{c^1} u_1(c^1), \quad \text{s.t.} (1-\delta) d_2 + \delta u_2(c^2) - u_2(c^1) \leq 0,$$
$$g_j(c^1) \leq 0, \quad j = 1, \ldots, k,$$

and

$$\max_{c^2} u_2\left(c^2\right), \quad \text{s.t.}\ (1-\delta)\, d_1 + \delta u_1\left(c^1\right) - u_1\left(c^2\right) \leq 0,$$
$$g_j\left(c^2\right) \leq 0, \quad j = 1, \ldots, k.$$

As mentioned in section 2.2.5, the Kuhn-Tucker first-order conditions are also sufficient conditions for a global maximum. Denote λ_0^1 and λ_j^1 as the Lagrange multipliers of the first maximization problem corresponding to the first constraint respectively the constraint involving the function g_j. Similar for the Lagrange multipliers λ_0^2 and λ_j^2 of the second maximization problem. Application of the Kuhn-Tucker first-order conditions to each problem separately yields two systems of first-order conditions given by

$$\begin{aligned}
&\frac{\partial u_1(c^1)}{\partial c_i} + \lambda_0^1 \frac{\partial u_2(c^1)}{\partial c_i} - \sum_{j=1}^{k} \lambda_j^1 \frac{\partial g_j(c^1)}{\partial c_i} = 0, \quad i = 1, \ldots, n, \\
&\lambda_0^1 \left[(1-\delta)\, d_2 + \delta u_2(c^2) - u_2(c^1)\right] = 0, \\
&\lambda_j^1 g_j(c^1) = 0, \quad j = 1, \ldots, k,
\end{aligned} \tag{3.18}$$

and

$$\begin{aligned}
&\frac{\partial u_2(c^2)}{\partial c_i} + \lambda_0^2 \frac{\partial u_1(c^2)}{\partial c_i} - \sum_{j=1}^{k} \lambda_j^2 \frac{\partial g_j(c^2)}{\partial c_i} = 0, \quad i = 1, \ldots, n, \\
&\lambda_0^2 \left[(1-\delta)\, d_1 + \delta u_1(c^1) - u_1(c^2)\right] = 0, \\
&\lambda_j^2 g_j(c^2) = 0, \quad j = 1, \ldots, k.
\end{aligned} \tag{3.19}$$

The generically unique pair of SPE contracts $\left(c^1, c^2\right)$ of proposition 3.14 are found by simultaneously solving these two systems of Kuhn-Tucker first-order conditions. Solving means finding the contracts c^1 and c^2 and all the Lagrange multipliers that satisfy (3.18) and (3.19). Note that in case the contract c^i, $i = 1, 2$, lies in the interior of the contract space C, then the well-known condition of equal marginal rates of substitution results, e.g., see section 2.2.5 for details.

Finally, if we compare the number of unknowns and equations that have to be solved with the number of variables and equations of (2.3), then we see that the number is doubled for the obvious reason that (2.3) has to be solved twice in proposition 3.14. From a computational point of view this makes the bargaining problem harder than the calculation of an arbitrary Pareto efficient contract.

3.5. Applications

Several economic applications are presented in this section. The first is the well-known textbook example of the 'divide a dollar' problem. The second application considers negotiations in a simple barter economy with two-goods and Cobb-Douglass utility functions.

3.5.1 Dividing a dollar

Two players negotiate how to divide the monetary benefits, being one dollar, generated by a costless 'investment' project they can engage in. An agreement is necessary before the project can be undertaken and no monetary benefits are collected during the negotiations. The contract $c = (c_1, c_2)$ specifies the division of the dollar and player i, $i = 1, 2$, receives a share of c_i. The contract space is given by $C = \{c \in \mathbb{R}_+^2 \mid c_1 + c_2 \leq 1\}$ and we allow for agreements in which the players agree to discard some part of the dollar.

Next, we additionally assume that each player is only concerned about his share of the dollar, that each player prefers a larger share to a lower share and that each player is risk averse. Formally, for each player i, $i = 1, 2$, there exists an *increasing* and *concave* function $\mu_i : [0, 1] \to \mathbb{R}$ such that $u_i(c) = \mu_i(c_i)$. Note that the utility function u_i satisfies assumption 2.4.[7] Without loss of generality the function μ_i is normalized meaning $\mu_i(0) = 0$. With respect to the exogenous disagreement point we assume that $d_i = \mu_i(0) = 0$. It is easy to see that a contract $c \in C$ is Pareto efficient if and only if $c_1 + c_2 = 1$.

The two maximization problems of proposition 3.14 that characterize the generically unique pair of SPE proposals $(c^1, c^2) = ((c_1^1, c_2^1), (c_1^2, c_2^2))$ can be written as

$$\max_{(c_1^1, c_2^1)} \mu_1\left(c_1^1\right), \quad \text{s.t.} \quad \mu_2\left(c_2^1\right) \geq \delta\mu_2\left(c_2^2\right), \quad c_1^1 + c_2^1 \leq 1$$

and

$$\max_{(c_1^2, c_2^2)} \mu_2\left(c_2^2\right), \quad \text{s.t.} \quad \mu_1\left(c_1^2\right) \geq \delta\mu_1\left(c_1^1\right), \quad c_1^2 + c_2^2 \leq 1,$$

where all nonnegativity constraints are neglected for simplicity. These two maximization problems are easy to solve. First, μ_i, $i = 1, 2$, is increasing implies that the proposed SPE division has to be efficient. Second, substitution of $c_j^i = 1 - c_i^i$, $i, j = 1, 2$, into player i's maximization problem yields

$$\max_{c_i^i} \mu_i\left(c_i^i\right), \quad \text{s.t.} \quad \mu_j\left(1 - c_i^i\right) \geq \delta\mu_j\left(c_j^j\right).$$

Since $\mu_i(c_i^i)$ is increasing in c_i^i whereas μ_j is decreasing in c_i^i we must have that the constraint is binding in the optimal c_i^i. So, we obtain that the generically unique pair of SPE contracts $(c^1, c^2) = ((c_1^1, 1 - c_1^1), (1 - c_2^2, c_2^2))$ are found by solving c_1^1 and c_2^2 in

$$\mu_1\left(1 - c_2^2\right) = \delta\mu_1\left(c_1^1\right) \quad \text{and} \quad \mu_2\left(1 - c_1^1\right) = \delta\mu_2\left(c_2^2\right)$$

From these equations it is easy to show that c_1^1 and c_2^2 are unique, because μ_i, $i = 1, 2$, is continuous and increasing combined with the uniqueness of (x^*, y^*) imply that there exist unique numbers c_1^1 and c_2^2 such that $\mu_1(c_1^1) = x_1^*$ and $\mu_2(c_2^2) = y_2^*$.

The system of two equations can be even further reduced. Since the function μ_i, $i = 1, 2$, is increasing it can be inverted, where μ_i^{-1} denotes its inverse function. Making use of the inverse functions implies that player 1's SPE share c_1^1 is the fixed point given by

$$c_1^1 = 1 - \mu_2^{-1}\left(\delta\mu_2\left(1 - \mu_1^{-1}\left(\delta\mu_1\left(c_1^1\right)\right)\right)\right). \tag{3.20}$$

A similar expression can be derived for player 2's SPE share c_2^2.

The following example illustrates the results thus far.

Example 3.23. *Consider the utility function μ_i, $i = 1, 2$, given by $\mu_i(c_i) = c_i^{\alpha_i}$, where $\alpha_i \in (0, 1]$ determines the Arrow-Pratt coefficient of absolute risk aversion $(1 - \alpha)/c_i^i$. Then $\mu_i^{-1}(z) = z^{1/\alpha_i}$ and (3.20) yield that*

$$c_1^1 = \frac{1 - \delta^{1/\alpha_2}}{1 - \delta^{(\alpha_1+\alpha_2)/(\alpha_1\alpha_2)}} \quad \textit{and} \quad c_2^2 = \frac{1 - \delta^{1/\alpha_1}}{1 - \delta^{(\alpha_1+\alpha_2)/(\alpha_1\alpha_2)}}.$$

The SPE shares depend upon the risk parameters α_1 and α_2. Player 1's SPE share c_1^1 is increasing in α_1 and decreasing in α_2. The economic implication is the following. Consider a heterogenous population of player 1's, each characterized by some $\alpha_1 \in (0, 1]$ who frequently negotiate over a dollar with one particular player 2, i.e., some fixed α_2. Then a relatively less risk averse player 1, i.e., a large value of α_1, does better against this particular player 2 than a relatively more risk averse player 1, i.e., a small value of α_1. So, risk aversion lowers the SPE share of the dollar. With respect to a heterogenous population of opponents it goes the other way around, the more risk averse the opponent is the higher a particular player 1's SPE share is.

Finally, symmetry is a special case, which means $\alpha_1 = \alpha_2 = \alpha$. Then

$$c_1^1 = c_2^2 = \frac{1 - \delta^{1/\alpha}}{1 - \delta^{2/\alpha}} > \frac{1}{2}$$

and player 1's first-mover advantage implies that this player obtains the largest share of the dollar. For risk-neutral players, i.e., $\alpha_1 = \alpha_2 = 1$, we obtain $c_1^1 = c_2^2 = \frac{1}{1+\delta} > \frac{1}{2}$. The latter expressions can be found in many textbooks on game theory.

Finally, in example 3.23 we saw that a relatively more risk averse player obtains a lower SPE share of the dollar. The general case of risk aversion in the context of divide a dollar is studied in Roth (1985). There it is shown that risk aversion negatively affects one own's SPE share. In this reference, the SPE share of player 1 is investigated for two different situations, one in which player 2 has the utility function u_2 and the other with a relatively more risk averse player 2 who has the utility given by $f(u_2(c))$, where f is a monotonically increasing and concave utility function. In example 3.23 the transformation f given by $f(z) = z^{\alpha_2'/\alpha_2}$, where $\alpha_2' < \alpha_2$, transforms the utility function $\mu_2(c_2) = c_2^{\alpha_2}$ into the utility function $f(\mu_2(c_2)) = c_2^{\alpha_2'}$ of a more risk averse opponent. The main result in Roth (1985) states that player 1's SPE share of the dollar if player 2 has a utility function u_2 is smaller than the SPE share player 1 would obtain if he negotiates with an opponent with the more risk averse utility function $f(u_2(c))$.

3.5.2 A barter economy

In this section we illustrate by means of a simple example how (3.18) and (3.19) can be applied to calculate the unique pair of SPE proposals in the classic Edgeworth box corresponding to a barter economy with two goods.

Consider a barter economy with two persons and two goods, labeled good 1 and 2. A contract $c = (c_1, c_2, c_3, c_4)$ specifies the quantities c_1 and c_2 of good 1 respectively good 2 player 1 obtains after trade and, similar, c_3 and c_4 denote player 2's quantities of good 1 respectively good 2 after trade. Before the negotiations start each player has some quantities of the goods in possession, called the initial endowments. Player 1's initial endowment $w^1 = (w_1^1, w_2^1)$ and player 2's initial endowment $w^2 = (w_1^2, w_2^2)$. The total amount of good j, $j = 1, 2$, is equal to $w_j^1 + w_j^2$. Furthermore, in order to derive simple analytical solutions we impose that player i, $i = 1, 2$, owns w units of good i and none of the other good, i.e., $w^1 = (w, 0)$ and $w^2 = (0, w)$. A contract is called an allocation in terms of a general equilibrium model, see e.g., Mas-Colell, Whinston, and Green (1995). Such contract is feasible if $c_1 + c_3 \leq w$, $c_2 + c_4 \leq w$ and all quantities in c are nonnegative. The set of feasible contracts C is given by

$$C = \left\{ c \in \mathbb{R}^4 \mid c_1 + c_3 \leq w,\ c_2 + c_4 \leq w,\ c_1, c_2, c_3, c_4 \geq 0 \right\}.$$

Each player only obtains utility from his own bundle of goods according to the Cobb-Douglas utility functions given by

$$u_1(c) = c_1^{\alpha} c_2^{\beta} \quad \text{and} \quad u_2(c) = c_3^{\alpha} c_4^{\beta},$$

where $\alpha + \beta \leq 1$ and $\alpha, \beta > 0$. Note that the Hessian of second-order derivatives of u_i, $i = 1, 2$, shows that this function is concave in c. The disagreement point d corresponds to the Cobb-Douglas utility levels corresponding to the initial endowments, i.e., $d_i = (w_1^i)^{\alpha}(w_2^i)^{\beta} = 0$, $i = 1, 2$.

In order to keep the exposition brief we discard the four nonnegativity constraints. Then substitution in (3.18) with $(1-\delta)d_i = 0, i = 1, 2, g_1(c) = c_1 + c_3$

and $g_2(c) = c_2 + c_4$ yields

$$\frac{\alpha}{c_1^1}u_1(c^1) - \lambda_1^1 = 0, \quad \frac{\beta}{c_2^1}u_1(c^1) - \lambda_2^1 = 0,$$
$$\lambda_0^1 \cdot \frac{\alpha}{c_3^1}u_2(c^1) - \lambda_1^1 = 0, \quad \lambda_0^1 \cdot \frac{\beta}{c_4^1}u_2(c^1) - \lambda_2^1 = 0,$$
$$\lambda_0^1\left[\delta\left(c_3^2\right)^\alpha\left(c_4^2\right)^\beta - \left(c_3^1\right)^\alpha\left(c_4^1\right)^\beta\right] = 0,$$
$$\lambda_1^1\left(c_1^1 + c_3^1 - w\right) = 0, \quad \lambda_2^1\left(c_2^1 + c_4^1 - w\right) = 0.$$

The four equations of the first and second row can be rewritten to obtain the standard result of equal marginal rates of substitution, i.e., $\frac{c_2^1}{c_1^1} = \frac{c_4^1}{c_3^1}$. Doing so means that we get rid of the three Lagrange multipliers λ_j^1, $j = 0, 1, 2$, in these equations. Furthermore, increasing utility functions and Pareto efficiency imply that all the constraints on the third and fourth row have to be binding.

Similar, substitution in (3.19) yields

$$\lambda_0^2 \cdot \frac{\alpha}{c_1^2}u_1(c^2) - \lambda_1^2 = 0, \quad \lambda_0^2 \cdot \frac{\beta}{c_2^2}u_1(c^2) - \lambda_2^2 = 0,$$
$$\frac{\alpha}{c_3^2}u_2(c^2) - \lambda_1^2 = 0, \quad \frac{\beta}{c_4^2}u_2(c^2) - \lambda_2^2 = 0,$$
$$\lambda_0^2\left[\delta\left(c_1^1\right)^\alpha\left(c_2^1\right)^\beta - \left(c_1^2\right)^\alpha\left(c_2^2\right)^\beta\right] = 0,$$
$$\lambda_1^2\left(c_1^2 + c_3^2 - w\right) = 0, \quad \lambda_2^2\left(c_2^2 + c_4^2 - w\right) = 0.$$

and the equations on the first two rows imply $\frac{c_2^2}{c_1^2} = \frac{c_4^2}{c_3^2}$ and all the constraints on the third and fourth row are binding.

Next, the four binding feasibility constraints can be rewritten as

$$c_3^1 = w - c_1^1, \quad c_4^1 = w - c_2^1, \quad c_3^2 = w - c_1^2, \quad c_4^2 = w - c_2^2$$

and we can simply eliminate c_3^1, c_4^1, c_3^2 and c_4^2 from the equations. Then, combining all the results thus far yields the much simpler system of equations in the four unknowns c_1^1, c_2^1, c_3^2 and c_4^2 given by

$$\frac{c_2^1}{c_1^1} = \frac{w - c_2^1}{w - c_1^1}, \quad \delta\left(c_3^2\right)^\alpha\left(c_4^2\right)^\beta - \left(w - c_1^1\right)^\alpha\left(w - c_2^1\right)^\beta = 0,$$
$$\frac{c_4^2}{c_3^2} = \frac{w - c_4^2}{w - c_3^2}, \quad \delta\left(c_1^1\right)^\alpha\left(c_2^1\right)^\beta - \left(w - c_3^2\right)^\alpha\left(w - c_4^2\right)^\beta = 0,$$

Solving for the four unknowns yields

$$c_1^1 = c_2^1 = c_3^2 = c_4^2 = \tfrac{w}{1+\delta^{\alpha+\beta}}$$

and, after substitution in the feasibility constraints, we obtain the unique pair of SPE proposals (c^1, c^2) given by

$$c^1 = \left(\frac{w}{1+\delta^{\alpha+\beta}}, \frac{w}{1+\delta^{\alpha+\beta}}, \frac{\delta^{\alpha+\beta} w}{1+\delta^{\alpha+\beta}}, \frac{\delta^{\alpha+\beta} w}{1+\delta^{\alpha+\beta}}\right),$$
$$c^2 = \left(\frac{\delta^{\alpha+\beta} w}{1+\delta^{\alpha+\beta}}, \frac{\delta^{\alpha+\beta} w}{1+\delta^{\alpha+\beta}}, \frac{w}{1+\delta^{\alpha+\beta}}, \frac{w}{1+\delta^{\alpha+\beta}}\right).$$

From this solution we see that player 1's first-mover advantage is translated in a larger bundle of the goods for this player.

3.6. Related literature

The alternating offer model was first proposed in Ståhl (1972) and, later but certainly independently, in Rubinstein (1982). The latter reference analyses the divide the dollar problem as discussed in section 3.5.1 The analysis of the bargaining problem in utility representation is also conducted in e.g., Binmore (1987c), chapter 7 in Van Damme (1991) and Okada (1991a). However, the simplest proof of lemma 3.2 is given in Houba (1993) and it requires less restrictive conditions than in the other references. These conditions are the assumptions made in chapter 2 with respect to the bargaining problem in utility representation. The proofs of uniqueness in Binmore (1987c), chapter 7 in Van Damme (1991) and Okada (1991a) are either more complicated or require an additional assumption or both. In Okada (1991a) it is assumed that the Pareto frontier of S^{IR} is piecewise linear and the proof is quite elaborate. In Binmore (1987c) and Van Damme (1991) it is additionally assumed that the Pareto frontier of the set of individually rational utility pairs is differentiable.

The translation of the unique pair of SPE utilities in proposition 3.14 into the generically unique pair of SPE contracts did not receive much attention in the literature. The only reference is Muthoo (1999), but he simply states that pair of SPE contracts is unique if and only if (3.17) has a unique solution in terms of contracts, which is not very informative. The framework of chapter 2 is very helpful here, because it provides sufficient conditions for generic uniqueness. The uniqueness results in Rubinstein (1982) and Hoel (1986) for the divide a dollar problem are derived under less restrictive assumptions than concavity of the utility functions. For models of barter economies with many individual traders we refer to the surveys in e.g., Osborne and Rubinstein (1990) and Gale (2000).

The proof of proposition 3.18 is based upon the modification of the method in Shaked and Sutton (1984) proposed by Fudenberg and Tirole (1991). The original method implicitly assumed that the proposing player always proposes an acceptable proposal, leaving out to investigate whether or not the proposing player is better off by doing so. The modification also checks the latter by explicitly allowing for unacceptable proposals. For the alternating offers this did not affect the earlier results. Later on, in chapter 8 on wage bargaining, we will encounter an example where this simple modification is indeed needed. In Binmore (1987c) an interesting variant of this method is proposed. For arbitrary $\hat{t} \in T$ he simply assumes that $E\left(h_1(\hat{t}), \hat{t}\right) = S^{IR}$ and then starts working backward in time until $E\left(h_1(0), 0\right)|_{E\left(h_1(\hat{t}),\hat{t}\right)=S^{IR}}$ as an argument of the set $E\left(h_1(\hat{t}), \hat{t}\right)$ is derived. Each step in this backward induction procedure follows the same logic as the method in Shaked and Sutton (1984). Finally, under the same assumptions as in this chapter, it is shown in Binmore (1987c) that the set $E\left(h_1(0), 0\right)|_{E\left(h_1(\hat{t}),\hat{t}\right)=S^{IR}}$ converges to the set E_1 of proposition 3.18 if $\hat{t}$ goes to infinity. For readers who are interested in cases in which the limit set does not consist of a single element we refer to e.g., Binmore (1987c) and Herrero (1989).

Notes

1 The numerator of x_1^* is positive if and only if $\alpha_2 > \delta(1-\alpha_1)$. Since $\alpha_2 \geq 1-\alpha_1 > \delta(1-\alpha_1)$ for all $\delta \in (0,1)$ the latter restriction is superfluous. Similar, the numerator of x_1^* is positive if and only if $\alpha_2 > \delta^{-1}(\delta - \alpha_1)$. The denominator is positive if and only if $\alpha_2 > \delta^2(1-\alpha_1)\left(\alpha_1 + \delta^2(1-\alpha_1)\right)^{-1}$. The right hand side of the latter restriction describes a convex function and its curve goes through the points $(0,1)$, $\left(\frac{\delta}{1+\delta}, \frac{\delta}{1+\delta}\right)$ and $(1,0)$. Finally, simply by application of the definition of a convex function, it can be shown that

$$\max\left\{\delta(1-\alpha_1), \delta^{-1}(\delta-\alpha_1)\right\} \geq \frac{\delta^2(1-\alpha_1)}{\alpha_1 + \delta^2(1-\alpha_1)}.$$

2 To see this for the function $\hat{V}_2^2(s^t)$, consider a sequence of utility pairs $\langle s^\ell \rangle_{\ell=0,1,\ldots,\infty}$, $s^\ell \in S$ and $s_1^\ell < q_1(V^1)$, that is convergent to some point (in S) on the line $s_1 = q_1(V^1)$. (For explanatory simplicity, we neglect the nongeneric case of sequences $\langle s^\ell \rangle_{\ell=0,1,\ldots,\infty}$ with $\lim_{\ell\to\infty} s_2^\ell = (1-\delta)d_2 + \delta V_2^2$.) Then $\hat{V}_2^2(s^\ell) = (1-\delta)d_2 + \delta V_2^2$ for all ℓ, whereas

$$\hat{V}_2^2(\lim_{\ell\to\infty} s^\ell) = \lim_{\ell\to\infty} s_2^\ell \neq (1-\delta)d_2 + \delta V_2^2.$$

Similar arguments show the discontinuity of the function $\hat{V}'^2$.

3 Note that the ε-Nash equilibrium concept, provided it is extended to allow for subgame perfectness, would allow for ε-optimal proposals $y^\varepsilon \in T_1'(V^1)$ such that $f_2\left(q_1(V^1)\right) - \varepsilon \leq y_2^\varepsilon < f_2\left(q_1(V^1)\right)$. However, the limit set of ε-optimal proposals in $T_1'\left(V^1\right)$ as ε goes to 0 still is the empty set. Furthermore, with respect to the set $T_1(V^1)$ the optimal proposal of lemma 3.7 remains optimal. So, the ε-Nash equilibrium concept does not yield new insights.

4 Note that these arguments would change if the set S^{IR} consists of a finite number of elements. Then $T_1'(V^1)$ is also finite and an optimal proposal exists, provided this set is nonempty. Moreover, there are two cases. Either $T_1'(V^1) = T_1(V^1)$ and rejecting at the threshold level does not yield a higher utility level, or $T_1'(V^1) \subset T_1(V^1)$. In the latter case the strategy $\sigma_1'((h_1(t), s^t))$ weakly dominates $\sigma_1((h_1(t), s^t))$ at the subgame $(h_1(t), s^t)$. However, thinking in terms of taking the limit with respect to grids on a given nonempty, compact and convex set $S^{IR} \in \mathbb{R}^2$ that become finer and finer until the whole set S^{IR} is obtained in the limit, then this weakly dominance of $\sigma_1'((h_1(t), s^t))$ over $\sigma_1((h_1(t), s^t))$ vanishes in the limit when the limit grid coincides with the set S^{IR}. So, the misunderstanding is correct for finite sets $S^{IR} \in \mathbb{R}^2$, but no longer for convex sets $S^{IR} \in \mathbb{R}^2$. In the latter cases, neither $\sigma_1'((h_1(t), s^t))$ weakly dominates $\sigma_1((h_1(t), s^t))$ nor vice versa.

5 This definition is very general and easy to apply in other contexts as well. For example, the literature on Industrial Organization contains models with a first-mover (dis)advantage. Also, two-player games in normal form have two strategically equivalent representations in extensive form and, obviously, $E_1 = E_2$ implies that no first-mover (dis)advantage exists as should be the only logical conclusion.

6 The following artificial example that is absolutely not related to the theory makes this point. Consider a bargaining problem (S, d) given by $S = \{s \in \mathbb{R}^2 \,|\, s_1 + s_2 \leq 1\}$ and $d = 0$. If $E_1 = \{(\frac{1}{2}, \frac{1}{2}\}$ and $E_2 = \{(\frac{1}{4}, \frac{1}{4})\}$, then player 1 has a first-mover advantage, but player 2 does not have a second-mover disadvantage.

7 The utility function $u_i : C \to \mathbb{R}$ is concave, because for all $c, c' \in C$ and $\lambda \in [0, 1]$ we have that

$$\begin{aligned} u_i(\lambda c + (1-\lambda)c') &= \mu_i(\lambda c_i + (1-\lambda)c_i') \geq \lambda\mu_i(c_i) + (1-\lambda)\mu_i(c_i') \\ &= \lambda u_i(c) + (1-\lambda)u_i(c'). \end{aligned}$$

Chapter 4

THE NASH PROGRAM

4.1. Introduction

In Nash (1953) two approaches for solving the bargaining problem were proposed. One of these approaches, called the *strategic* approach, explicitly considers the bargaining process and the players strategic opportunities as an essential part of any theory of negotiations. Bargaining procedure 2.14 and the analysis conducted in chapter 3 is illustrative for the strategic approach. In contrast, the other approach, called the *axiomatic* approach, is not concerned with specifying the bargaining process at all. Instead in the axiomatic approach the solution concept of bargaining problems is defined by a list of certain mathematical properties, called *axioms*, that it is required to satisfy. In a manner of speaking, the axiomatic approach treats the bargaining procedure as a 'production function' on the domain of all bargaining problems that mechanically produces a bargaining solution as its output for every bargaining problem being its input. The axiomatic approach, which originates with the work of Nash (1950), pre-dates the strategic approach and for this historical incident the standard definition of the bargaining problem lacks the description of a bargaining procedure.

In this chapter we will restrict ourself to introducing *Nash's bargaining solution*, which is first proposed in Nash (1950) and one of possibly many axiomatic bargaining solutions. It will turn out that this axiomatic bargaining solution is related to the unique equilibrium of the bargaining procedure proposed in Nash (1953) and to the unique SPE derived in chapter 3. The proposed procedure is a game in normal form and the players' strategies are called demands and, hence, its name *Nash's demand game*.[1] The relationship between the axiomatic and strategic approach is one of the major results in bargaining theory. Already it was Nash himself who realized that axiomatic solution concepts should be firmly rooted in a noncooperative, strategic bargaining framework. Unless some sensible strategic bargaining model is able to support an axiomatic solution by an equilibrium strategy profile, the plausibility of the formulated axioms remain in doubt. In particular, the research agenda for investigating on the one hand axiomatic bargaining solutions and on the other hand equilibrium outcomes of strategic bargaining procedures has already been put forward in Nash (1953) and is appropriately called the *Nash program.*

This chapter is organized as follows. The first section contains the analysis of Nash's bargaining solution given the imposed list of axioms underlying it. The second section is devoted to relating the equilibrium outcomes of Nash's demand game and the equilibrium outcomes of the alternating offer procedure to Nash's axiomatic solution. The third section clarifies the relation between the axiomatic and strategic approach.

4.2. Nash's bargaining solution

An axiomatic bargaining solution for a class of bargaining problems satisfies a list of several 'desirable' mathematical properties, called *axioms*, which seem intuitively appealing for any solution of the bargaining problem. Mathematically speaking, an axiomatic solution for a class of bargaining problems is a mapping with this class of problems as its domain and the set of feasible outcomes for each bargaining problem as its image and this axiomatic solution satisfies the list of stated axioms. The *Nash bargaining solution* (NBS), which was first proposed in Nash (1950), is the unique mapping that satisfies four such axioms.

In this section Nash's four axioms are first stated in terms of the bargaining problem in utility representation and Nash's axiomatic bargaining solution is characterized. Then the results are reformulated in terms of contracts and utility functions and some methodological problems are addressed. Since some parts of the methodological problems are very technical and detailed we advise readers not familiar with axiomatic bargaining theory to skip sections 4.2.3 and 4.2.7 on first reading.

4.2.1 Utility representation

Nash's bargaining solution is derived from four axioms that are introduced in this section. As in Nash (1950) we consider bargaining problems in utility representation.

Consider the whole class of bargaining problems in utility representation (S, d), meaning that all convex sets $S \subset \mathbb{R}^2$ and disagreement points $d \in S$ are considered. Mathematically speaking, an axiomatic solution for this class of bargaining problems is a function

$$B : \{ (S,d) | \ S \subset \mathbb{R}^2 \text{ is a convex set, } d \in S \} \to S$$

that satisfies a list of axioms.

The first axiom imposed is the so-called *efficiency* axiom which states that the axiomatic bargaining solution should be Pareto efficient. In Nash (1950) this axiom is motivated as follows. If both players are rational, then they will not agree upon a Pareto inefficient utility pair $s \in S$ because they will realize that they both can do better than s and, therefore, they will continue the negotiations to reach a better agreement.

Axiom 4.1. *$B(S,d)$ is Pareto efficient.*

The second axiom imposed is the *symmetry* axiom which states that the only significant differences between the two players correspond to the shape of the set S, i.e., the players' utility functions, and the disagreement point d. In Nash (1950) symmetry is explained as 'equal bargaining skills', while in Nash (1953) this point of view is regarded as misleading. In the latter article it is argued that assuming complete information and rational players is equivalent to assuming that the players have 'equal bargaining abilities'. Before we state the symmetry axiom, we first define a *symmetric* bargaining problem as a problem (S,d) satisfying $d_1 = d_2$ and $(s_1, s_2) \in S$ if and only if $(s_2, s_1) \in S$ for all $(s_1, s_2) \in S$.

Axiom 4.2. *If (S,d) is symmetric, then $B_1(S,d) = B_2(S,d)$.*

The third axiom imposed is referred to as *invariance of affine transformations* and it states that any affine transformation of the bargaining problem (S,d) will cause the same transformation of the bargaining solution $B(S,d)$. Formally, for given $\alpha \in \mathbb{R}^2_+$ and $\beta \in \mathbb{R}^2$ the associated *affine transformation* $\left(\hat{S}(\alpha,\beta), \hat{d}(\alpha,\beta)\right)$ of (S,d) is defined as

$$\hat{S}(\alpha,\beta) = \{\hat{s} \in \mathbb{R}^2 \mid \exists s \in S : \hat{s}_i = \alpha_i s_i + \beta_i, \quad i = 1,2\}$$

and $\hat{d}_i(\alpha,\beta) = \alpha_i d_i + \beta_i$, $i = 1,2$.

The axiom of invariance of affine transformations is as follows.

Axiom 4.3. *If $\left(\hat{S}(\alpha,\beta), \hat{d}(\alpha,\beta)\right)$ is an affine transformation of (S,d), then $B\left(\hat{S}(\alpha,\beta), \hat{d}(\alpha,\beta)\right) = (\alpha_1 B_1(S,d) + \beta_1, \alpha_2 B_2(S,d) + \beta_2)$.*

The latter axiom is often used to normalize the bargaining problem. The bargaining problem (S,d) is said to be *normalized* if $d = 0$ and $S^{IR} \subseteq [0,1]^2$. Every bargaining problem can be normalized by applying the affine transformation $\alpha_i = \frac{1}{s_i^{\max} - d_i} > 0$ and $\beta_i = \frac{-d_i}{s_i^{\max} - d_i}$, $i = 1,2$.

The fourth axiom imposed is referred to as *independence of irrelevant alternatives* which states that the bargaining solution does not change if alternatives other than $B(S,d)$ are removed from the set S. It is easy to verify that the optimal solution of any optimization problem satisfies this axiom. Without going into details, we already mention that the converse is also true, namely this axiom induces that $B(S,d)$ corresponds to the solution of an optimization problem.

Axiom 4.4. *If $T \subset S$ and $B(S,d) \in T$, then $B(T,d) = B(S,d)$.*

The following theorem was first derived in Nash (1950) and it is stated without proof.

Theorem 4.5. *The unique solution that satisfies axioms 4.1-4.4 is the solution function $N(S,d)$ given by*

$$N(S,d) = \arg\max_{s \in S^{IR}} (s_1 - d_1)(s_2 - d_2). \tag{4.1}$$

The function $N(S,d)$ is the Nash bargaining solution (NBS) and this function is the only function that satisfies the four axioms above. Furthermore, none of these four axioms can be suspended in theorem 4.5, e.g., Osborne and Rubinstein (1990), Peters (1992) and Roth (1979). However, in these references it is shown that axiom 4.1 in this theorem can be replaced by the following axiom.

Axiom 4.6. *$B(S,d)$ is strictly individually rational, i.e., $B(S,d) > d$.*

Replacing the efficiency axiom by the latter axiom changes theorem 4.5 into the next theorem.

Theorem 4.7. *The unique function $B(S,d)$ that satisfies axioms 4.2-4.4 and 4.6 is the function $N(S,d)$ given in (4.1).*

The following example shows that in symmetric bargaining problems the NBS is easy to compute without solving maximization problem (4.1) if axiom 4.1 and 4.2 are combined.

Example 4.8. *Consider example 3.5 once more. The bargaining problem (S,d) features $d = 0$ and S is equal to the convex hull of $(1,0)$, $(0,1)$, $(0,0)$ and (α,α), $\alpha \in \left(\frac{1}{2},1\right)$. Thus, this bargaining problem is symmetric and axiom 4.2 imposes that $N_1(S,d) = N_2(S,d)$. Combined with axiom 4.1 we immediately obtain $N(S,d) = (\alpha,\alpha)$. The NBS can also be found by solving (4.1), i.e.,*

$$\max s_1 s_2, \quad s.t \quad s_2 + \tfrac{1-\alpha}{\alpha} s_1 \leq 1, \quad s_2 + \tfrac{\alpha}{1-\alpha} s_1 \leq \tfrac{\alpha}{1-\alpha}, \quad -s_1, -s_2 \leq 0.$$

Since the objective function $s_1 s_2$ is a strictly quasiconcave function and the set S is a convex set we have that the first-order Kuhn-Tucker conditions are necessary and sufficient conditions for a maximum at (α,α). We leave the verification to the reader.

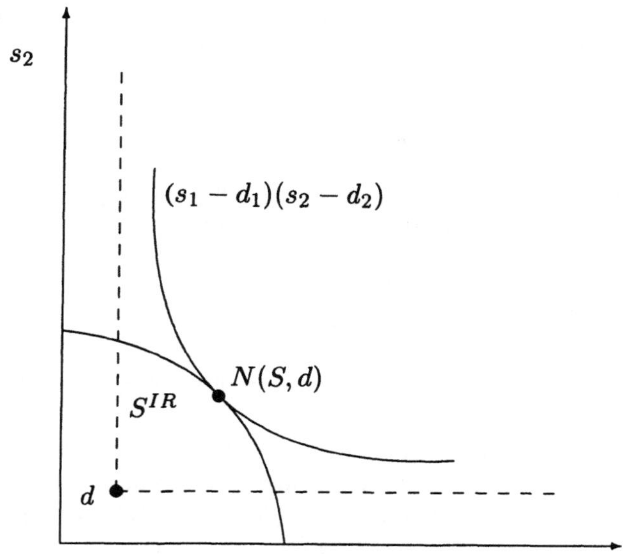

Figure 4.1. Nash's bargaining solution $N(S, d)$ is the unique point on the Pareto frontier of S^{IR} where the Nash product curve through this point is tangent to the Pareto frontier.

4.2.2 Two geometrical properties

In this section two geometrical properties of the NBS are discussed. The first geometrical property is simply the standard geometrical property of any optimization problem in the two-dimensional space: $N(S, d)$ is equal to the unique point on the Pareto frontier of S^{IR} where the Nash product curve through the NBS is tangent to the Pareto frontier of S^{IR}. Since the set S is convex and the Nash product is both strictly increasing and strictly quasiconcave in $s \in \mathbb{R}^2$, it follows that any arbitrary Nash product curve intersects the Pareto frontier of S^{IR} at most twice. However, there is a unique point on the Pareto frontier where the Nash product curve through this point is tangent to the Pareto frontier of S^{IR}, see figure 4.1 for an illustration. This figure also illustrates that the NBS satisfies axiom 4.6, i.e., strict individual rationality, as stated in theorem 4.7. This follows because $s_i = d_i$ for at least one player $i = 1, 2$ yields a Nash product of 0, whereas any $s > d$ in S yields a positive Nash product and, therefore, the maximum Nash product is also positive.

The second geometrical property states: The slope of the line through d and $N(S, d)$ is equal to minus the value (or the absolute value) of the slope of the line through $N(S, d)$ that is tangent to the Nash product curve through $N(S, d)$. In case the Pareto frontier can be described by a differentiable function f_2 we can rewrite the first-order conditions of (4.1) to obtain

$$\frac{f_2(s_1) - d_2}{s_1 - d_1} = -f_2'(s_1).$$

In case f_2 is not differentiable in $N_1(S, d)$, then the slope lies in between the left and right derivative of f_2 in $N_1(S, d)$.[2] Formally,

$$-f_2'^{-}(s_1) \leq \frac{f_2(s_1) - d_2}{s_1 - d_1} \leq -f_2'^{+}(s_1),$$

where $f_i'^{-}$ and $f_i'^{+}$ denote the left and right derivative of f_i, respectively. Figure 4.2 illustrates this second property.

In the general case of bargaining procedure 2.14 the players can influence the disagreement point d through their disagreement actions and, therefore, it is of interest to study how a change in d affects $N(S, d)$. The following proposition states a necessary and sufficient condition for a shift from d to d' that does not harm player 1, i.e., $N_1(S, d') \geq N_1(S, d)$. Consequently, since the NBS is Pareto efficient, such shift does not benefit player 2. The proposition also states a necessary and sufficient condition such that the shift from d to d' is beneficial for player 1 at the expense of player 2. Needless to say, that a shift from d' to d has opposing effects.

Proposition 4.9. *Let $d' \in S$ be located on or below the line through d and $N(S, d)$. Then $N_1(S, d') \geq N_1(S, d)$ and $N_2(S, d') \leq N_2(S, d)$. Moreover, both inequalities are strict if and only if*

$$\frac{N_2(S, d) - d_2'}{N_1(S, d) - d_1'} > -f_2'^{+}(N_1(S, d)).$$

Corollary 4.10. *If f_2 is differentiable in $N_1(S, d)$, then the condition for strict inequalities is satisfied if d' lies below the line through d and $N(S, d)$.*

The results of the last proposition can be applied to the subclass of bargaining problems (S, d) with $d \in S$ in order to define and characterize the set of disagreement points $d \in S$ that all yield a NBS that is equal to some $s^* \in S^{PE}$. Loosely speaking, we keep the set S fixed while varying the disagreement point $d \in S$ and we are interested in the set of disagreement points such that $N(S, d) = s^*$. Denote the latter set as $D(s^*) \subseteq S$. Reformulation of proposition 4.9 means that $N(S, d) = s^*$ if and only if the slope of the line

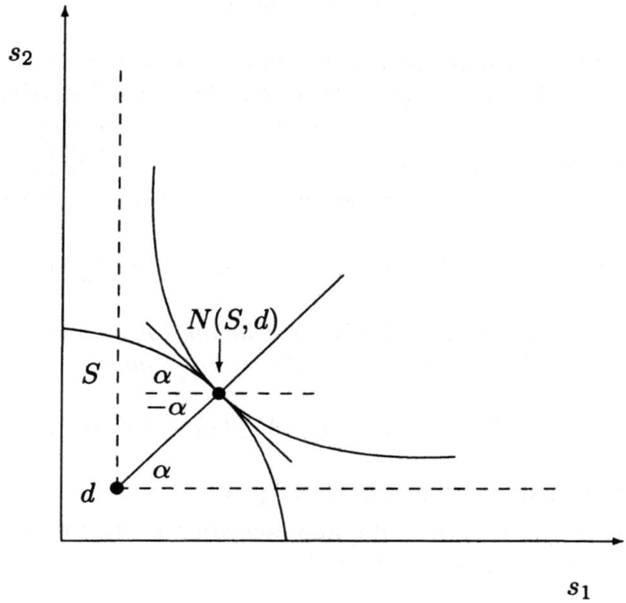

Figure 4.2. The slope of the line through d and $N(S,d)$ is equal to the absolute value of the slope of the line through $N(S,d)$ that is tangent to the Nash product curve through $N(S,d)$.

through $d \in S$ and $N(S,d) = s^*$ lies in between the left and right derivative of f_2 in s_1^*, i.e.

$$-f_2^{\prime -}(s_1^*) \leq \frac{s_2^* - d_2}{s_1^* - d_1} \leq -f_2^{\prime +}(s_1^*).$$

Rewriting these inequalities into the functional form of a line in $\mathbb{R}^2$ yields

$$D(s^*) = \left\{ d \in S \,\middle|\, s_2^* - f_2^{\prime +}(s_1^*)(d_1 - s_1^*) \leq d_2 \leq s_2^* - f_2^{\prime -}(s_1^*)(d_1 - s_1^*) \right\}.$$

Moreover, if f_2 is differentiable in s_1^*, then $D(s^*)$ is the intersection of the line $d_2 = s_2^* - f_2'(s_1^*)(d_1 - s_1^*)$ with the set S. Since $d \in D(s^*)$ implies that $N(S,d) = s^*$ we can regard the set $D(s^*)$ as the inverse of the function $N(S,d)$ with respect to disagreement points $d \in S$. Obviously, the set $D(s^*)$ is a convex set, e.g., Jansen and Tijs (1981).

The next example illustrates how to derive the set $D(s^*)$.

Example 4.11. *Consider example 4.8. Recall that $N(S,d) = (\alpha,\alpha)$. The function f_2 is not differentiable in $s_1 = \alpha$, because*

$$-f_2'^{-}(\alpha) = \tfrac{1-\alpha}{\alpha} \leq 1 \leq -f_2'^{+}(\alpha) = \tfrac{\alpha}{1-\alpha}.$$

The tangent line of the Nash product curve through (α,α) has a slope with absolute value of 1 and this slope lies in between the absolute values of the left and right derivative of f_2 in $s_1 = \alpha$. Thus, the first graphical property is satisfied. Also the second geometrical property is satisfied, because $\frac{N_1(S,d)-d_1}{N_2(S,d)-d_2} = 1$ lies in between the absolute values of the left and right derivatives. Furthermore,

$$D((\alpha,\alpha)) = \left\{ d \in \mathbb{R}^2_+ \;\middle|\; \alpha + \tfrac{\alpha}{1-\alpha}(d_1 - \alpha) \leq d_2 \leq \alpha + \tfrac{1-\alpha}{\alpha}(d_1 - \alpha) \right\}.$$

See figure 4.3 for an illustration of this set. Similar, for s^* on the line piece $s_2 = 1 - \frac{1-\alpha}{\alpha}s_1$, $s_1 \in [0,\alpha)$, we have $f_2'(s_1^*) = -\frac{1-\alpha}{\alpha}$ and

$$D(s^*) = \left\{ d \in \mathbb{R}^2_+ \mid d_2 = 1 - 2\tfrac{1-\alpha}{\alpha}s_1^* + \tfrac{1-\alpha}{\alpha}d_1 \right\}.$$

4.2.3 Bargaining in the contract space

In the axiomatic approach the utility representation is often taken as one of the primitives. Or, to quote from Nash (1950),

> We shall regard two anticipations (i.e., lotteries, eds.) which have the same utility for any utility function corresponding to either individual as equivalent so that the graph (i.e., the set S, eds.) becomes a complete representation of the essential features of the situation.

This means that all bargaining problems with contract space C, utility functions u_i, $i = 1.2$, and exogenous disagreement outcome that have the same utility representation (S,d) are treated as being equal. In Roemer (1988) a reconstruction of the axiomatic approach is undertaken that treats the bargaining problem as its primitive. The reconstructed axioms are less restrictive, which we will not further discuss. One remark is appropriate, the subject in section 4.2.3 is very technical and only of interest to those readers that are interested in advanced matters.

For explanatory reasons we assume that there exists a $c^d \in C$ such that $d_i = u_i(c^d)$, $i = 1,2$. The class of bargaining problems as introduced in chapter 2 is then given by $\{(C, \{u_i\}_{i=1,2}, c^d)\}$, meaning that all bargaining problems that satisfy the assumptions made in section 2.2 are considered as its domain. Since we cannot rule out the nongeneric multiplicity of Pareto efficient contracts (see section 2.2.7), an axiomatic solution for this class of bargaining problems is, mathematically speaking, a mapping or correspondence

$$B : \{(C, \{u_i\}_{i=1,2}, c^d)\} \rightarrow C$$

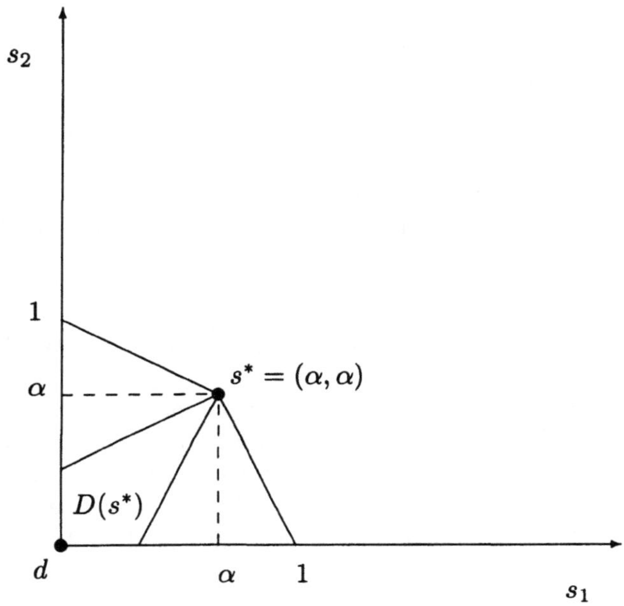

Figure 4.3. The set $D(s^*)$ for $s^* = (\alpha, \alpha)$ in example 4.11.

that satisfies a list of axioms. Note that we abuse notation here by using the symbol B twice, but from now on we will always state the arguments of B in order to specify the exact context.

Similar as before, the first axiom is the so-called *efficiency* axiom .

Axiom 4.12. *Each element of* $B(C, \{u_i\}_{i=1,2}, c^d)$ *is Pareto efficient.*

The second axiom is a reformulation of the *symmetry* axiom , i.e., the symmetric outcome is a bargaining solution if the bargaining problem is symmetric. This requires that we first define a symmetric bargaining problem. In words, a symmetric bargaining problem means that there are no significant differences between the two players with respect to the set C, the players' utility functions, and the disagreement point. In order to make this simple statement formal we follow Rubinstein, Safra, and Thomson (1992), who first introduce the notion of a symmetry function for bargaining problems and then define symmetric bargaining problems as those bargaining problems that admit such symmetry function. The symmetry function is defined as follows.

Definition 4.13. *The function $\phi : C \to C$ is a symmetry function relative to the bargaining problem $(C, \{u_i\}_{i=1,2}, c^d)$ if it satisfies the following three properties:*

i) $\phi(c^d) = c^d$,

ii) $\phi(c) = c'$ *if and only if* $\phi(c') = c$ *for all* $c \in C$ *and*

iii) $u_i(c) \geq u_i(c')$ *if and only if* $u_i(\phi(c)) \geq u_i(\phi(c'))$ *for all* $c \in C$.

Every bargaining problem that admits a symmetry function is said to be a symmetric bargaining problem. The following axiom is the symmetry axiom.[3]

Axiom 4.14. *If the bargaining problem $(C, \{u_i\}_{i=1,2}, c^d)$ is symmetric, then*

$$\{c \in C | \ \arg\max_{c \in C} \tfrac{1}{2}u_1(c) + \tfrac{1}{2}u_2(c)\} \subseteq B(C, \{u_i\}_{i=1,2}, c^d).$$

The third axiom translates the *invariance of affine transformations* and states that affine transformations of the utility functions do not change the physical contracts in $B\left(C, \{u_i\}_{i=1,2}, c^d\right)$. Formally, for given $\alpha_i \in \mathbb{R}_+$ and $\beta_i \in \mathbb{R}$, $i = 1, 2$, the affine transformation $\hat{u}_i : C \to \mathbb{R}$ of u_i is defined as $\hat{u}_i(c) = \alpha_i u_i(c) + \beta_i$ for all $c \in C$. The axiom of invariance of affine transformations is as follows.

Axiom 4.15. *If $\hat{u}_i$, $i = 1, 2$, is an affine transformation of u_i, then*

$$B(C, \{\hat{u}_i\}_{i=1,2}, c^d) = B(C, \{u_i\}_{i=1,2}, c^d).$$

Note that c^d does not shift, only the utilities attached to it, i.e., $\alpha_i u_i(c^d) + \beta_i$.

Finally, axiom 4.4 is replaced by an axiom called *consistency across dimensions*. The rationale behind this axiom is one of sequential allocation consistency. Let us illustrate this by means of an example. Consider a two-person barter economy with two goods and the second good does not enter player 1's utility function. Then negotiations can be handled in either of two ways. Either the players simultaneously negotiate over both goods or, alternatively, player 2 first receives the aggregate endowment of the second good and then the players negotiate over the division of the first good. The axiom imposes that both ways of handling the negotiations should not affect the final division of the first good across the players. So, the bargaining solution for the bargaining problem with the reduced dimension must describe the allocation of the goods preferred by both players.

Of course, the barter economy is special because the restrictions upon $c_k \in [0, w_k]$, $k = 1, 2$, are not linked with each other. In for example the divide a dollar problem of section 3.5.1 the variable c_i, $i = 1, 2$, does not enter in the opponent's utility function but $c_1 + c_2 \leq 1$ links c_1 and c_2 and prevents us from lowering the dimensions. Similarly, production in a barter economy is likely to link the players' consumption goods. Generally, links between the variables in C may prevent reducing the dimension unless removing the dimension or variable c_k does not affect the constraints on other variables.

All these considerations are made formal as follows. First, player i, $i = 1, 2$, is said not to derive utility from c_k, $k = 1, \dots, n$, if his utility function u_i is constant in c_k, i.e., the partial derivative $\frac{\partial u_i(c)}{\partial c_k} = 0$ for all $c \in C$. Next, we deal with possible links in the contract space. Suppose, after suitable renumbering, that player i, $i = 1, 2$, for some $m_i \in \{1, \dots, n-1\}$ does not derive utility from the variables c_k, $k = m_i + 1, \dots, n$, while his opponent does. The question is, do there for some $\ell_i \in \{m_i + 1, \dots, n\}$ exist two sets, denoted as C^{ℓ_i} and $C^{n-\ell_i}$, such that $C^{\ell_i} \subset \mathbb{R}^{\ell_i}$, $C^{n-\ell_i} \subset \mathbb{R}^{n-\ell_i}$ and

$$C = \left\{ (c^{\ell_i}, c^{n-\ell_i}) \in \mathbb{R}^n \;\middle|\; c^{\ell_i} \in C^{\ell_i}, c^{n-\ell_i} \in C^{n-\ell_i} \right\}.$$

If the answer is affirmative, then we say that C can be *reduced* by $n - \ell_i$ dimensions to dimension ℓ_i.

Finally, we specify utility functions on C^{ℓ_i}. If player i, $i = 1, 2$, is the player who does not derive utility from the removed variables, this is simply the same utility function as on C. With respect to his opponent, player j, $j = 1, 2$ and $j \neq i$, the value of the variables removed matters. Following Roemer (1988) we take as a reference point the values of c_k, $k = \ell_i + 1, \dots, n$, associated with $B(C, \{u_i\}_{i=1,2}, c^d)$, which is generically a singleton. Denote $(\hat{c}^{\ell_i}, \hat{c}^{n-\ell_i})$ as one of the elements in $B(C, \{u_i\}_{i=1,2}, c^d)$. Then we define the utility functions $u_i^{\ell_i} : C^{\ell_i} \to \mathbb{R}$ and $u_j^{\ell_i} : C^{\ell_i} \to \mathbb{R}$ as

$$u_i^{\ell_i}(c^{\ell_i}) = u_i(c^{\ell_i}, \hat{c}^{n-\ell_i}) \quad \text{and} \quad u_j^{\ell_i}(c^{\ell_i}) = u_j(c^{\ell_i}, \hat{c}^{n-\ell_i}).$$

With respect to the disagreement *utilities* in the ℓ_i-dimensional bargaining problem a similar transition has to be made and this yields $u_i(c^{\ell_i,d}, \hat{c}^{n-\ell_i})$ and $u_j(c^{\ell_i,d}, \hat{c}^{n-\ell_i})$ as the players' disagreement utilities. Unfortunately, since the bargaining solution requires a specification in terms of the disagreement *point* in the contract space C^{ℓ_i} some cumbersome notation is needed. The disagreement point c^d is to be replaced by $c^{\ell_i,d} = (c_1^d, \dots, c_{\ell_i}^d, \hat{c}^{n-\ell_i}) \in C$, because the utility functions $u_i^{\ell_i}$ require this. We can now state the consistency axiom across dimensions.[4]

Axiom 4.16. *Let, for some $m_i \in \{1, \dots, n-1\}$, player i, $i = 1, 2$, be the player who does not derive utility from c_k, $k = m_i + 1, \dots, n$, let it be possible to reduce the set C to dimension $\ell_i \geq m_i$ and let $(\hat{c}^{\ell_i}, \hat{c}^{n-\ell_i}) \in B(C, \{u_i\}_{i=1,2}, c^d)$. Then only if $u_j(c^{\ell_i,d}, \hat{c}^{n-\ell_i}) = d_j$ we impose $\hat{c}^{\ell_i} \in B(C^{\ell_i}, \{u_i^{\ell_i}\}_{i=1,2}, c^{\ell_i,d})$.*

The following theorem was first derived in Roemer (1988) and it is stated without proof.

Theorem 4.17. *The unique correspondence $B(C, \{u_i\}_{i=1,2}, c^d)$ that satisfies axioms 4.12-4.16 is the correspondence $N : \{C, \{u_i\}_{i=1,2}, c^d\} \to C$ given by*

$$N(C, \{u_i\}_{i=1,2}, c^d) = \left\{ \arg \max_{c \in C^{IR}} \, (u_1(c) - d_1)\,(u_2(c) - d_2) \right\}. \qquad (4.2)$$

Moreover, generically this correspondence consists of a single element.

Once more we mention that the Nash product has no economic meaning, which is discussed in section 4.2.6 of this chapter.

The following example illustrates the NBS in the contract space. Moreover, it also illustrates the nongeneric nonuniqueness in NBS contracts.

Example 4.18. *Consider the bargaining problem of example 2.10 once more, i.e.*

$$C = \left\{ c \in \mathbb{R}^3_+ \mid c_1 + c_2 + c_3 = 1 \right\}, \quad d = 0$$

and for $\alpha \in (0,1)$ the players' utility functions are given by

$$u_1(c) = c_1 + \alpha c_2 \quad \textit{and} \quad u_2(c) = \alpha c_2 + c_3.$$

This bargaining problem induces the same Pareto frontier as the bargaining problem in utility representation (S,d) of example 4.8 and we immediately obtain

$$N(S,d) = \begin{cases} \left(\frac{1}{2}, \frac{1}{2}\right), & \text{if } \alpha \le \frac{1}{2}, \\ (\alpha, \alpha), & \text{if } \alpha \ge \frac{1}{2}. \end{cases}$$

Similar as in example 3.16 the NBS contracts c^* solve

$$c_1^* + \alpha c_2^* = N_1(S,d) \quad \text{and} \quad \alpha c_2^* + c_3^* = N_2(S,d).$$

Then for $\alpha < \frac{1}{2}$ and $\alpha > \frac{1}{2}$ we obtain the unique NBS contract $c^* = \left(\frac{1}{2}, 0, \frac{1}{2}\right)$, respectively, $c^* = (0,1,0)$. However, multiplicity arises for the nongeneric case $\alpha = \frac{1}{2}$, where every contract $c^* = (\gamma_1, \gamma_2, \gamma_3)$ given by

$$\gamma_1 = \tfrac{1}{2} - \tfrac{1}{2}\gamma_2, \quad \gamma_3 = \tfrac{1}{2} - \tfrac{1}{2}\gamma_2, \quad \gamma_2 \in [0,1]$$

is a NBS contract. So, similar as for alternating offers, we obtain uniqueness for the NBS in the utility representation and generic uniqueness in terms of contracts.

4.2.4 Computation of axiomatic contracts

In many economic applications standard optimization methods can be applied, because it is additionally assumed that the utility functions are twice differentiable and the contract space C can be represented by a finite number of twice differentiable and quasiconvex functions as in (2.1). In this section we impose the latter assumptions and translate the maximization problem of theorem 4.17 into the associated Kuhn-Tucker first-order conditions that can be solved with standard numerical methods.

In (2.1) the contract space C is represented by k twice-differentiable and quasiconvex functions $g_j : C \to \mathbb{R}$, $j = 1, \ldots, k$, i.e.

$$C = \{c \in \mathbb{R}^n \mid g_j(c) \leq 0,\ j = 1, \ldots, k\}.$$

Then the maximization problem stated in theorem 4.17 can be rewritten as the nonlinear programming problem given by

$$\max_c (u_1(c) - d_1)(u_2(c) - d_2), \quad \text{s.t.} \quad \begin{aligned} g_j(c) &\leq 0, \quad j = 1, \ldots, k, \\ d_i - u_i(c) &\leq 0, \quad i = 1, 2. \end{aligned}$$

The objective function is quasiconcave, which can be easily seen as follows. Similar as in consumer theory we have that a monotone increasing transformation of the objective function does not change the optimum expressed in the contract space. The transformation $f : \mathbb{R} \to \mathbb{R}$ given by $f(z) = \ln(z)$ is such a transformation and yields $\ln(u_1(c) - d_1) + \ln(u_2(c) - d_2)$ as the objective function. Since the function u_i, $i = 1, 2$, is concave we have that $\ln u_i$ is concave. So, the latter objective function is quasiconcave and the domain is convex. Since the logarithmic form and the Nash-product form induce the same indifference curves in $\mathbb{R}^n$ it then follows that the Nash product curves are quasiconcave. Then, similar as in section 2.2.5, the Kuhn-Tucker first-order conditions are also sufficient conditions for a global maximum.

Denote λ_j, $j = 1, \ldots, k$, as the Lagrange multipliers of the maximization problem corresponding to the constraint involving the function g_j. Similar, μ_i, $i = 1, 2$, is the Lagrange multiplier for player i's individual rationality constraint. Application of the Kuhn-Tucker first-order conditions yields

$$\begin{aligned} &(u_2(c) - d_2)\frac{\partial u_1(c)}{\partial c_\ell} + (u_1(c) - d_1)\frac{\partial u_2(c)}{\partial c_\ell} + \qquad\qquad (4.3) \\ &-\sum_{j=1}^{k} \lambda_j \frac{\partial g_j(c)}{\partial c_\ell} - \mu_1 \frac{\partial u_1(c)}{\partial c_\ell} - \mu_2 \frac{\partial u_2(c)}{\partial c_\ell} = 0, \qquad \ell = 1, \ldots, n, \\ &\lambda_j g_j(c) = 0, \qquad j = 1, \ldots, k, \\ &\mu_i(u_i(c) - d_i) = 0, \qquad i = 1, 2. \end{aligned}$$

Since the NBS is strict individually rational it immediately follows that in the optimum $\mu_1 = \mu_2 = 0$, which reduces the system to a smaller system of $n + k$ equations and same number of variables.

According to the efficiency axiom the generically unique NBS contract is Pareto efficient. Nevertheless, we investigate how the system of $k + n$ variables and the same number of equations relate to system (2.3) that characterizes Pareto efficient contracts and that features the extra variable λ_0 and the extra equation $\lambda_0 \left(-u_2(c) + \overline{u}_2\right) = 0$. Let c^* and $\lambda_1^*, \ldots, \lambda_k^*$ solve (4.3). Then the number $u_2\left(c^*\right) - d_2 > 0$ and we can divide all $k + n$ equations by this number. The latter yields a system of equations that has much resemblance with (2.3). Indeed, a solution for (2.3) can now be given simply by choosing

$$
\begin{aligned}
&c = c^*, && \overline{u}_2 = u_2\left(c^*\right), && \qquad (4.4)\\
&\lambda_0 = \tfrac{u_1(c^*)-d_1}{u_2(c^*)-d_2} > 0, && \lambda_j = \tfrac{\lambda_j^*}{u_2(c^*)-d_2}, && j = 1, \ldots, k.
\end{aligned}
$$

It can be easily checked that the latter solution solves (2.3). So, this is an alternative way to show that the generically unique NBS contract is Pareto efficient.

4.2.5 Two critical remarks

In this section we formulate two points of critique with respect to Nash's bargaining solution as discussed in the previous sections. The first point concerns the lack of a proper economic interpretation of the Nash product. The second point has to do with the meaning of the invariance of affine transformations axiom in Nash (1950), which is in our view not properly understood in the literature. This point is made by reformulating the latter axiom as an axiom that states that the bargaining solution in terms of the physical contract should be independent of the utility representation of the players' preferences and, then, we show its implications. A resolution to both points of critique is proposed in e.g., Rubinstein, Safra, and Thomson (1992), which is postponed to the next two sections.

The first point of critique concerns the notion of the Nash product that served us so well in characterizing Nash's bargaining solution. One of the primitives of the bargaining problem are the players' utility functions and these are economically well defined and understood. The problem is that the (Nash) product of two utility functions implicitly induces a complete, transitive, continuous and quasiconcave preference relation over the contract space C, but this preference relation is not specified as being part of the bargaining problem. Therefore, the economic meaning of the product of two utility functions is unclear. Of course, one could argue that the Nash product represents a social welfare function but for a theory of bargaining this is rather awkward since each of the negotiating players is concerned with his own well being while the resulting outcome can be mistakenly seen as if they jointly maximize a social welfare function. Therefore, the Nash product should only be seen as a mathematical construct that is useful in characterizing the NBS but that lacks an economic foundation. In the next section an alternative definition for the NBS is introduced that is based upon the primitives of the bargaining problem only and that therefore does not rely on the Nash product (except for its characterization).

The second point of critique concerns the meaning of the invariance of affine transformations axiom, which we believe is the view Nash (1950) had in mind. First, in Nash (1950) the contract space is a simplex of lotteries with arbitrary dimensions and the players' utility functions are expected utility functions. This interpretation follows from this seminal article, where the second section is entirely devoted to the theory of expected utility over lotteries. This point of view is further supported by the following two quotes from the same article:

> In a bargaining situation one anticipation (i.e., lottery, eds.) is especially distinguished; this is the anticipation of no cooperation between the bargainers. It is natural, therefore, to use utility functions for the two individuals which assign the number zero to this anticipation. This leaves each individual's utility function determined only up to multiplication by a positive real number. Henceforth, any utility functions used shall be understood to be so chosen.

and further on:

> Of course, the graph (i.e., the set S, eds.) is only determined up to changes of scale since the utility functions are not completely determined.

So, the assumption of expected utility functions over the simplex of lotteries coincides with the reconstructed axiom 4.15 and is the same as axiom 4.3 in terms of the utility representation.

The true interpretation of axiom 4.15 is that the bargaining solution in terms of the physical contract should not depend on the particular form of the utility functions that are supposed to represent the players' preferences. Since expected utility functions are unique up to an affine transformation, see e.g., chapter 6 in Mas-Colell, Whinston, and Green (1995), such affine transformations must not influence the physical bargaining outcome. However, in general C is not a simplex of lotteries and the players' preferences may not correspond to utility functions that are unique up to an affine transformation. For example, in a standard barter economy the players' have complete, transitive, convex and continuous preferences over bundles of goods and the associated utility functions are unique up to a monotone transformation. Example 4.21 shows that the NBS division in the latter context depends upon the utility representation chosen and this can be considered as an undesirable property of a bargaining solution.

It would be natural to replace the invariance of affine transformations axiom by an axiom that states that the physical bargaining solution should be independent of the utility representation of the players' preferences. However, doing so would deprive economics of its most popular axiomatic bargaining solution as we will argue in the remainder of this section.

First, we state an alternative axiom that reflects the interpretation of axiom 4.15 that John Nash intended in our view, which we will call the axiom of *invariance of utility representation.*

Axiom 4.19. *If the utility functions u_i and u_i', $i = 1, 2$, both represent player i's preferences on C, then $B(C, \{u_i\}_{i=1,2}, c^d) = B(C, \{u_i'\}_{i=1,2}, c^d)$.*

How this axiom effects any bargaining solution depends upon the players' preferences. For the class of additively separable utility functions on C, i.e., utility functions of the form $u(c) = \sum_{k=1}^{n} u_k(c_k)$, it is known that these functions are unique up to an affine transformation, see e.g., p. 99 in Mas-Colell, Whinston, and Green (1995). This class includes the special case of expected utility functions on a simplex of lotteries. For additive separable utility functions axiom 4.19 requires that the physical bargaining solution is independent of affine transformations similar as in axiom 4.15 and these utility functions automatically satisfy this condition. In the context of a barter economy, the utility functions are unique up to a monotone transformation $f : \mathbb{R} \to \mathbb{R}$ and the axiom imposes $B(C, \{u_i\}_{i=1,2}, c^d) = B(C, \{f(u_i)\}_{i=1,2}, c^d)$, which is more demanding than axiom 4.15. But for this class of utility functions we obtain an impossibility result: If axiom 4.19 replaces axiom 4.15 in theorem 4.17, then the NBS does no longer satisfy all four axioms. Since the NBS was already the only surviving solution in theorem 4.17 this simply means that we also lose the only candidate for the solution, i.e., an impossibility. The following theorem formally states this result.

Theorem 4.20. *Let each player's preferences belong to the class of preferences that can be represented by utility functions that are unique up to some monotone transformation including all affine transformations.*

i) If each player's utility function is unique up to an affine transformation, then the unique correspondence $B(C, \{u_i\}_{i=1,2}, c^d)$ that satisfies axioms 4.12, 4.14, 4.16 and 4.19 is the NBS correspondence $N(C, \{u_i\}_{i=1,2}, c^d)$.

ii) If one of the player's utility function does not satisfy the condition stated under i), then no correspondence $B(C, \{u_i\}_{i=1,2}, c^d)$ exists that satisfies axioms 4.12, 4.14, 4.16 and 4.19, i.e., $B(C, \{u_i\}_{i=1,2}, c^d) = \emptyset$.

The important message is that there does not exist a NBS contract if the axiom of invariance of utility representation replaces the invariance of affine transformations axiom, unless both player's preference relation can be represented by utility functions that are unique up to an affine transformation such as for example the class of additively separable utility functions. The latter class of preferences includes expected utility (as originally proposed in Nash (1950)) and discounting in intertemporal economic problems of a one-dimensional stream of a per-period variable. Since in most economic applications preferences are considered that do not correspond to this restricted class, this theorem deprives economics of its most popular axiomatic bargaining solution unless a resolution is found, which is proposed in the following two sections. Note that the same argument in principle applies to every axiomatic bargaining solution that has the invariance of affine transformations as one of its axioms, such as for example the solution in Kalai and Smorodinsky (1975).

The concluding example illustrates that the NBS need not satisfy axiom 4.19.

Example 4.21. *Consider the barter economy of section 3.5.1, i.e.*

$$C = \left\{c \in \mathbb{R}^4 \mid c_1 + c_3 \leq w,\ c_2 + c_4 \leq w,\ c_1, c_2, c_3, c_4 \geq 0\right\},$$

$c^d = (w, 0, 0, w)$ *and*

$$u_1(c) = c_1^\alpha c_2^\beta \quad \text{and} \quad u_2(c) = c_3^\alpha c_4^\beta,$$

where $\alpha + \beta \leq 1$ *and* $\alpha, \beta > 0$. *Application of (4.3) yields*

$$N(C, \{u_i\}_{i=1,2}, c^d) = \left(\tfrac{1}{2}w, \tfrac{1}{2}w, \tfrac{1}{2}w, \tfrac{1}{2}w\right).$$

Next, each agent's utility function in the classical Edgeworth box is unique up to a monotone transformation. One particular transformation is $f(z) = \sqrt{z}$. *Let* $u_1'(c) = f(u_1(c)) = c_1^{\alpha/2} c_2^{\beta/2}$. *It can then easily be verified that the unique NBS allocation is equal to*

$$N(C, \{u_1', u_2\}, c^d) = \left(\tfrac{1}{3}w, \tfrac{1}{3}w, \tfrac{2}{3}w, \tfrac{2}{3}w\right) \neq N(C, \{u_i\}_{i=1,2}, c^d).$$

This shows that the axiomatic bargaining solution in terms of the negotiated allocation critically depends upon player 1's utility representation of his complete, transitive, (quasi)concave and continuous preference relation. General equilibrium theory treats both cases of the Edgeworth box as being equal, whereas the NBS does not. We regard the latter as an undesirable property which questions the validity of the theory.

4.2.6 A reinterpretation

In Rubinstein, Safra, and Thomson (1992) an alternative definition for the Nash bargaining solution is proposed that does not refer to the Nash product. The rationale behind this definition is that players' attitudes towards risk matter. This risk has to be introduced first and is understood to arise from some kind of (unspecified) bargaining process. In this section the equivalence between this alternative definition and Nash's bargaining solution in case of expected utility functions on the enlarged domain is established.

As mentioned, the players' attitudes towards risk matter but in order to have an interesting and applicable economic theory the contract space should not be restricted to a simplex of lotteries. A simple way out is to enlarge the domain of the preference relations to $\hat{C} = C \times [0,1]$ and introduce utility functions on it as follows. For explanatory simplicity we assume that there exists a contract $c^d \in C$ such that $u_i(c^d) = d_i$, $i = 1,2$, that is taken as the disagreement point. Then for every $p \in [0,1]$ we define the lottery $pc + (1-p)c^d$ as the two-prize lottery in which the contract c has probability p and the disagreement outcome c^d has the complementary probability of $1 - p$.[5] Obviously, every $c \in C$ corresponds to such two-prize lottery with $p = 1$, i.e., $(c, 1) \in \hat{C}$. We define $\hat{c}^d = (c^d, 1)$. Next, player i, $i = 1,2$, is assumed to have the utility function $\hat{u}_i : \hat{C} \rightarrow \mathbb{R}$ over these two-prize lotteries given by $\hat{u}_i(c,p)$ such that $\hat{u}_i(c,1) = u_i(c)$. Note that expected utility with respect to p corresponds to $\hat{u}_i(c,p) = pu_i(c)+(1-p)u_i(c^d)$ and $\hat{u}_i$ is unique up to an affine transformation.

The alternative definition in Rubinstein, Safra, and Thomson (1992) is best understood as follows. Suppose $c^* \in C$ is the contract that currently lies on the negotiation table and consider a contract $c \in C$ such that $\hat{u}_1(c,1) > \hat{u}_1(c^*,1)$, i.e., $u_1(c) > u_1(c^*)$. So, player 1 could raise his objections against c^* and as an alternative propose c instead. Now suppose that every time one of the players objects against the proposal currently on the table the players face some risk $p \in [0,1]$ of a permanent breakdown after which each player receives his disagreement utility. Then player 1 will only raise his objections and propose c if the utility $\hat{u}_1(c,p) > \hat{u}_1(c^*,1) = u_1(c^*)$. If player 2 would give in to player 1 by accepting c, then c^* can never be an outcome of the negotiations. So, if c^* is a candidate for being the bargaining outcome of our theory, then player 2 has to successfully raise his counter objections against c. The idea of the definition is that player 2 raises c^* as his counter objection against c and is willing to face the same risk to put c^* back on the table again, which would make player 1's effort fruitless. Player 2 is only willing to do so if $\hat{u}_2(c^*,p) \geq \hat{u}_2(c,1) = u_2(c)$. In a manner of speaking, if player 1 has the objection c against c^* then player 2 should have the counter objection c^* against c. The definition requires that each player should have a counter objection against any objection the opponent might have independent of the probability p. So, each of the players should be able to successfully defend c^* against every possible objection against it. The following definition formalizes this interpretation.

Definition 4.22. *The contract $c^* \in C$ is a NBS if for all $p \in [0,1]$ and for all $c \in C$ and $i,j = 1,2$, $j \neq i$, it holds that if $\hat{u}_i(c,p) > \hat{u}_i(c^*,1)$ then $\hat{u}_j(c^*,p) \geq \hat{u}_j(c,1)$.*

For the special case of utility functions $\hat{u}_i$, $i = 1,2$, that represent expected utility preferences upon $\hat{C}$ the alternative definition for the NBS c^* is equivalent to the maximization of the Nash product.

Theorem 4.23. *Let* $\hat{u}_i(c,p) = pu_i(c) + (1-p)u_i(c^d)$, $i = 1,2$, *for all* $c \in C$, *then the contract* c^* *satisfies definition 4.22 if and only if*

$$c^* = \arg\max_{c \in C^{IR}} \left(u_1(c) - d_1\right)\left(u_2(c) - d_2\right).$$

Note that expected utility in p corresponds to $\hat{u}_i(c,p) = pu_i(c) + (1-p)u_i(c^d)$ and that then $\hat{u}_i$ is unique up to an affine transformation. The latter means that we have defined a domain for which it is natural that $\alpha_i u_i(c) + \beta_i$ represents the same preference relation as u_i, whereas a (nonlinear) monotone transformation of u_i yields a different preference relation. It is this particular interpretation that is consistent with axiom 4.15.

Finally, for the special case of expected utility functions u_i over a simplex of lotteries C we do not need the extension of the domain to $\hat{C}$. The reason is that for all $p \in [0,1]$ the compound lottery $pc + (1-p)c^d$, $i = 1,2$, is equivalent to the simple lottery $c' \in C$ given by $c'_k = pc_k + (1-p)c^d_k$, $k = 1, \ldots, n$, and therefore

$$\hat{u}_i(c,p) = \hat{u}_i(c',1) = u_i(c'), \quad i = 1,2.$$

4.2.7 Alternative axioms for Nash's bargaining solution

As already mentioned, the class of bargaining problems considered in Nash (1950) feature a contract space of lotteries and players that have expected utility preferences. In Rubinstein, Safra, and Thomson (1992) the latter domain is extended to allow for nonexpected utility preferences and these preferences do not satisfy the axiom of invariance of affine transformations. So, in order to extend axiomatic bargaining theory the latter axiom is removed. Also the axiom of independence of irrelevant alternatives is dropped, because it is often criticized in the literature. Instead, one new axiom is proposed that replaces the latter two axioms on the domain of nonexpected utility preferences. In Rubinstein, Safra, and Thomson (1992) it is then shown that replacing the two old axioms by this new axiom on the domain of nonexpected utility functions uniquely characterizes the alternative definition of the Nash bargaining solution stated in definition 4.22, provided the domain of preferences considered is restricted by imposing two additional assumptions. In this section these three axioms are discussed and the main result in Rubinstein, Safra, and Thomson (1992) is stated. However, nowhere in the latter reference there is any hint to the discussion leading to theorem 4.23 . Therefore, we investigate whether or not the new axiom automatically implies that axiom 4.19 of invariance of utility representation is also satisfied and we provide an affirmative answer. This means that the critique of the previous section no longer applies to the new axiomatization of the NBS. Similar as for section 4.2.3, the discussion in this section is very technical and only of interest to those readers that are really interested in such advanced matters.

A bargaining solution in terms of a contract in C in case utility functions are taken on the enlarged domain $\hat{C}$ is the correspondence

$$B : \left\{ \left(\hat{C}, \{\hat{u}_i\}_{i=1,2}, \hat{c}^d \right) \right\} \to C$$

and is denoted as $B(\hat{C}, \{\hat{u}_i\}_{i=1,2}, \hat{c}^d)$. All axioms are formulated in terms of $\hat{u}_i$, $i = 1, 2$.

The first two axioms of the three axioms proposed in Rubinstein, Safra, and Thomson (1992) are the efficiency and the symmetry axiom, which now read.

Axiom 4.24. *Each element of $B(\hat{C}, \{\hat{u}_i\}_{i=1,2}, \hat{c}^d)$ is Pareto efficient.*

Axiom 4.25. *If $(\hat{C}, \{\hat{u}_i\}_{i=1,2}, \hat{c}^d)$ admits a symmetry function $\phi : C \to C$, then $\phi\left(B(\hat{C}, \{\hat{u}_i\}_{i=1,2}, \hat{c}^d)\right) = B(\hat{C}, \{\hat{u}_i\}_{i=1,2}, \hat{c}^d)$, i.e., all NBS contracts are invariant under the symmetry function.*

These two axioms state a property of the bargaining solution for each bargaining problem in isolation. Recall that the axioms of independence of irrelevant alternatives and invariance of affine transformation link different bargaining problems. In Rubinstein, Safra, and Thomson (1992) the latter two axioms are replaced by just one single axiom. There is a choice in varying the domain of the preferences (as is the case in axiom 4.16) or the preferences itself (as in axiom 4.15) or both. The third axiom fixes the domain $\hat{C}$ and is concerned with linking bargaining problems that have different preferences.

In the following axiom player i, $i = 1, 2$, with utility function u_i' instead of u_i is less eager to raise his objections but with respect to his counter objection c^* his preferences are identical to $\hat{u}_i$ and, therefore, his behaviour does not change. So, player i is still able to defend c^* against every objection of his opponent, but raises less objections himself. Then this change in preference should not affect the bargaining outcome.

Axiom 4.26. *Let $B(\hat{C}, \{\hat{u}_i\}_{i=1,2}, \hat{c}^d) = c^*$ and $\hat{u}_i' : \hat{C} \to \mathbb{R}$ such that*
i) for all $c \in C$ such that $\hat{u}_i(c, 1) \geq \hat{u}_i(c^*, 1)$:

$$\hat{u}_i(c, p) = \hat{u}_i(c^*, 1) \Rightarrow \hat{u}_i'(c, p) \leq \hat{u}_i'(c^*, 1),$$

ii) for all $c \in C$ such that $\hat{u}_i(c, 1) \leq \hat{u}_i(c^*, 1)$:

$$\hat{u}_i(c, 1) = \hat{u}_i(c^*, q) \Rightarrow \hat{u}_i'(c, 1) = \hat{u}_i'(c^*, q).$$

Then $B(\hat{C}, \{\hat{u}_i'\}_{i=1,2}, \hat{c}^d) = B(\hat{C}, \{\hat{u}_i\}_{i=1,2}, \hat{c}^d)$.

The latter axiom is weaker than the original axioms 4.12 and 4.14 it replaces, e.g., Rubinstein, Safra, and Thomson (1992). The following proposition shows that this axiom is weaker than the invariance of utility representation axiom. Of course, only for the enlarged domain of bargaining problems.

Proposition 4.27. *Let $\{\hat{C}, \{\hat{u}_i\}_{i=1,2}, \hat{c}^d\}$ be the class of bargaining problems. Then axiom 4.26 implies axiom 4.19.*

Imposing the three axioms simultaneous is not sufficient to guarantee generic uniqueness of the NBS of definition 4.22 on the whole domain of nonexpected utility preferences. However, there is still an interesting subdomain for which generic uniqueness (this includes existence) can be shown, which requires two additional assumptions upon the nonexpected utility preferences as stated in the following theorem. For a proof we refer to Rubinstein, Safra, and Thomson (1992).[6]

Theorem 4.28. *Let $\hat{u}_i$, $i = 1, 2$, be increasing in p and strong conditionally certainty equivalent. Then the generically unique bargaining solution $c^* \in C$ that satisfies axioms 4.24-4.26 is the generically unique NBS c^* of definition 4.22.*

Finally, for the special case in which C is a simplex of lotteries and u_i, $i = 1, 2$, are (non)expected utility functions we do not need the extension $\hat{C}$ of the domain. The reason is that the compound lottery $pc + (1-p)c^d$, $i = 1, 2$, is equivalent to its reduced lottery $c' \in C$ and therefore $\hat{u}_i(c, p) = u_i(c')$, $i = 1, 2$.

4.2.8 Alternative axiomatic solutions

As we have seen Nash's bargaining solution is the only bargaining solution that satisfies a list of originally four axioms and we already mentioned that the efficiency axiom could be replaced by the strictly individual rationality axiom. Important issues are: Does each axiom reflect some economic intuition and why are we interested in a particular list of axioms? As a critique to some of these four axioms several alternative lists of axioms have been proposed and, consequently, alternative axiomatic solutions have been derived, see e.g., Kalai and Smorodinsky (1975), Kalai (1977), Roth (1979), Osborne and Rubinstein (1990).

The most well-known critique is found in Kalai and Smorodinsky (1975) and it concerns the independence of irrelevant alternatives axiom, i.e., axiom 4.4. In short, consider two bargaining problems in utility representation, denoted by (S, d) and (T, d), that have the same disagreement point d. Suppose that these two problems are such that i) the endpoints $(d_1, s_2^{\max})$ and $(s_1^{\max}, d_2)$ of the Pareto frontier of S^{IR} are also the endpoints of the Pareto frontier of the set of individually rational utilities in T and ii) the Pareto frontier of the set T has shifted outwards away from the disagreement point. So, $S^{IR} \subset T^{IR}$ and for each $s \in S^{IR}$ there exists a $t \in T^{IR}$ such that $t \geq s$. Then the bargaining solution B satisfies the *monotonicity* axiom if $B(T, d) \geq B(S, d)$ for any pair of such bargaining problems.

The rationale of this axiom is that the bargaining problem (T, d) is more favourable to both players in terms of feasible utility pairs if compared to (S, d)

and if this is the case then the bargaining solution should accrue the incremental benefits of shifting from (S, d) to (T, d) to both of the players. Formulated in this way, this monotonicity axiom sounds reasonable. However, examples can be constructed for which the NBS fails the monotonicity axiom, e.g., Kalai and Smorodinsky (1975), and this is traced back to the independence of irrelevant alternatives axiom. Replacing the latter axiom by the monotonicity axiom yields an alternative bargaining solution, which is nowadays named after Kalai and Smorodinsky (1975). In Roemer (1988) the monotonicity axiom is reformulated as a monotonicity of resources. If, in terms of a barter economy, the initial endowments or resources increase, then the bargaining solution should be such that both players obtain more of the resource.

However, the Kalai-Smorodinsky solution is also invariant of affine transformations and, therefore, a similar version of theorem 4.20 can be easily derived. In Rubinstein, Safra, and Thomson (1992) also a new axiomatization of the latter bargaining solution is given for the domain of nonexpected utility preferences. In this new axiomatization the monotonicity axiom and the invariance of affine transformations axiom are simultaneously replaced by a single axiom on the domain $\hat{C}$. Similar as in proposition 4.27 we note that, without giving any further details, that the new axiom always implies the invariance of utility representation axiom (on this domain).

This brief discussion illustrates a drawback of the axiomatic approach. Apparently, there are many conflicting lists of axioms for solutions to satisfy that lead to many possible axiomatic solutions (more than just the two mentioned in this chapter). But the axiomatic approach does not give us any guidance to choose among these competing lists. Another limitation of the axiomatic approach is that the axioms are thought to represent the outcome of some kind of bargaining process and by nature of the axiomatic approach this bargaining process is treated as a black box. In such a bargaining process strategic considerations will play a role in the final outcome and it is not obvious how strategic considerations can be translated into axioms.

4.3. Strategic bargaining and Nash's bargaining solution

In this section the relation between the strategic approach of chapter 3and the axiomatic approach discussed in the previous section is investigated. First, the bargaining procedure proposed in Nash (1953), called Nash's demand game, is investigated and its unique Nash equilibrium after equilibrium selection is related to Nash's bargaining solution. Then we argue that Nash's demand game should be seen as the reduced normal-form game of some unspecified extensive form and alternating offers is a natural candidate. Next, the unique SPE of alternating offers is also related to the Nash's bargaining solution by showing that player 1's and player 2's SPE proposal both converge to Nash's bargaining solution as the time between bargaining rounds vanishes.

4.3.1 Nash's demand game

Nash's demand game, first introduced in Nash (1953), is a game in normal form for the bargaining problem in utility representation (S, d) and the players' strategies are called demands. Hence, the name *Nash's demand game*. The players state simultaneously and independently of each other a *minimum* demand in terms of utility for which this player is willing to settle the bargaining problem. The demand of player i, $i = 1, 2$, is denoted as x_i, $x_i \in \mathbb{R}$. If the players' demands are feasible, i.e., $x \in S$, then both demands can be fulfilled and it is said that the players agree and player i, $i = 1, 2$, obtains x_i. If both demands are not feasible, i.e., $x \notin S$, then both demands cannot be fulfilled and it is said that players disagree and player i, $i = 1, 2$, gets d_i. In order to state the players' utility function in this game we define the discontinuous indicator function $i : S \to \{0, 1\}$ as

$$i\,(x) = \begin{cases} 1, & \text{if } x \in S, \\ 0, & \text{if } x \notin S. \end{cases}$$

Thus, $i\,(x) = 1$ if x is feasible and $i\,(x) = 0$ otherwise. So, player i's utility as a function of x is given by $d_i + i\,(x)\,(x_i - d_i)$. The formal definition of Nash's demand game is as follows.

Bargaining Procedure 4.29. *Nash's demand game is the game in normal form given by $N = \{1, 2\}$, player i's demand $x_i \in \mathbb{R}$ and player i's utility function $d_i + i\,(x)\,(x_i - d_i)$, $i = 1, 2$.*

The standard equilibrium concept for games in normal form is the Nash equilibrium. In Nash (1953) it is shown that the set of non-trivial pure strategy Nash equilibria S^{NE} is equal to the set of utility pairs that are both individually rational and Pareto efficient, i.e.

$$S^{NE} = \{\, x \in S^{IR} \mid x_2 = f_2(x_1),\ x_1 \in [d_1, s_1^{\max}] \,\}.$$

Thus, for the Nash equilibrium $x^* \in S^{NE}$ it holds that the demands are feasible and player i, $i = 1, 2$, obtains the utility x_i^*. It is easy to verify that every pair of demands in S^{NE} is a Nash equilibrium. For instance, if player 1 would demand more, i.e., $x_1 > x_1^* = f_1\,(x_2^*)$, then the demands are no longer feasible and player 1 obtains $d_1 < x_1^*$. Similar, player 1's demand $x_1 < f_1\,(x_2^*)$ implies that $(x_1, x_2^*) \in S$ but player 1's utility $x_1 < x_1^*$. So, given player 2's demand x_2^*, player 1's demand $x_1^* = f_1\,(x_2^*)$ maximizes player 1's discontinuous utility function. For completeness we mention that there exists a large set of trivial Nash equilibria in which player i demands $x_i^* \geq s_i^{\max}$, $i = 1, 2$, and the players fail to reach an agreement. We leave the verification to the reader.

In Nash (1953) a selection method to reduce the multiplicity of Nash equilibria is proposed that is called in the terminology of modern game theory the

H-essential equilibrium concept, e.g., see Van Damme (1991). Without going into the details of this equilibrium concept, the main idea is to slightly perturb the demand game in such a way that the players are slightly uncertain about the Pareto efficient frontier of S^{IR}. It is without loss of generality to assume that $S = S^{IR}$. For the parameter $\varepsilon > 0$ the probability $p_\varepsilon(x)$, for $x - d \in \mathbb{R}^2_+$, denotes the probability that both demands are feasible and, consequently, $1 - p_\varepsilon(x)$ is the probability that both demands are not feasible. Following Van Damme (1991) we assume that the function $p_\varepsilon : \{x - d \in \mathbb{R}^2_+\} \to [0,1]$ is given by

$$p_\varepsilon(x) = \begin{cases} 1, & \text{if } \gamma(x) \leq 1, \\ e^{-\varepsilon^{-1}(\gamma(x)-1)^2}, & \text{if } \gamma(x) > 1, \end{cases}$$

where $\gamma(x) = \min\{\gamma | x = d + \gamma(s - d), s \in S^{IR}\}$ is a measure for the distance from $x \geq d$ to the Pareto frontier of S^{IR}. Like a balloon the set $\{d + \gamma(s - d) | s \in S^{IR}\}$ proportionally inflates with the amount of 'air' γ relative to some kind of origin d and $\gamma(x)$ is the minimum amount of air necessary for some point $s \in S^{IR}$ to touch x. So, the line through x and d intersects $S^{PE} \cap S^{IR}$. If s^* represents the latter intersection point, then $x - d = \gamma(x)(s^* - d)$. By definition, $\gamma(x) \leq 1$ if $x \in S$ and $\gamma(x) > 1$ if $x \notin S$ and $x \geq d$.

Player i's utility function in the perturbed Nash's demand game is the expected utility function given by $d_i + p_\varepsilon(x)(x_i - d_i)$ and, since, $\lim_{\varepsilon \downarrow 0} p_\varepsilon(x) = i(x)$ for all $x \geq d$ the game of procedure 4.29 is the limit of the perturbed game as ε goes to 0.

Bargaining Procedure 4.30. *Nash's perturbed demand game is the game in normal form given by* $N = \{1, 2\}$, *player* i*'s demand* $x_i \in \mathbb{R}$ *and player* i*'s utility function* $d_i + p_\varepsilon(x_1, x_2)(x_i - d_i)$, $i = 1, 2$.

The following theorem states that the modified version of Nash's demand game admits a unique Nash equilibrium and that the pair of associated Nash equilibrium demands converges to the NBS $N(S, d)$ in the limit as the uncertainty vanishes.

Theorem 4.31. *For* $\varepsilon > 0$ *the bargaining procedure 4.30 admits a unique Nash equilibrium* $x(\varepsilon)$ *characterized by* $x(\varepsilon) = d + \frac{1}{2}(1 + \sqrt{1 + 4\varepsilon})(N(S, d) - d)$. *Moreover,* $x(\varepsilon) \to N(S, d)$ *as* ε *goes to* 0.

Corollary 4.32. *Nash's demand game of bargaining procedure 4.29 admits a unique* H*-essential equilibrium* x^* *given by* $x^* = N(S, d) \in S^{NE}$.

The following example illustrates the previous theorem and can be regarded as a less elaborated proof for the simple case of a linear Pareto frontier.

Example 4.33. *Consider the bargaining problem* (S, d) *given by*

$$S = \{s \in \mathbb{R}^2_+ \mid s_1 + s_2 \leq 1\} \quad \text{and} \quad d = 0.$$

For bargaining procedure 4.29 we have that every $x^* \in \{(s_1, 1-s_1) \,|s_1 \in [0,1]\}$ *is a non-trivial Nash equilibrium with agreement upon the pair of utilities* x^*. *The set of trivial Nash equilibria resulting in disagreement is given by* $\{s \in \mathbb{R}^2_+ \,|s \geq (1,1)\}$.
Consider bargaining procedure 4.30. Then $\gamma((x_1,x_2)) = x_1 + x_2$. *The expected utility* $p_\varepsilon(x)x_i$ *is given by*

$$p_\varepsilon(x)x_i = \begin{cases} x_i, & \text{if } x \in S, \\ e^{-\varepsilon^{-1}(x_1+x_2-1)^2}x_i, & \text{if } x \notin S. \end{cases}$$

Player i's best response solves the first-order conditions, i.e.

$$\tfrac{\partial p_\varepsilon(x)x_i}{\partial x_i} = \begin{cases} 1, & \text{if } x_i \leq 1-x_j, \\ \varepsilon^{-1}p_\varepsilon(x)\left[2x_i(1-x_1-x_2)+\varepsilon\right], & \text{if } x_i > 1-x_j. \end{cases}$$

Obviously, the derivative is not equal to 0 *for* $x_i \leq 1 - x_j$. *So, player i's best response satisfies* $x_i > 1 - x_j$. *Then*

$$\tfrac{\partial p_\varepsilon(x)x_1}{\partial x_1} - \tfrac{\partial p_\varepsilon(x)x_2}{\partial x_2} = 2\varepsilon^{-1}p_\varepsilon(x)(x_1-x_2)(1-x_1-x_2) = 0,$$

i.e., $x_1 = x_2$ *because* $x_1 + x_2 > 1$. *Substitution of* $x_2 = x_1$ *into the first-order condition for* $i = 1$ *yields* $2x_1(1-2x_1) + \varepsilon = 0$, *which in turn leads to the unique Nash equilibrium*

$$x_1(\varepsilon) = x_2(\varepsilon) = \tfrac{1}{4} + \tfrac{1}{4}\sqrt{(1+4\varepsilon)} = \tfrac{1}{2} + \tfrac{1}{2}\sqrt{(1+4\varepsilon)}N_i(S,d).$$

Obviously, $\lim_{\varepsilon\downarrow 0} x_i(\varepsilon) = N_i(S,d)$.

4.3.2 Interpretation of demands

Nash's demand game is often mistaken for a static game due to the formulation as a game in normal form, but this was certainly not the intention John Nash had in mind. In Nash (1953), (p. 129), it is said

> By beginning with a space of mixed strategies instead of talking about a sequence of moves, etc., we presuppose a reduction of the strategic potentialities of each player to the normal form.

So, in Nash (1953) the demands are interpreted as representing the player's strategies in some unspecified dynamic bargaining process. For finite games in extensive-form with perfect recall this point of view can be justified by referring to Kuhn's theorem, e.g., see Fudenberg and Tirole (1991) or Mas-Colell, Whinston, and Green (1995), which states that the behavioural strategies[7] in any finite game in extensive-form with perfect recall are equivalent to the mixed strategies in the corresponding reduced game in normal form. This reduced game is a game in normal form in which the players choose their mixed strategies simultaneously and independently of each other. Therefore, Kuhn's theorem states that every Nash equilibrium in the extensive form corresponds to a Nash equilibrium in the reduced normal form and vice versa. When John Nash wrote his seminal articles on bargaining games the extensive form was not yet fully understood and he had to resort to a bargaining game in normal form. Reinhard Selten's work on subgame perfect equilibria and trembling-hand perfect Nash equilibria, e.g., see Selten (1975), paved the way to analyze games in extensive form.

We pursue this insight a little further. For finite games in extensive form with perfect recall and perfect information it is known that the set of subgame perfect equilibria coincides with the set of trembling-hand perfect (Nash) equilibria in the agent normal form, e.g., see Van Damme (1991), Fudenberg and Tirole (1991) or Mas-Colell, Whinston, and Green (1995). The alternating offer model can be approximated by a finite game in extensive form by taking a fine grid on S^{IR} or $S^{IR} \cap S^{PE}$ and by truncating the infinite horizon procedure at round t_f, $t_f \in T$ for large t_f. So, the subgame perfect equilibrium of the approximated game corresponds to the trembling-hand perfect equilibrium of the associated agent normal form.

Infinite games arise if either proposals belong to a continuous space S^{IR} or $t_f = \infty$ or both, as is the case in alternating offers. For infinite games in general the equivalence between the two equilibrium concepts is not so clear. However, in Carlsson (1991) it is shown that the unique trembling-hand-perfect Nash equilibrium of Nash's demand game in procedure 4.29 corresponds to the pair of demands given by the NBS, similar as in corollary 4.32. These results hint at that the reduced normal form associated with the extensive form game of some bargaining procedure is strategically equivalent to Nash's demand game and that this extensive form has the players agree upon the NBS in every SPE, i.e., uniqueness in SPE utilities.

To conclude, let us briefly demonstrate how alternating offers can be reduced to Nash's demand game. First, restrict attention to Markov strategies, which we justify by referring to the results in chapter 3 where the unique SPE features such strategies. Since each Markov strategy is uniquely described by its history-independent proposal and a history independent set of acceptable proposals T_i, $i = 1, 2$, we already have reduced the players' strategy spaces and player 1 chooses the pair (x, T_1), where $x \in S$ and $T_1 \subseteq S$, whereas player 2 chooses the pair (y, T_2), $y \in S$ and $T_2 \subseteq S$. In order to obtain Nash's demand game we have to impose an additional restriction upon the strategy spaces that relates the players' threshold levels to their proposals: Player i, $i = 1, 2$, chooses $z_i \in \mathbb{R}$ and the pair (z_1, z_2) induces the players' Markov strategies given by

$$x = (z_1, f_2(z_1)), \; y = (f_1(z_2), z_2), \; T_i = \{s \in S \mid s_i \geq z_i\}, \; i = 1, 2. \tag{4.5}$$

So, we impose that player i's demand z_i, $i = 1, 2$, coincides with this player's threshold level. The associated outcome (z_1, z_2) of this reduced normal form is given by

$$\begin{cases} x, & \text{if } (z_1, z_2) \in S, \\ d, & \text{if } (z_1, z_2) \notin S. \end{cases}$$

For completeness we mention that the results in Nash (1953) with respect to non-trivial Nash equilibria in this reduced normal form directly translate to every $(z_1, z_2) \in S^{PE}$ being a Nash equilibrium in the extensive form of alternating offers, which is indeed the case as is shown in Rubinstein (1982). The result in Carlsson (1991) then implies that $(z_1, z_2) = N(S, d)$ corresponds to the unique trembling-hand perfect equilibrium of this reduced normal form. Note that our additional restriction rules out the SPE strategies of theorem 3.19, because $(z_1, z_2) = (x_2^*, y_1^*) \notin S^{IR}$ and the threshold level of player i, $i = 1, 2$, is equal to the present value $(1 - \delta)d_i + \delta z_i$.

4.3.3 Convergence in alternating offers

In this section it is established that for the bargaining problem in utility representation both the SPE proposals x^* and y^* of theorem 3.19 converge to the NBS $N(S, d)$ as the time between bargaining rounds vanishes, i.e., as Δ goes to 0. For alternating offers this is equivalent to $\delta = e^{-r\Delta}$ goes to 1 and this means that the exogenous risk of breakdown vanishes, i.e., $1 - \delta$ goes to 0. In terms of discounting the impatience of the players vanishes. The convergence result for contracts is postponed to section 4.3.4.

In order to express that the unique pair of SPE proposals of theorem 3.19 depends upon δ we write $x^*(\delta)$ and $y^*(\delta)$. The convergence, as δ goes to 1, of these SPE proposals to the NBS $N(S, d)$ should not come as a surprise. First, from $x_2^*(\delta) = (1 - \delta) d_2 + \delta y_2^*(\delta)$ it is immediately clear that if both $x^*(\delta)$ and $y^*(d)$ converge, then these two points must have a common limit point, i.e., $\lim_{\delta \to 1} x^*(\delta) = \lim_{\delta \to 1} y^*(\delta)$. Next, $x^*(\delta)$ and $y^*(\delta)$ lie on the same Nash product curve and for all $\delta < 1$ this curve intersects the Pareto frontier twice. If $x^*(\delta)$ and $y^*(\delta)$ have a common limit, then this means that the two distinct intersection points converge to each other and coincide in the limit. But then, in the limit, the Nash product curve through the common limit point must be tangent to the Pareto frontier. So, by the first geometrical property, the only candidate for the limit that qualifies is the NBS. The following theorem makes this statement precise and, in its proof, the existence of the common limit is also established.

Theorem 4.34. *Let $(x^*(\delta), y^*(\delta))$ the unique pair of SPE proposals of theorem 3.19. Then* $\lim_{\delta \to 1} x^*(\delta) = \lim_{\delta \to 1} y^*(\delta) = N(S, d)$.

Corollary 4.35. *Player 1's first-mover advantage vanishes in the limit as δ goes to* 1.

In section 3.4.3 it was shown that the proposing player has a first-mover advantage and that this arises from the fact that rejecting a proposal is costly to the responding player. As δ goes to 1 the latter costs decrease and, in the limit, these costs simply vanish.

The convergence result of theorem 4.34 can also be illustrated with the second geometrical property of the NBS of section 4.2.2. Given the unique pair of SPE proposals $(x^*(\delta), y^*(\delta))$ construct the points $(x_1^*(\delta), y_2^*(\delta)) \notin S^{IR}$ and $(y_1^*(\delta), x_2^*(\delta)) \in S^{IR}$ from $x^*(\delta)$ and $y^*(\delta)$. Then the line through d and $(x_1^*(\delta), y_2^*(\delta))$ contains the point $(x^*(\delta), x_2^*(\delta))$, which is illustrated in figure 4.4. This follows from

$$\frac{x_2^*(\delta) - d_2}{y_1^*(\delta) - d_1} = \frac{\delta\left(y_2^*(\delta) - d_2\right)}{\delta\left(x_1^*(\delta) - d_1\right)} = \frac{y_2^*(\delta) - d_2}{x_1^*(\delta) - d_1},$$

where the first equality is derived after applying twice the result that the responding player is indifferent between accepting the SPE proposal or rejecting it. Next, application of the same indifference property twice in calculating the slope of the line through $y^*(\delta)$ and $x^*(\delta)$ yields

$$\frac{x_2^*(\delta) - y_2^*(\delta)}{x_1^*(\delta) - y_1^*(\delta)} = \frac{x_2^*(\delta) - d_2 - \left(y_2^*(\delta) - d_2\right)}{x_1^*(\delta) - d_1 - \left(y_1^*(\delta) - d_1\right)} = -\frac{y_2^*(\delta) - d_2}{x_1^*(\delta) - d_1}.$$

So, the slope of the line through d, $\left(y_1^*(\delta), x_2^*(\delta)\right)$ and $\left(x_1^*(\delta), y_2^*(\delta)\right)$ is minus the slope of the line through $x^*(\delta)$ and $y^*(\delta)$. So, if $s^* \in S^{PE}$ is the common limit point then as δ goes to 1 the line through $y^*(\delta)$ and $x^*(\delta)$ becomes tangent to the Pareto frontier in the common limit point and

$$\lim_{\delta \to 1} \frac{x_2^*(\delta) - d_2}{y_1^*(\delta) - d_1} = \frac{f_2(s_1^*) - d_2}{s_1^* - d_1}$$

and

$$\lim_{\delta \to 1} -\frac{x_2^*(\delta) - y_2^*(\delta)}{x_1^*(\delta) - y_1^*(\delta)} \in \left(f_2^{\prime -}(s_1^*), f_2^{\prime +}(s_1^*)\right),$$

which is exactly the second geometrical property of the NBS.

In the previous section we argued how alternating offers could be reduced to a game in normal form that was strategically equivalent to Nash's demand game. In Nash (1953) it is argued that

> Of course, one cannot represent all possible bargaining devices as moves in the noncooperative game. The negotiation process must be formalized and restricted, but in such a way that each participant is still able to utilize all the essential strengths of his position.

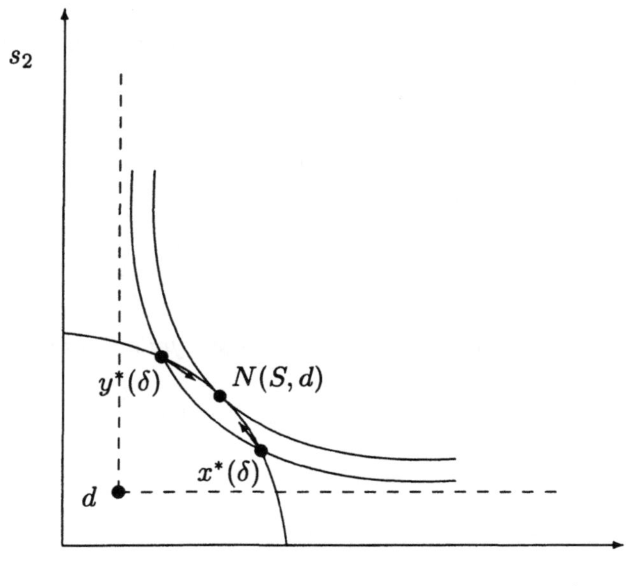

Figure 4.4. The property of convergence of the SPE proposals $x^*(\delta)$ and $y^*(\delta)$ of theorem 3.19 to Nash's bargaining solution $N(S,d)$.

In the light of theorem 3.19 the restriction of the strategy space to Markov strategies only can be justified as an innocuous restriction, but the additional restriction in (4.5) is certainly not since it requires in any non-trivial Nash equilibrium that $(z_1, z_2) \in S^{PE} \cap S^{IR}$, which rules out $z_1 = x_1^*(\delta)$ and $z_2 = y_2^*(\delta)$ as a pair of Nash equilibrium demands. So, an alternative restriction should be imposed to link the unique SPE to the trembling-hand perfect Nash equilibrium of Nash's demand game and the analysis in chapter 3 simply suggests the following: For some quadruple $(z_1, z_2, \alpha_1, \alpha_2)$ the players' strategies in the reduced normal form are given by

$$x = (z_1, f_2(z_1)), \; y = (f_1(z_2), z_2), \; T_i = \{s \in S \mid s_i \geq \alpha_i\}, \; i = 1, 2.$$

This means that $\alpha_i = (1-\delta)d_i + \delta z_i$ is no longer ruled out as a threshold rule, whereas the additional restriction $\alpha_i = z_i$ rules out such threshold rule. So, this version of Nash's demand game features strategies $(z_i, \alpha_i) \in \mathbb{R}^2, i = 1, 2$, and is rich enough to admit $(z_1, z_2, \alpha_1, \alpha_2) = (x_1^*, y_2^*, q_1(x^*), p_2(y^*))$ as one of the nontrivial Nash equilibria. Although we will not pursue this line of research any further, it is our conjecture that this generalized Nash's demand game restores the relation between the unique SPE of the alternating offers model and the trembling-hand perfect equilibrium of this reduced normal form. Of course, if one adds trembles to the reduced normal form, then one must take into account that all outcomes $\left(1-\delta^t\right)d+\delta^t x$, t even, and $\left(1-\delta^t\right)d+\delta^t y$, t odd, come to play a role in the Nash equilibrium analysis in order to capture the analysis off the equilibrium path.

We conclude this section by providing an example.

Example 4.36. *Consider examples 3.5 and 4.8 once more, i.e., $d=0$ and for $\alpha\in\left[\frac{1}{2},1\right)$ the piecewise linear function f_1 is given by*

$$f_1\left(s_2\right)=\begin{cases}1-\frac{1-\alpha}{\alpha}s_2, & \text{if } s_2\in[0,\alpha],\\ \frac{\alpha}{1-\alpha}-\frac{\alpha}{1-\alpha}s_2, & \text{if } s_2\in(\alpha,1].\end{cases}$$

Then convergence to $N(S,d)=(\alpha,\alpha)$ as δ goes to 1 follows from

$$x^*\left(\delta\right)=\left(\tfrac{\alpha}{(1-\delta)\alpha+\delta},\tfrac{\delta\alpha}{(1-\delta)\alpha+\delta}\right)\to(\alpha,\alpha),$$

and

$$y^*\left(\delta\right)=\left(\tfrac{\delta\alpha}{(1-\delta)\alpha+\delta},\tfrac{\alpha}{(1-\delta)\alpha+\delta}\right)\to(\alpha,\alpha).$$

This illustrates the main result of theorem 4.34.

4.3.4 Convergence in the contract space

In this section the results of the previous section are translated in terms of convergence in the contract space. Then the convergence result is briefly illustrated in terms of the Kuhn-Tucker first-order conditions stated in section 3.4.4 and 4.2.4.

The next theorem reformulates the convergence result in theorem 4.34. Formally, the pair of SPE proposals $(c^1(\delta),c^2(\delta))$ form a upper-semi continuous correspondence of δ that is generically a continuous function. We leave the technical details to the reader.

Theorem 4.37. *Let $\left(c^1(\delta),c^2(\delta)\right)$ be the generically unique pair of SPE proposals of proposition 3.15. Then*

$$\lim_{\delta\to1}c^1(\delta)=\lim_{\delta\to1}c^2(\delta)=N(C,\{u_i\}_{i=1,2},c^d).$$

This convergence result should be interpreted in terms of the reformulated bargaining problem in section 4.2.6. To see this, recall from section 2.3.4 that $U_i(p(\sigma);\delta)=(1-\delta^{t(\sigma)})d+\delta^{t(\sigma)}u_i(c^{t(\sigma)}(t(\sigma)))$ and that δ can be interpreted as a probability. Then, in terms of section 4.2.6, we may interpret every path $p(\sigma)$ as the pair $(c^{t(\sigma)}(t(\sigma)),\delta^{t(\sigma)})\in\hat{C}$ and write $U_i(p(\sigma);\delta)=$

$\hat{u}_i(c^{t(\sigma)}(t(\sigma)), \delta^{t(\sigma)})$, where $\hat{u}_i$ represents the expected utility function of section 4.2.6 with probability $p = \delta^{t(\sigma)}$. So, the interpretation of δ, $\delta \in [0,1)$, as a probability is consistent with the assumption of expected utility functions underlying both the bargaining problem and the axiomatic Nash's bargaining solution, i.e., axiom 4.3.

Finally, we illustrate theorem 4.37 in terms of the Kuhn-Tucker first-order conditions that characterize the SPE proposals $c^1(\delta)$ and $c^2(\delta)$ in (3.18), respectively, (3.19) and the NBS $N(C, \{u_i\}_{i=1,2}, c^d\}$ as characterized in (4.3). We neglect the individual rationality constraints and assume for explanatory simplicity that the SPE contracts are unique for all $\delta \in (0,1)$. Denote $\hat{c}^1$, $\hat{c}^2$, $\hat{\lambda}^1_j$ and $\hat{\lambda}^2_j$, $j = 0, 1, \ldots, k$, as the limit values of $c^1(\delta)$, $c^2(\delta)$, $\hat{\lambda}^1_j(\delta)$ and $\hat{\lambda}^2_j(\delta)$ as δ goes to 1. Substitution of $\delta = 1$ and these limit values into the Kuhn-Tucker first-order conditions (3.18), respectively, (3.19) means that the limit values solve

$$\begin{aligned}
\frac{\partial u_1(\hat{c}^1(\delta))}{\partial c_i} + \hat{\lambda}^1_0(\delta)\frac{\partial u_2(\hat{c}^1)}{\partial c_i} - \sum_{j=1}^{k} \hat{\lambda}^1_j \frac{\partial g_j(\hat{c}^1)}{\partial c_i} &= 0, \qquad i = 1, \ldots, n,\\
\hat{\lambda}^2_0 \frac{\partial u_1(\hat{c}^2)}{\partial c_i} + \frac{\partial u_2(\hat{c}^2)}{\partial c_i} - \sum_{j=1}^{k} \hat{\lambda}^2_j \frac{\partial g_j(\hat{c}^2)}{\partial c_i} &= 0, \qquad i = 1, \ldots, n,\\
\hat{\lambda}^1_0 \left(u_2\left(\hat{c}^2\right) - u_2\left(\hat{c}^1\right)\right) &= 0,\\
\hat{\lambda}^2_0 \left(u_1\left(\hat{c}^1\right) - u_1\left(\hat{c}^2\right)\right) &= 0,\\
\hat{\lambda}^1_j g_j\left(\hat{c}^1\right) &= 0, \qquad j = 1, \ldots, k,\\
\hat{\lambda}^2_j g_j\left(\hat{c}^2\right) &= 0, \qquad j = 1, \ldots, k.
\end{aligned}$$

The equations on the third and fourth row imply that $\hat{c}^1 = \hat{c}^2$ and, thus, as before that $c^1(\delta)$ and $c^2(\delta)$ have a common limit value. Then substitution of $\hat{c}^2 = \hat{c}^1$ into the second row implies that the first and second row yield

$$\hat{\lambda}^2_0 = \left(\hat{\lambda}^1_0\right)^{-1}, \quad \hat{\lambda}^2_j = \left(\hat{\lambda}^1_0\right)^{-1} \hat{\lambda}^1_j, \quad j = 1, \ldots, k.$$

So, the number of variables and equations can be reduced by a half and this simply yields (4.3). The latter yields the NBS characterized in (4.4) and, hence,

$$\hat{c}^1 = c^*, \quad \hat{\lambda}^1_0 = \frac{u_2(c^*) - d_2}{u_1(c^*) - d_1}, \quad \hat{\lambda}^1_j = \lambda^*_j, \; j = 1, \ldots, k.$$

The following example illustrates the convergence of the SPE contracts to the NBS contract. This example also illustrates that there is convergence of sets in case of nongeneric nonuniqueness.

Example 4.38. *Consider examples 3.16 and 4.18 once more, i.e.*

$$C = \left\{ c \in \mathbb{R}^3_+ \mid c_1 + c_2 + c_3 = 1 \right\}, \, c^d = 0$$

and for $\alpha \in (0,1)$

$$u_1(c) = c_1 + \alpha \quad \text{and} \quad u_2(c) = \alpha c_2 + c_3.$$

As before, we distinguish $\alpha < \frac{1}{2}$, $\alpha > \frac{1}{2}$ and $\alpha = \frac{1}{2}$.
First, $\alpha < \frac{1}{2}$. Then the convergence to $N(C, \{u_i\}_{i=1,2}, c^d) = (1,0,1)$ as δ goes to 1 follows from

$$c^1(\delta) = \left(\tfrac{1}{1+\delta}, 0, \tfrac{\delta}{1+\delta}\right) \to (1,0,1), \quad c^2(\delta) = \left(\tfrac{\delta}{1+\delta}, 0, \tfrac{1}{1+\delta}\right) \to (1,0,1).$$

Second, $\alpha > \frac{1}{2}$. Then $N(C, \{u_i\}_{i=1,2}, c^d) = (0,1,0)$ and the convergence as δ goes to 1 follows from

$$c^1(\delta) = \left(\tfrac{(1-\delta)\alpha}{(1-\delta)\alpha+\delta}, \tfrac{\delta}{(1-\delta)\alpha+\delta}, 0\right) \to (0,1,0)$$

and

$$c^2(\delta) = \left(0, \tfrac{\delta}{(1-\delta)\alpha+\delta}, \tfrac{(1-\delta)\alpha}{(1-\delta)\alpha+\delta}\right) \to (0,1,0).$$

Third, $\alpha = \frac{1}{2}$, i.e., the case of nongeneric multiplicity. Recall that $c^1(\delta) = (\gamma_1, \gamma_2, \gamma_3)$ is given by

$$\gamma_1 = \tfrac{1}{1+\delta} - \tfrac{1}{2}\gamma_2, \quad \gamma_3 = \tfrac{\delta}{1+\delta} - \tfrac{1}{2}\gamma_2, \quad \gamma_2 \in \left[0, \tfrac{2\delta}{1+\delta}\right]$$

and that by symmetry, $c^2(\delta) = (\gamma_3, \gamma_2, \gamma_1)$, $\gamma_2 \in \left[0, \frac{2\delta}{1+\delta}\right]$. Then letting δ go to 1 means that these two sets coincide in the limit and that the common limit set is characterized by

$$\gamma_1 = \tfrac{1}{2} - \tfrac{1}{2}\gamma_2, \quad \gamma_3 = \tfrac{\delta}{2} - \tfrac{1}{2}\gamma_2, \quad \gamma_2 \in [0,1].$$

Since the latter set is equal to the set of NBS contracts characterized in example 4.18 this illustrates theorem 4.37.

4.4. The two approaches are complementary

The axiomatic and the strategic approach to the bargaining problem are two different approaches and yet the results obtained in the previous section indicate that both approaches can produce similar predictions. An important question is whether or not some specific interpretation should be given to the results that can be derived in both approaches. If not, then one adopts the view that these results are simply mathematical artifacts of the two approaches which should not receive any significance. This point of view does not conform with the popular view in bargaining theory.

Already in Nash (1953) the significance of the relation between the two approaches is stressed and it is even argued that both approaches are complementary, because

> each approach helps to justify and clarify the other.

Indeed the fact that the axiomatic Nash solution approximates the SPE agreement reached in the strategic alternating offer model implies that the axioms underlying the NBS are approximately justified for this strategic bargaining procedure, provided the risk of breakdown is sufficiently small.

The axioms underlying the NBS can also be used to interpret the unique SPE outcome of the alternating offer model with a positive risk of breakdown. The unique SPE outcome in which players immediately agree upon an efficient agreement is in accordance with axiom 4.1 and delay in the alternating offer model is inefficient. If the bargaining problem is symmetric, then $y_2^*(\delta) = x_1^*(\delta)$ in terms of utilities and $c^2(\delta) = \phi(c^1(\delta))$, where ϕ is the symmetry function of section 4.2.3. However, the risk $\delta < 1$ introduces an asymmetry between the proposing and the responding player that is reflected by the first-mover advantage and only in the limit where the players are perfectly patient symmetry is restored. As mentioned before, the independence of irrelevant alternatives axiom imposes some kind of maximization problem and, so, proposition 3.14 implies that the players ignore irrelevant alternatives if we would shrink the contract space C without removing the contracts $c^1(\delta)$ and $c^2(\delta)$. The discussion on the monotonicity axiom in section 4.2.8 than hints at that the SPE proposals violate the monotonicity axiom.

Despite the nice results obtained thus far the use of either approach is rather limited. The axiomatic approach is limited, because many lists of conflicting axioms can be thought of and, hence, many axiomatic solutions exist. Furthermore, it is not obvious how strategic considerations should be translated into axioms. At this point the strategic approach can be of use. First, the strategic considerations can be modelled in a strategic model and then the corresponding equilibrium outcome can indicate which axiom fits the strategic considerations. Also in selecting among axiomatic solutions the strategic approach can play a role by requiring that any sound axiomatic solution should correspond to the equilibrium of some strategic bargaining model of which the bargaining procedure represents the essential characteristics.

Also the use of the strategic approach has its limitations. One limitation is that the number of bargaining procedures one can think of seems to be unlimited and, therefore, it is not possible to analyze all these bargaining procedures. Furthermore, from standard game theory it is known that equilibria, whether Nash or subgame perfect, are extremely sensitive to the precise rules of the game and, hence, the equilibria in the strategic bargaining models are also sensitive to the specification of the bargaining procedure. This implies that even small changes which can be regarded as irrelevant can have a dramatic impact on the equilibrium outcome of the strategic bargaining model. The sensitivity of the results obtained thus far with respect to changes in the underlying assumptions is postponed to chapter 5 and 6.

Given the limitations to both approaches these should be regarded as complementary in the following way: If the outcomes of an axiomatic solution do not agree with the outcomes of *any* bargaining procedure, then the axiomatic solution does not capture the strategic relevance in any bargaining procedure and this solution is likely to be of little relevance. Conversely, a bargaining procedure that agrees with *no* axiomatic solution does not correspond to any axiomatic intuition and one may expect that such a bargaining procedure does not have a sound intuitive interpretation. Therefore, such a bargaining procedure is of limited applicability. These arguments stress the importance of finding mutually reinforcing axiomatic solutions and equilibria of strategic bargaining procedures.

4.5. Related Literature

For an extensive treatment of axiomatic bargaining theory we refer to e.g., Roth (1979) and Peters (1992). Theorem 4.5 is a standard result in bargaining theory and a proof can be found in, e.g., Binmore (1987a), Van Damme (1991), Kaneko (1980), Nash (1950), Meyerson (1991), Osborne and Rubinstein (1990), Roth (1979) and Peters (1992). In Jansen and Tijs (1981) every axiomatic bargaining solution with the property that the set $D(s^*)$ introduced in section 4.2.2 is a convex set is called *regular* and the NBS is therefore regular.

The reconstruction of axiomatic bargaining theory is based upon Roemer (1988), where the bargaining problem is defined as a two-person barter economy with an arbitrary number of goods (without properly specifying the initial endowments, although these seem implicitly be taken as the worst outcome). The results extend to the class of bargaining problems introduced in chapter 2.2, see e.g., page 5 in Roemer (1988). The reformulation in section 4.2.3 is our own, although the idea of the symmetry function is taken from Rubinstein, Safra, and Thomson (1992). The latter reference contains a valuable discussion of the methodological problems. The analysis in Roemer (1988) is not restricted to the NBS, but also to several other axiomatic solutions, such as Kalai and Smorodinsky (1975), egalitarian, monotone utility path and proportional. Of all these alternative solution concepts only the Kalai-Smorodinsky solution is reconstructed in Rubinstein, Safra, and Thomson (1992). Axiom 4.19, i.e., independence of utility representation, is based upon our own reinterpretation of Nash (1950) and theorem 4.20 and proposition 4.27 are our own.

For a discussion on bargaining problems, arbitrage problems, social welfare functions and fair bargains we refer to Peters (1992). In Mariotti (1999) the independence of irrelevant alternatives axiom is replaced by an axiom that embodies the Suppes-Sen principle of a fair bargain. The resulting Nash bargaining solution should be regarded as a fair bargaining scheme instead of a bargaining solution.

For alternative assumptions on the function $p_\varepsilon(x)$ in Nash's perturbed demand game and their consequences for convergence to the NBS see e.g., Binmore (1987b), Van Damme (1991) and Osborne and Rubinstein (1990). In Carlsson (1991) it is first shown that the trembling-hand Nash equilibrium of Nash's demand game converges to the NBS. An evolutionary underpinning of the NBS in Nash's demand game can be found in chapter 8 of Young (1998). The convergence of the SPE proposals was first derived in Binmore, Rubinstein, and Wolinsky (1986). Related discussions can also be found in Van Damme (1991), Okada (1991a) and Osborne and Rubinstein (1990). For a procedure that lasts only two rounds and that exactly implements the NBS see e.g., Rubinstein, Safra, and Thomson (1992).

Nash's solution is not the only axiomatic solution that corresponds to some kind of bargaining procedure. In Moulin (1984) a bargaining procedure is proposed that justifies the axiomatic solution in Kalai and Smorodinsky (1975). In chapter 9 of Young (1998) this axiomatic solution is derived in an evolutionary game-theoretic context.

The role of risk aversion in the divide a dollar problem is first investigated in Kihlstrom, Roth, and Schmeidler (1991). Later analyses of risk aversion were conducted in the space of lotteries under the assumption of expected utilities, see e.g., Roth and Rothblum (1982) for a partial analysis assuming a riskless disagreement outcome and Safra, Zhou, and Zilcha (1990) for the complete analysis. We refer to Hanany and Safra (2000) for the characterization of the Nash bargaining solution for nonexpected utilities over lotteries. The divide a dollar problem with (nonexpected) decision weight utility functions is treated in Houba, Tieman, and Brinksma (1998). Bargaining problems with nonexpected utility functions represented by the Choquet integral are considered in Ok and Zhou (2000).

For convergence results of nonexpected utility functions featuring loss aversion we refer to Shalev (1997). In general, a tough technical problem with nonexpected utility functions is that the stationarity underlying the utility functions U_i, $i = 1, 2$, introduced in section 2.3.4 is lost and one has to solve for equilibria in a nonstationary model.

Notes

1 Actually, Nash's demand game is the second stage of Nash's variable threat game, which is a two-stage game with endogenous disagreement actions. The full game will be analyzed in chapter 7.

2 Consider the function $f_i : \mathbb{R} \to \mathbb{R}$, $i = 1, 2$. The left derivative $f_i'^-$ of f_i in s_j, $j = 1, 2$ and $j \neq i$, is defined as $f_i'^-(s_j) = \lim_{z \uparrow s_j} \frac{\partial f_i(z)}{\partial z}$. Similar, the right derivative $f_i'^+$ of f_i in s_j is defined as $f_i'^+(s_j) = \lim_{z \downarrow s_j} \frac{\partial f_i(z)}{\partial z}$.

3 In Rubinstein, Safra, and Thomson (1992) the symmetry axiom is stated as follows: If the bargaining problem admits a symmetry function, then all NBS contracts are a fixed point of the symmetry function, i.e.,

$$\phi\left(B(C, \{u_i\}_{i=1,2}, c^d)\right) = B(C, \{u_i\}_{i=1,2}, c^d).$$

Here we state the symmetry axiom as proposed in Roemer (1988).

4 In Roemer (1988) the 'only if' conditions also includes $u_i(c^{\ell_i,d}, \hat{c}^{n-\ell_i}) = d_i$. Since the removed variables do not enter player i's utility function the latter is trivially satisfied and therefore superfluous.

5 In Rubinstein, Safra, and Thomson (1992) the domain is even further enlarged because all finite lotteries on C are part of the domain and preferences are defined over these lotteries.

6 The domain in Rubinstein, Safra, and Thomson (1992) consists of all finite lotteries on C, which is larger than the domain considered here. Formulated for all lotteries strong conditionally certainty equivalence (CCE*) has much resemblance with the independence axiom, e.g., see chapter 6 in Mas-Colell, Whinston, and Green (1995), but is less restrictive in the sense that it requires that one of the three lotteries in the independence axiom is a degenerated lottery. For our special case CCE* coincides with homogeneity, which means that:

$$\hat{u}_i(c, 1) \geq \hat{u}_i(c', p) \Rightarrow \hat{u}_i(c, \alpha) \geq \hat{u}_i(c', \alpha p), \quad \alpha \in [0, 1].$$

7 Behavioural strategies are defined as mixed strategies on the set of pure actions in each decision node.

Chapter 5

COMPREHENSIVE BARGAINING PROBLEMS

5.1. Introduction

How far can the assumptions made in chapter 2 be relaxed without devastating the results obtained in chapter 3 and 4? And if results qualitatively change, what new insights does this yield? These are the main questions addressed in this and the next chapter. This chapter is solely devoted to the consequences of utility functions that are quasiconcave instead of concave, whereas the next chapter contains a diversity of topics, including nonstationary bargaining situations and alternative bargaining procedures. Variable or endogenous threats are postponed to the second part of this book and the chapters thereafter. Thus, we still assume a *fixed* disagreement point.

Recall that assumption 2.4 imposes concavity upon the players' utility functions over contracts and, according to proposition 2.12, this suffices to ensure that the corresponding bargaining problem in utility representation has a convex set of available utility pairs. If the utility functions are continuous and quasiconcave, then the associated bargaining problem in utility representation becomes strongly comprehensive meaning that the set of available utility pairs may no longer be convex.

The first step is to define strongly comprehensive bargaining problems in utility representation. The remainder of this chapter then resembles much of the order in which the analysis was conducted in chapter 3 and 4: We first characterize the set of MPE strategies and, second, the set of SPE strategies. Then the axiomatic approach for these bargaining problems is studied and two generalizations of the NBS are introduced, which are then in a final step related to MPE and SPE strategies. The translation in terms of equilibrium contracts is briefly discussed. Finally, one major result of this chapter is that quasiconcave utility functions are not sufficient to guarantee uniqueness of the SPE strategy and we end this chapter with discussing sufficient conditions for uniqueness.

In this chapter, despite our efforts to avoid unnecessary duplications with respect to previous chapters, new techniques and concepts such as equilibrium switching and two generalizations of Nash's bargaining solution have to be introduced. The reader should also be aware that several of the techniques introduced in this chapter as well as many of the results obtained return in later chapters.

5.2. Comprehensive bargaining problems

In this section the notion of strong comprehensiveness is introduced for bargaining problems in utility representation.

The starting point is to relax assumption 2.4 and to allow for *continuous* utility functions $u_i : \mathbb{R}^n \rightarrow \mathbb{R}$ that are *quasiconcave*. So, concave utility functions become a special case. For the latter case of bargaining problems proposition 2.12 states that the set S of available utility pairs in the corresponding bargaining problem in utility representation (S, d) is a convex set. For quasiconcave utility functions the set S is connected but need no longer be convex.[1] Moreover, given assumptions 2.2, 2.7 and 2.9 Pareto efficient contracts have been characterized in (2.2). Also for quasiconcave utility functions does maximization problem (2.2) yield generically a unique solution and the formal arguments are similar as those in section 2.2.7. Moreover, player 1's maximum attainable utility is a continuous function of $\overline{u}_2$ and strictly decreasing in $\overline{u}_2$. This implies that for quasiconcave utility functions the Pareto frontier of the set S in $\mathbb{R}^2$ still is a decreasing curve in terms of s_1.

The property that the Pareto frontier of S can be described by a decreasing function is closely related to the notion of strong comprehensiveness, which is defined as follows. Consider the bargaining problem (S, d), where $S \subset \mathbb{R}^2$ is a connected set. We first discuss the weaker notion of comprehensiveness. Let for a point $s \in S^{IR}$ the set $R(s) \subset \mathbb{R}^2$ be given by $R(s) = \{ s' \in \mathbb{R}^2 \,|\, d \leq s' \leq s \}$, which is obviously a rectangle. A bargaining problem is called comprehensive if for all $s \in S^{IR}$ it holds that $R(s) \subseteq S^{IR}$. The notion of *strongly comprehensive* strengthens this idea and additionally requires that the rectangle $R(s)$ belongs to the *interior* of the set S^{IR} (except of course the upper-right corner s if it lies on the Pareto frontier of S^{IR}). Recall that we assumed that the bargaining problem is essential, i.e., there exists an $s \in S$ such that $s > d$. Then the formal definition is as follows.

Definition 5.1. *The set S^{IR} is strongly comprehensive if for each $s \in S^{IR}$ and all $s' \neq s$ such that $d < s' \leq s$ it holds that $s' \in Int\, S^{IR}$.*

Definition 5.1 rules out that there exists a rectangular in S^{IR} such that its upper-horizontal or right-vertical edge describes some part of the Pareto frontier.

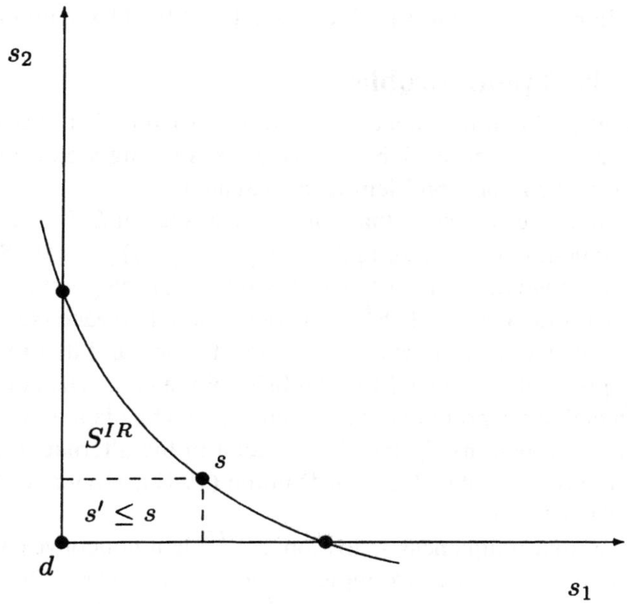

Figure 5.1. The set S^{IR} of the bargaining problem (S, d) is strongly comprehensive.

Equivalently, the Pareto frontier does not contain horizontal or vertical parts. The latter implies that the function f_i, $i = 1, 2$, as defined in section 2.2.8, that describes the Pareto frontier does not make any jumps, i.e., is not discontinuous. Furthermore, the function f_i, $i = 1, 2$, cannot be increasing, because then (take $i = 2$) we would arrive at the contradiction that a (small) part of the Pareto frontier $\{s \in S \mid s_2 = f_2(s_1)\}$ would belong to the interior of S^{IR}. These simple insights imply the next proposition.

Proposition 5.2. *If the set S^{IR} is strongly comprehensive, then the functions f_1 and f_2 that describe the Pareto frontier are continuous and strictly decreasing. Moreover, if S^{IR} is a convex set, then it is also strongly comprehensive.*

This proposition means that the notion of strong comprehensiveness is less restrictive than imposing S^{IR} is a convex set, but it is still sufficient to guarantee nice properties for the functions describing the Pareto frontier. The difference between S^{IR} is convex versus S^{IR} is (only) strongly comprehensive can be summarized by whether or not the function f_j, $j = 1, 2$, is concave.

5.3. Markov perfect equilibrium

In this section MPE strategies are characterized. As in chapter 3, each MPE is related to a fixed point problem. A novel feature is that for a generic class of strongly comprehensive bargaining problems multiplicity of fixed points occurs.

5.3.1 The fixed point problem

The fixed point problem introduced in section 3.3.1 remains to play an important role in case the set of available utility pairs is strongly comprehensive. In this section this fixed point problem is investigated.

In lemma 3.2 it was established that convexity of the set S^{IR} is a sufficient condition for uniqueness of the fixed point $(x, y) = (p(y), q(x))$. Similar as in section 3.3.1 we have that for every fixed point (x, y) the points x and y lie on the same Nash product curve. If S^{IR} is a convex set, then each Nash product curve intersects the Pareto frontier of S^{IR} at most twice and this insight was exploited in the proof of lemma 3.2 to establish uniqueness. The convexity of the set S^{IR} is crucial to the proof and it can therefore not be dispensed.[2] Similar, the concavity of the functions f_1 and f_2 is crucial in the alternative proofs in e.g., Binmore (1987c), chapter 7 in Van Damme (1991), Houba and Bennett (1997) and Okada (1991a).

Does this mean that uniqueness is lost once S^{IR} is a nonconvex set? This question is difficult to answer as example 5.3 below illustrates. This example shows that strong comprehensiveness of S^{IR} is not a sufficient condition to guarantee the uniqueness of a fixed point. Furthermore, this example cannot be dismissed, because the area in the parameter space where multiplicity of the fixed points arises has positive measure meaning that multiplicity is *generic*. Moreover, the area (in the parameter space) featuring multiplicity touches the area of bargaining problems with a convex set S^{IR}. Nevertheless, example 5.3 also shows that there are strongly comprehensive bargaining problems (for which S^{IR} is not a convex set) that still admit a unique fixed point and that this case is also generic. In general, the literature on bargaining has not yet produced more than the partial answer given in chapter 3: The set S^{IR} is a convex set is a sufficient condition for uniqueness of the fixed point. In our view the research agenda for finding sufficient conditions for either uniqueness or multiplicity should be conducted in terms of the contract space C and utility functions u_i, $i = 1, 2$, and is postponed to section 5.6.2.

Finally, we mention that for each fixed point (x, y) of $p \times q$ (even in case of multiplicity) the properties stated in lemma 3.3 still hold.

Example 5.3. *Let* $\alpha_1, \alpha_2 \in (0, 1)$. *Consider the bargaining problem* (S, d) *with* $d = 0$ *and the strictly decreasing Pareto frontier of* $S = S^{IR}$ *consisting of the two linear parts passing through the points* $(0, 1)$, (α_1, α_2) *and* $(1, 0)$. *Then* $S = S^{IR} = \{ s \in \mathbb{R}^2_+ \mid s_2 \leq f_2(s_1) \}$ *is strongly comprehensive and the* *function* $f_i : [0,1] \to [0,1]$, $i = 1,2$, *is given by*

$$f_1(s_2) = \begin{cases} 1 - \frac{1-\alpha_1}{\alpha_2} s_2, & \textit{if } s_2 \in [0, \alpha_2], \\ \frac{\alpha_1}{1-\alpha_2} - \frac{\alpha_1}{1-\alpha_2} s_2, & \textit{if } s_2 \in (\alpha_2, 1], \end{cases}$$

and

$$f_2(s_1) = \begin{cases} 1 - \frac{1-\alpha_2}{\alpha_1} s_1, & \textit{if } s_1 \in [0, \alpha_1], \\ \frac{\alpha_2}{1-\alpha_1} - \frac{\alpha_2}{1-\alpha_1} s_1, & \textit{if } s_1 \in (\alpha_1, 1]. \end{cases}$$

Note that S^{IR} *fails to be a convex set for* $\alpha_1 + \alpha_2 < 1$. *Furthermore, example 3.5 corresponds to* $\alpha_1 = \alpha_2 = \alpha \geq \frac{1}{2}$. *Since* $y_1 < x_1$ *by lemma 3.3 we have the following three cases:* $y_1^* < \alpha_1 < x_1^*$, $y_1^* < x_1^* < \alpha_1$ *and* $\alpha_1 < y_1^* < x_1^*$.
1. Consider $y_1^* < \alpha_1 < x_1^*$, *or* $x_2^* < \alpha_2 < y_2^*$. *Then (3.1) becomes*

$$\begin{aligned} x_2 &= \delta y_2, & x_1 &= 1 - \tfrac{1-\alpha_1}{\alpha_2} x_2, \\ y_1 &= \delta x_1, & y_2 &= 1 - \tfrac{1-\alpha_2}{\alpha_1} y_1, \end{aligned}$$

and solving this system of equations yields the unique solution

$$\begin{aligned} x^* &= \left(\frac{\alpha_1(\alpha_2 - \delta(1-\alpha_1))}{\alpha_1\alpha_2 - \delta^2(1-\alpha_1)(1-\alpha_2)}, \frac{\delta\alpha_2(\alpha_1 - \delta(1-\alpha_2))}{\alpha_1\alpha_2 - \delta^2(1-\alpha_1)(1-\alpha_2)} \right), \\ y^* &= \left(\frac{\delta\alpha_1(\alpha_2 - \delta(1-\alpha_1))}{\alpha_1\alpha_2 - \delta^2(1-\alpha_1)(1-\alpha_2)}, \frac{\alpha_2(\alpha_1 - \delta(1-\alpha_2))}{\alpha_1\alpha_2 - \delta^2(1-\alpha_1)(1-\alpha_2)} \right). \end{aligned}$$

All numerators and denominators are positive if and only if

$$\alpha_2 > \delta(1-\alpha_1), \quad \alpha_2 > \delta^{-1}(\delta - \alpha_1), \quad \alpha_2 > \frac{\delta^2(1-\alpha_1)}{\alpha_1 + \delta^2(1-\alpha_1)} \tag{5.1}$$

and the latter restriction is superfluous.[3] *Furthermore,* $y_1^* \leq \alpha_1 \leq x_1^*$ *imposes the restrictions* $\alpha_1 \geq \frac{\delta}{1+\delta}$ *and* $\alpha_2 \geq \frac{\delta}{1+\delta}$, *which are more restrictive than the restrictions in (5.1). Next, all numerators and denominators are negative if and only if*

$$\alpha_2 < \delta(1-\alpha_1), \alpha_2 < \delta^{-1}(\delta - \alpha_1), \alpha_2 < \frac{\delta^2(1-\alpha_1)}{\alpha_1 + \delta^2(1-\alpha_1)} \tag{5.2}$$

and the latter restriction is superfluous.[4] *A negative denominator means that* $y_1^* \leq \alpha_1 \leq x_1^*$ *imposes the restrictions* $\alpha_1 \leq \frac{\delta}{1+\delta}$ *and* $\alpha_2 \leq \frac{\delta}{1+\delta}$, *which are more restrictive than the restrictions in (5.2). So, the expressions for* (x^*, y^*) *are consistent with* $y_1^* \leq \alpha_1 \leq x_1^*$ *if and only if either* $\alpha_i \leq \frac{\delta}{1+\delta}$, $i = 1,2$, *or* $\alpha_i \geq \frac{\delta}{1+\delta}$, $i = 1,2$.
2. Consider $y_1^* < x_1^* < \alpha_1$, *or* $\alpha_2 < x_2^* < y_2^*$. *Then (3.1) becomes*

$$\begin{aligned} x_2 &= \delta y_2, & x_1 &= \tfrac{\alpha_1}{1-\alpha_2} - \tfrac{\alpha_1}{1-\alpha_2} x_2, \\ y_1 &= \delta x_1, & y_2 &= 1 - \tfrac{1-\alpha_2}{\alpha_1} y_1, \end{aligned}$$

which yields the unique solution

$$x^* = \left(\tfrac{\alpha_1}{(1+\delta)(1-\alpha_2)}, \tfrac{\delta}{1+\delta}\right), \quad y^* = \left(\tfrac{\delta\alpha_1}{(1+\delta)(1-\alpha_2)}, \tfrac{1}{1+\delta}\right)$$

These expressions are consistent with $y_1^* < x_1^* < \alpha_1$ if and only if $\alpha_2 \leq \frac{\delta}{1+\delta}$, which is the same restriction as before.
3. Consider $\alpha_1 < y_1^* < x_1^*$, or $x_2^* < y_2^* < \alpha_2$. Similar as the previous cases, (3.1) becomes

$$x_2 = \delta y_2, \qquad x_1 = 1 - \tfrac{1-\alpha_1}{\alpha_2} x_2,$$
$$y_1 = \delta x_1, \qquad y_2 = \tfrac{\alpha_2}{1-\alpha_1} - \tfrac{\alpha_2}{1-\alpha_1} y_1,$$

and the unique solution is given by

$$x^* = \left(\tfrac{1}{1+\delta}, \tfrac{\delta\alpha_2}{(1+\delta)(1-\alpha_1)}\right), \quad y^* = \left(\tfrac{\delta}{1+\delta}, \tfrac{\alpha_2}{(1+\delta)(1-\alpha_1)}\right),$$

which is consistent with $\alpha_1 < y_1^* < x_1^*$ if and only if $\alpha_1 \leq \frac{\delta}{1+\delta}$.
The restrictions for each of the three cases can be easily represented in the (α_1, α_2)-space. The lines $\alpha_1 = \frac{\delta}{1+\delta}$ and $\alpha_2 = \frac{\delta}{1+\delta}$ divide the square $(0,1)^2$ into four smaller squares. For $\alpha_i \geq \frac{\delta}{1+\delta}$, $i = 1, 2$, the first case applies, whereas $\alpha_1 \geq \frac{\delta}{1+\delta}$ and $\alpha_2 \leq \frac{\delta}{1+\delta}$ imply the second case. The third case corresponds to $\alpha_1 \leq \frac{\delta}{1+\delta}$ and $\alpha_2 \geq \frac{\delta}{1+\delta}$. For these three smaller squares $p \times q$ admits a unique fixed point. However, for $\alpha_i \leq \frac{\delta}{1+\delta}$, $i = 1, 2$, all three cases apply simultaneously, meaning that $p \times q$ admits three fixed points. Furthermore, as δ goes to 1, the point $\left(\delta(1+\delta)^{-1}, \delta(1+\delta)^{-1}\right)$ converges to $\left(\frac{1}{2}, \frac{1}{2}\right)$. The latter point lies on the border of the area $\alpha_1 + \alpha_2 \geq 1$, i.e., the area of parameter values for which S^{IR} is a convex set. This means that for each $(\alpha_1, \alpha_2) < \left(\frac{1}{2}, \frac{1}{2}\right)$ and for all $\delta > \max\left\{\frac{\alpha_1}{1-\alpha_1}, \frac{\alpha_2}{1-\alpha_2}\right\}$ the function $p \times q$ admits three fixed points. Thus, the multiplicity does not vanish as δ goes to 1. Finally, multiplicity is generic, because the associated area has positive measure $\delta^2(1+\delta)^{-2} \in \left(0, \frac{1}{4}\right)$. Note that uniqueness is also generic, because this corresponds to an area with positive measure $1 - \delta^2(1+\delta)^{-2} > \frac{3}{4}$.

5.3.2 MPE in utility representation

In this section we show that for strongly comprehensive bargaining problems in utility representation each MPE is related to the fixed point problem (3.1) and vice versa.

The characterization of MPE strategies follows similar steps as explained in section 3.3.3-3.3.5 and many of the results obtained there are still valid. The reason is that in this section in terms of the functions f_i, $i = 1, 2$, only the assumption of concavity is dropped and this property of f_i was not used to prove many of the earlier results. The next proposition is a generalization of proposition 3.12 and relates MPE strategies one-to-one to the fixed points of (3.1).

Proposition 5.4. *If S^{IR} is strictly comprehensive, then Markov strategies form a MPE if and only if $(x,y) = (p(y), q(x))$ and the sets of acceptable proposals given by*

$$T_1(x) = \{s \in S |\ s_1 \geq q_1(x)\}, \quad T_2(y) = \{s \in S |\ s_2 \geq p_2(y)\}.$$

Moreover, $(x,y) = (p(y), q(x))$ implies that $x \in T_2(y)$ and $y \in T_1(x)$.

Proposition 5.4 states that there is a one-to-one relation between MPE strategies and the fixed points of (3.1). This means that the uniqueness of MPE strategies can be reformulated as when does the function $p \times q$ admit a unique fixed point (x^*, y^*). The latter problem was addressed in the previous section.

In case (3.1) admits multiple fixed points there exist multiple MPE strategies and the players have to coordinate on the same equilibrium. The Nash equilibrium as well as the SPE concept simply assumes that in equilibrium this coordination problem does not arise and that both players correctly anticipate which equilibrium is going to be played, see e.g. the text books on game theory Fudenberg and Tirole (1991), Meyerson (1991) and Osborne and Rubinstein (1994). Since game theory has not yet solved this problem, we simply characterize all the MPE or SPE and regard this topic as a game theoretic problem rather than a problem specific to bargaining theory.

We conclude this section with an example that states the MPE strategies for the bargaining problem in example 5.3 in case there are multiple fixed points.

Example 5.5. *Consider the bargaining problem of example 5.3 once more and additionally assume $\alpha_1, \alpha_2 < \frac{1}{2}$ and $\delta > \max\left\{\frac{\alpha_1}{1-\alpha_2}, \frac{\alpha_2}{1-\alpha_1}\right\}$. Then the function $p \times q$ admits three fixed points and there is one pair of MPE strategies related to each fixed point. The pair of MPE strategies for the first case of example 5.3 is given by*

$$\sigma_1(h_t) = \begin{cases} \textit{proposes} & \left(\frac{\alpha_1(\alpha_2-\delta(1-\alpha_1))}{\alpha_1\alpha_2-\delta^2(1-\alpha_1)(1-\alpha_2)}, \frac{\delta\alpha_2(\alpha_1-\delta(1-\alpha_2))}{\alpha_1\alpha_2-\delta^2(1-\alpha_1)(1-\alpha_2)}\right), \\ \textit{responds yes} & \textit{if } s_1(t) \geq \frac{\delta\alpha_1(\alpha_2-\delta(1-\alpha_1))}{\alpha_1\alpha_2-\delta^2(1-\alpha_1)(1-\alpha_2)}, \end{cases}$$

and

$$\sigma_2(h_t) = \begin{cases} \textit{proposes} & \left(\frac{\delta\alpha_1(\alpha_2-\delta(1-\alpha_1))}{\alpha_1\alpha_2-\delta^2(1-\alpha_1)(1-\alpha_2)}, \frac{\alpha_2(\alpha_1-\delta(1-\alpha_2))}{\alpha_1\alpha_2-\delta^2(1-\alpha_1)(1-\alpha_2)}\right), \\ \textit{responds yes} & \textit{if } s_2(t) \geq \frac{\delta\alpha_2(\alpha_1-\delta(1-\alpha_2))}{\alpha_1\alpha_2-\delta^2(1-\alpha_1)(1-\alpha_2)}. \end{cases}$$

Similar, the MPE strategies for the second case of example 5.3 are given by

$$\sigma_1(h_t) = \begin{cases} \textit{proposes} & \left(\frac{\alpha_1}{(1+\delta)(1-\alpha_2)}, \frac{\delta}{1+\delta}\right), \\ \textit{responds yes} & \textit{if } s_1(t) \geq \frac{\delta\alpha_1}{(1+\delta)(1-\alpha_2)}, \end{cases}$$

and

$$\sigma_2(h_t) = \begin{cases} \textit{proposes} & \left(\frac{\delta\alpha_1}{(1+\delta)(1-\alpha_2)}, \frac{1}{1+\delta}\right), \\ \textit{responds yes} & \textit{if } s_2(t) \geq \frac{\delta}{1+\delta}. \end{cases}$$

Finally, the third case of example 5.3 implies

$$\sigma_1(h_t) = \begin{cases} \textit{proposes} & \left(\frac{1}{1+\delta}, \frac{\delta\alpha_2}{(1+\delta)(1-\alpha_1)}\right), \\ \textit{responds yes} & \textit{if } s_1(t) \geq \frac{\delta}{1+\delta}, \end{cases}$$

and

$$\sigma_2(h_t) = \begin{cases} \textit{proposes} & \left(\frac{\delta}{1+\delta}, \frac{\alpha_2}{(1+\delta)(1-\alpha_1)}\right), \\ \textit{responds yes} & \textit{if } s_2(t) \geq \frac{\delta\alpha_2}{(1+\delta)(1-\alpha_1)}. \end{cases}$$

5.4. Subgame perfect equilibrium

In this section the set of SPE utility pairs is characterized in two steps. First, the upper and lower bounds on this set are related to the MPE utilities and, next, the set is characterized. In order to derive the latter we introduce the technique of equilibrium switching. Recall that the unique SPE characterized in chapter 3 featured immediate agreement. In case of multiple fixed points equilibria without this property can be derived, i.e., with delay in reaching an agreement, and we will do so in section 5.4.4. These equilibria involve equilibrium switching.

5.4.1 Bounds for SPE utilities

In section 3.4.1 the method in Shaked and Sutton (1984) was introduced as a powerful method that could be applied to show that the set of SPE utility pairs consists of a unique element. In this section this method is again applied in order to derive bounds upon the sets of SPE utility pairs.

The method of Shaked and Sutton exploited the fact that the set of SPE utility pairs is history independent up to round t, $t \in \mathbb{N}$, being even or odd, as was shown in lemma 3.17. Since the proof of the latter lemma is independent of the shape of the set S^{IR} this result still applies. Adopting the same notation, we denote $E_j \subseteq S^{IR}$, $j = 1, 2$, as the set of SPE utility pairs at the start of any round t where player j is the proposing player. As before, $\underline{s}_i^j$, respectively, $\overline{s}_i^j$, $i, j = 1, 2$, denotes the largest lower and smallest upper bound upon player i's

SPE utilities in the set E_j. The following proposition relates these bounds to the fixed points of $p \times q$.

Proposition 5.6. *If S^{IR} is strictly comprehensive, then*

$$\begin{aligned}
\underline{s}_1^1 &= \min\{x_1 \,|(x,y) = (p(y), q(x))\}, & \overline{s}_2^1 &= f_2(\underline{s}_1^1),\\
\overline{s}_1^1 &= \max\{x_1 \,|(x,y) = (p(y), q(x))\}, & \underline{s}_2^1 &= f_2(\overline{s}_1^1),\\
\underline{s}_2^2 &= \min\{y_2 \,|(x,y) = (p(y), q(x))\}, & \overline{s}_1^2 &= f_1(\underline{s}_2^2),\\
\overline{s}_2^2 &= \max\{y_2 \,|(x,y) = (p(y), q(x))\}, & \underline{s}_1^2 &= f_1(\overline{s}_2^2).
\end{aligned}$$

Moreover, $(\underline{s}_1^1, \overline{s}_2^1) = p(\underline{s}_1^2, \overline{s}_2^2)$ *and* $(\overline{s}_1^2, \underline{s}_2^2) = q(\overline{s}_1^1, \underline{s}_2^1)$.

This proposition states that each player's SPE utilities in E_j, $j = 1, 2$, are bounded from below by what this player considers as the worst possible MPE in the set of MPE utilities. So, with respect to the set $E_j \subseteq S^{IR}$, $j = 1, 2$, it follows that

$$E_j \subseteq \left\{ s \in \mathbb{R}^2 \;\middle|\; s_1 \geq \underline{s}_1^j,\ s_2 \geq \underline{s}_2^j,\ s_2 \leq f_2(s_1) \right\},$$

meaning that the set E_j in the two-dimensional space is bounded from below by a vertical and a horizontal line, whereas it is bounded from above by the Pareto frontier of S^{IR}. The intersection of the horizontal (or vertical) line with the Pareto frontier corresponds to a pair of MPE utilities and, therefore, belongs to the set E_j, $j = 1, 2$. This in turn implies that each player's highest possible SPE utility is bounded from above by his most favourable MPE utility. Moreover, player i's highest possible SPE utility corresponds to his opponent's worst possible SPE utility and vice versa.

In example 5.3 we saw that multiplicity of fixed points of $p \times q$ cannot be ruled out. Nevertheless, proposition 5.6 reinforces the earlier result of proposition 3.18: If the function $p \times q$ admits a unique fixed point, then obviously $\underline{s}_i^j = \overline{s}_i^j$, $i, j = 1, 2$, meaning that the alternating offers has a unique pair of SPE strategies given by the unique pair of MPE strategies.

Corollary 5.7. *If (x^*, y^*) is the unique fixed point of $p \times q$, then $E_1 = \{x^*\}$ and $E_2 = \{y^*\}$.*

Although the bounds of proposition 5.6 are informative in case the fixed point of $p \times q$ is unique, we still have not characterized the shape of the set E_j, $j = 1, 2$, in case of multiplicity, which is postponed for the next sections.

The next example illustrates proposition 5.6.

Example 5.8. *Consider the bargaining problem of example 5.3 and 5.5 once more and additionally assume $\alpha_1, \alpha_2 < \frac{1}{2}$ and $\delta > \max\left\{\frac{\alpha_1}{1-\alpha_2}, \frac{\alpha_2}{1-\alpha_1}\right\}$. Then* *application of proposition 5.6 implies that*

$$\underline{s}_1^1 = \frac{\alpha_1}{(1+\delta)(1-\alpha_2)}, \qquad \overline{s}_2^1 = f_2\left(\underline{s}_1^1\right) = \frac{\delta}{1+\delta},$$
$$\overline{s}_1^1 = \frac{1}{1+\delta}, \qquad \underline{s}_2^1 = f_2\left(\overline{s}_1^1\right) = \frac{\delta\alpha_2}{(1+\delta)(1-\alpha_1)},$$
$$\underline{s}_2^2 = \frac{\alpha_2}{(1+\delta)(1-\alpha_1)}, \qquad \overline{s}_1^2 = f_1\left(\underline{s}_2^2\right) = \frac{1}{1+\delta},$$
$$\overline{s}_2^2 = \frac{1}{1+\delta}, \qquad \underline{s}_1^2 = f_1\left(\overline{s}_2^2\right) = \frac{\delta\alpha_1}{(1+\delta)(1-\alpha_2)}.$$

So, with respect to the set E_j, $j = 1,2$, it follows that

$$E_1 \subseteq \left\{ s \in \mathbb{R}^2 \,\middle|\, s_1 \geq \frac{\alpha_1}{(1+\delta)(1-\alpha_2)}, s_2 \geq \frac{\delta\alpha_2}{(1+\delta)(1-\alpha_1)}, s_2 \leq f_2(s_1) \right\},$$
$$E_2 \subseteq \left\{ s \in \mathbb{R}^2 \,\middle|\, s_1 \geq \frac{\delta\alpha_1}{(1+\delta)(1-\alpha_2)}, s_2 \geq \frac{\alpha_2}{(1+\delta)(1-\alpha_1)}, s_2 \leq f_2(s_1) \right\}.$$

5.4.2 Equilibrium switching

In this section the idea of equilibrium switching is introduced and translated to alternating offers. This powerful idea returns several times in later chapters, but here we explain the general idea behind it mainly for the simple case of strategies that induce immediate agreement.

Consider a strongly comprehensive bargaining problem (S, d) that admits multiple MPE strategies. Let the pair of strategies σ feature immediate agreement upon $s^0(\sigma) = s^*$ for some $s^* \in S$. Immediate agreement upon s^* is an SPE outcome if $i)$ player 1 does not have an incentive to propose some other $s^0 \neq s^* \in S$, $ii)$ player 2 does not have an incentive to reject s^* and $iii)$ the strategies σ specify SPE behaviour from every subgame $h_1(1) = (s^0, no)$ onwards. The third requirement is easy to fulfil if the players follow a pair of MPE strategies from round $t = 1$ onward.

The key idea of *equilibrium switching* is not to have a history independent choice of one particular MPE for all $h_1(1) \in H_1(1)$, but rather to let the MPE strategies to be followed from round $t = 1$ onward depend upon the history $h_1(1)$. For example, the players start the MPE strategies associated with $\left(\underline{s}_1^1, \overline{s}_2^1\right)$ of proposition 5.6 at round $t = 1$ if $h_1(1) \neq (s^*, no)$ and the players start the MPE strategies associated with $\left(\overline{s}_1^2, \underline{s}_2^2\right)$ after the history $h_1(1) = (s^*, no)$. So, the main difference with Markov strategies is that here the value functions $V_i^2(h_1(1), 1)$ for $h_1(1) = (s^0, no)$ are history dependent and *discontinuous* in s^0 at s^*. Furthermore, dependent upon the exact history these value functions switch between two MPE strategies, which is called equilibrium switching.

In Abreu (1988) the idea of equilibrium switching was introduced for infinitely repeated games, but its methodology is generally applicable. In order to be of use, equilibrium switching strategies should possess a 'stick and carrot' character, meaning that if player i, $i = 1,2$, deviates from the specified strategies this player will be punished, i.e., hit with the stick, whereas player i is rewarded, i.e., earned the carrot, if he complies to the specified strategies. In order to fully exploit the idea of stick and carrot each player's punishment should be as severe as possible. To see this, consider some mild punishment that is sufficient to deter a player from deviating of the prescribed strategies. All punishment that are more severe also deter this player from deviating and we can restrict attention to the most severe punishment. Within a game theoretic context, a player's most severe punishment coincides with the worst possible SPE strategies for this player.

Application of these insights to alternating offers is simple. First, proposition 5.6 implies that each player's worst (best) SPE utility coincides with this player's worst (best) MPE utility. So, MPE utility pairs can be used as worst punishments in equilibrium switching and these are already characterized in proposition 5.6. Player 1's worst punishment in terms of a pair of MPE utilities is $\left(\underline{s}_1^1, \overline{s}_2^1\right)$ and, similar, player 2's worst punishment is given by $\left(\overline{s}_1^2, \underline{s}_2^2\right)$. From these utility pairs we observe that bargaining games are special in the sense that each player's worst equilibrium punishment coincides with his opponent's highest equilibrium utility that is attainable in the negotiations and vice versa. This means that each player's opponent has a strong incentive to carry out the equilibrium punishment.

Finally, note that the discontinuous value functions stated at the start of this section feature the stick and carrot character. Furthermore, each player's punishment is the most extreme within the set of MPE utilities. The associated pair of strategies σ is given in table 5.1, where the players start in the initial state named IA (immediate agreement) and go to the absorbing state i, $i = 1, 2$, immediately after player i deviated from the prescribed actions in the initial state. For more details on tables representing strategies we refer to section 2.3.9.

5.4.3 SPE with equilibrium switching

For explanatory reasons the strategies involving equilibrium switching and immediate agreement are analyzed in this section in order to derive the equilibrium conditions upon agreements. The results are derived with dynamic programming. The general case of SPE strategies featuring delay is left for the next section.

Consider the pair of strategies σ represented by table 5.1, which feature equilibrium switching and immediate agreement upon $s^0(\sigma) = s^*$ for some $s^* \in S$. Then this pair of strategies forms a SPE if and only if *i*) player 1 does not have an incentive to propose a different proposal $s^0 \in S$, $s^0 \neq s^*$, *ii*) player 2 does not have an incentive to reject s^* and *iii*) the strategies σ specify SPE behaviour from every subgame $h_1(1) = \left(s^0, no\right)$ onwards. As explained before, equilibrium switching automatically takes care of the third requirement and can therefore be disregarded in the analysis of this section.

The aim of the analysis is to derive conditions upon s^* such that yes maximizes player 2's utility. Recall that given the strategies σ player 2's value function at information set $h_1(1) = (s^0, no)$ is given by

$$V_2^2\left(\left(s^0, no\right), 1\right) = \begin{cases} (1-\delta)\, d_2 + \delta \underline{s}_2^2, & \text{if } s^0 = s^*, \\ (1-\delta)\, d_2 + \delta \overline{s}_2^2, & \text{if } s^0 \neq s^*. \end{cases}$$

Then application of dynamic programming implies that yes maximizes player 2's utility at information set $h_1(1) = (s^0, no)$ if and only if

$$yes = \arg \max_{\{yes,no\}} \begin{cases} s_2^0, & \text{if } yes, \\ (1-\delta)\, d_2 + \delta \underline{s}_2^2, & \text{if } no \text{ and } s^0 = s^*, \\ (1-\delta)\, d_2 + \delta \overline{s}_2^2, & \text{if } no \text{ and } s^0 \neq s^*. \end{cases}$$

There are two cases: $s^0 = s^*$ and $s^0 \neq s^*$. In the former case yes is optimal for player 2 if $s_2^* \geq (1-\delta)\, d_2 + \delta \underline{s}_2^2 = \underline{s}_2^1$ and, similarly, if $s_2^0 \geq (1-\delta)\, d_2 + \delta \overline{s}_2^2 = \overline{s}_2^1$ for the latter case.[5] Thus, immediate agreement upon s^* imposes the equilibrium condition $s_2^* \geq \underline{s}_2^1$ and the associated optimal σ_2 is given by

$$\sigma_2\left(s^0, 0\right) = \begin{cases} yes, & \text{if } s^0 = s^* \text{ (and } s_2^* \geq \underline{s}_2^1\text{)}, \\ yes, & \text{if } s^0 \neq s^* \text{ and } s_2^0 \geq \overline{s}_2^1, \\ no, & \text{if } s^0 \neq s^* \text{ and } s_2^0 < \overline{s}_2^1. \end{cases}$$

This solves $ii)$ of the equilibrium requirements above.

The analysis continues by deriving conditions upon s^* for the requirement $i)$. Player 1's value function $V_1^1(h_1(0), 0)$ at the start of the bargaining game associated with the strategies σ, the equilibrium condition $s_2^* \geq \underline{s}_2^1$ and player 2's optimal response is given by

$$V_1^1\left(h_1(0), 0\right) = \begin{cases} s_1^*, & \text{if } s^0 = s^* \text{ (and } s_2^* \geq \underline{s}_2^1\text{)}, \\ s_1^0, & \text{if } s^0 \neq s^* \text{ and } s_2^0 \geq \overline{s}_2^1, \\ (1-\delta)\, d_1 + \delta \underline{s}_1^2, & \text{if } s^0 \neq s^* \text{ and } s_2^0 < \overline{s}_2^1. \end{cases}$$

Similar as before, we are interested in equilibrium conditions upon s^* such that

$$s^* = \arg \max_{s^0 \in S} V_1^1\left(h_1(0), 0\right). \tag{5.3}$$

This requirement implies two conditions. First, for all $s^0 \in S$ such that $s_2^0 \geq \overline{s}_2^1$ it must hold that $s_1^0 \leq s_1^*$. Second, for all $s^0 \in S$ such that $s_2^0 < \overline{s}_2^1$ it must hold that $(1-\delta)\, d_1 + \delta \underline{s}_1^2 \leq s_1^*$. Consider the first condition. Since f_1 is strictly decreasing we have that $s_2^0 \geq \overline{s}_2^1$ implies that $s_1^0 \leq f_1\left(s_2^0\right) \leq f_1\left(\overline{s}_2^1\right) = \underline{s}_1^1$. The latter means that s^0 has to lie to the left of or on the line that is the

State		IA	1	2
Player 1	proposes	s^*,	$(\underline{s}_1^1, \overline{s}_2^1)$,	$(\overline{s}_1^1, \underline{s}_2^1)$,
	responds *yes*	if $s_1^t \geq s_1^*$,	if $s_1^t \geq \underline{s}_1^2$,	if $s_1^t \geq \overline{s}_1^2$,
Player 2	proposes	s^*,	$(\underline{s}_1^2, \overline{s}_2^2)$,	$(\overline{s}_1^2, \underline{s}_2^2)$,
	responds *yes*	if $s_2^t \geq s_2^*$,	if $s_2^t \geq \overline{s}_2^1$,	if $s_2^t \geq \underline{s}_2^1$,
Transitions		Go to state 1 if either $s^t \neq s^*$, t even, or 1 rejects $s_1^t = s^*$, t odd. Go to state 2 if either $s^t \neq s^*$, t odd, or 2 rejects $s^t = s^*$, t even.	absorbing	absorbing

Table 5.1. An example of strategies that induce immediate agreement upon s^* in every subgame and a player's deviation is followed by this player most severe punishment, where $(\overline{s}_1^i, \underline{s}_2^i)$ and $(\underline{s}_1^i, \overline{s}_2^i)$, $i = 1, 2$, are characterized in proposition 5.6.

lower bound upon E_1. So, any deviation $s^0 \neq s^*$ that will be accepted by player 2 yields a utility of at most player 1's worst possible SPE utility. Such deviation would certainly be profitable in case $s_1^* < \underline{s}_1^1$, because then player 1's deviation $s^0 = (\underline{s}_1^1, \overline{s}_2^1)$ is accepted by player 2. Since the SPE concept rules out profitable one-stage deviations we must have that immediate agreement upon s^* is weakly preferred to such deviation, which yields the condition $s_1^* \geq \underline{s}_1^1$. It turns out that the latter restriction makes the second condition superfluous, because $(1-\delta)\,d_1 + \delta \underline{s}_1^2 = (1-\delta)\,d_1 + \delta p_1(\underline{s}_1^1, \overline{s}_2^1) < \underline{s}_1^1$.

Summarizing, we have derived the following proposition.

Proposition 5.9. *The strategies σ represented by table 5.1 are SPE strategies featuring immediate agreement upon s^* at round $t(\sigma) = 0$ if and only if $s_1^* \geq \underline{s}_1^1$ and $s_2^* \geq \underline{s}_2^1$. Moreover, $E_1 = \{s \in S^{IR} \mid s_1^* \geq \underline{s}_1^1,\ s_2^* \geq \underline{s}_2^1\}$.*

The intuition for why these strategies σ work is as follows. If player 1 does not propose $s^0 = s^*$, then this player is punished while the agreement s^* has the feature of a reward, i.e., the carrot, provided the latter is preferred to being punished. Similarly, player 2 is forced to accept s^*, because rejecting means that player 2 is punished. Player 2 will be persuaded to accept s^* if doing so is not worse than being punished. Note that this proposition also applies if the function $p \times q$ admits a unique fixed point (x^*, y^*). Then $(\underline{s}_1^1, \underline{s}_2^1) = x^*$ lies on the Pareto frontier and $E_1 = \{x^*\}$ as in proposition. Moreover, the only strategy of table 5.1 that is SPE necessarily features $s^* = x^*$. The obvious reason is that equilibrium switching is trivial in the sense that both player 1's and player 2's worst possible SPE coincide and, therefore, player 2's value function $V_2^2((s^0, no), 1)$ becomes continuous in $s^0 = s^*$, because $\underline{s}_2^2 = \overline{s}_2^2 = y_2^*$.

Proposition 5.9 implies that the agreement s^* need not be Pareto efficient meaning that the players do not reap all of the potential benefits available to them. However, the fear of being punished, that is build in the strategies, prevents the players to propose a Pareto efficient contract that is better for both players when compared to s^*.

5.4.4 SPE with delay

In this section we want to go one step beyond deriving the set of SPE utility pairs at even bargaining rounds and derive the set of SPE *payoff relevant* outcome paths of the bargaining game for subgames starting at even and odd rounds.

All equilibrium utilities in the set E_j, $j = 1, 2$, are uniquely determined by some payoff relevant path, which consists of the agreement $s^{t(\sigma)}(\sigma) \in S^{IR}$ and the round $t(\sigma) \in \mathbb{N}$ of its conclusion. So, equilibrium utilities have the form $\left(1 - \delta^{t(\sigma)}\right) d + \delta^{t(\sigma)} s^{t(\sigma)}(\sigma)$. Delay in reaching an agreement occurs if $t(\sigma) > 0$. The characterization of all payoff relevant paths consists of deriving equilibrium conditions upon $t(\sigma) \in \mathbb{N}$ and $s^{t(\sigma)}(\sigma) \in S^{IR}$ such that $\left(1 - \delta^{t(\sigma)}\right) d + \delta^{t(\sigma)} s^{t(\sigma)}(\sigma) \in E_j$, $j = 1, 2,$. The idea is again to introduce suitable strategies that feature equilibrium switching and then derive equilibrium conditions for these strategies.

The strategies σ that are investigated are represented by table 5.2 and these strategies extend the strategies of table 5.1 by adding one column as the initial state. These modified strategies implement the agreement $s^{t(\sigma)}(\sigma) = s^*$ at round $t(\sigma)$ and the players disagree until round $t(\sigma) \in \mathbb{N}$. In order to sustain such an outcome path, players must first make proposals until round $t(\sigma)$ that are rejected. We regard $(s_1^{\max}, d_2)$ and $(d_1, s_2^{\max})$ as natural candidates for these proposals.[6] They have the flavour that the proposing player demands what in Kalai and Smorodinsky (1975) is called his utopia utility in S^{IR} and is not willing to settle for anything less. In case player i, $i = 1, 2$, would make an offer that leaves less to himself than $s_i^{\max}$, then this player is immediately punished because the strategies switch to this player's worst possible SPE in state i. So, similar as before, equilibrium switching is applied as a device to prevent players from deviating from σ. At round $t(\sigma)$ the strategies move to state IA (of immediate agreement) of the second column, which is the first column of table 5.1, including the same state transition. The behaviour in state IA describes agreement upon $s^{t(\sigma)}(\sigma) = s^*$. For completeness, we mention that the last two columns, represent the state i, $i = 1, 2$, meaning that player i is punished and these are identical to the states in table 5.1.

Deriving equilibrium conditions upon s^* and $t(\sigma)$ is similar as in the previous section and, therefore, deferred to the appendix.

State		Initial
Player 1	proposes	$(s_1^{\max}, d_2)$, always
	responds *yes*	if $s_1^t \geq q_1\left(s_1^{\max}, d_2\right)$,
Player 2	proposes	$(d_1, s_2^{\max})$, always,
	responds *yes*	if $s_2^t \geq p_2\left(d_1, s_2^{\max}\right)$,
Transitions		Go to state IA of table 5.1 if $t \geq t(\sigma^*)$. Go to state 1 of table 5.1 if $s_1^t < s_1^{\max}$, t even. Go to state 2 of table 5.1 if $s_2^t < s_2^{\max}$, t odd.

Table 5.2. The initial state of the SPE strategies σ of proposition 5.10, where the states IA, 1 and 2 refer to the columns of table 5.1. If s^* and $t(\sigma)$ satisfy the conditions stated in proposition 5.10, then these strategies are SPE and induce the agreement s^* after a delay of $t(\sigma)$ rounds.

Proposition 5.10. *The strategies of table 5.2 are SPE if and only if*

$$\left(1-\delta^{t(\sigma)}\right) d+\delta^{t(\sigma)} s^* \geq \left(\underline{s}_1^1, \underline{s}_2^1\right). \tag{5.4}$$

Moreover, $E_j=\left\{s \in S^{IR} \,\middle|\, s_1^* \geq \underline{s}_1^j,\ s_2^* \geq \underline{s}_2^j\right\} \subset S^{IR}$, $j=1,2$.

Equilibrium condition (5.4) implies that the maximum delay of any SPE and the associated SPE agreement, denoted as $\hat{t}$ and $\hat{s}$, are found by solving

$$\max_{t \in \mathbb{R}, s \in \mathbb{R}^2} t, \quad \text{s.t.} \quad -\left(1-\delta^t\right) d_i-\delta^t s_i \leq -\underline{s}_i^1, \qquad i=1,2, \tag{5.5}$$

$$s_2 \leq f_2\left(s_1\right),$$

where we treat t as a continuously variable for explanatory simplicity. It is easy to see that the maximum delay is finite, because $\underline{s}_i^1 > d_i$, $i=1,2$, and $t=\infty$ would make the right hand side of the i-th constraint equal to d_i independent of $s \in S$. Furthermore, for a continuous variable t, the Pareto efficiency constraint should be binding in any optimum $\left(\hat{t}, \hat{s}\right)$. If not, then $\hat{s}_2 < f_2\left(\hat{s}_1\right)$ leaves some room to slightly increase both variables s_1 and s_2 such that all three constraints are still satisfied and, doing so, creates some extra slack in the first two constraints which can then be exploited to slightly increase t also, which would contradict that $\left(\hat{t}, \hat{s}\right)$ is an optimum. So, at the optimum $\left(\hat{t}, \hat{s}\right)$ the final agreement $\hat{s}$ should be Pareto efficient.

Example 5.11. *Consider the bargaining problem of example 5.3 once more and additionally assume* $\alpha_1=\alpha_2=\alpha<\frac{1}{2}$ *and* $\delta>\frac{\alpha}{1-\alpha}$. *Combining the results of example 5.8 and proposition 5.10 imply that both* $\subseteq$ *in the example are* $=$. *Next, both* E_1 *and* E_2 *are strict subsets of* S^{IR}, *because* $d=0$ *and all lower bounds are positive. Furthermore, making use of* $\hat{s}_2=f_2\left(\hat{s}_1\right)$ *and the symmetry of* *the problem yields* $\hat{s} = (\alpha, \alpha)$ *and* $\hat{t} = -(\ln \delta)^{-1} \ln(1+\delta)(1-\alpha) > 0$ *solve maximization problem (5.5).*

5.5. Nash program

This section deals with the Nash program for strongly comprehensive bargaining problems. First, the axiomatic approach is investigated and two generalizations for the NBS are introduced. Next, the relation between the axiomatic and strategic approach is investigated, which means that we study whether or not the sets of MPE and SPE utility pairs of the alternating offers model converge to these two generalized axiomatic solutions as the time between bargaining rounds vanishes.

5.5.1 Generalized Nash's solutions

In this section two extensions of the axiomatic Nash's bargaining solution for strongly comprehensive bargaining problems are briefly discussed. These generalizations return in section 5.5.1 where the limit set of MPE and SPE utility pairs are characterized.

The extension of Nash's solution to the class of strongly comprehensive bargaining problems (S, d) follows a similar route as in chapter 4, although some technicalities have to be addressed. The first is that multiple solutions cannot be avoided and, therefore, mathematically speaking any axiomatic solution for the class of strongly comprehensive bargaining problems is a correspondence

$$B : \{(S,d) \mid S^{IR} \text{ is strongly comprehensive}\} \to S.$$

The first generalization was proposed in Kaneko (1980). In this reference Nash's original four axioms were reformulated in terms of correspondences and an additional axiom was introduced. The latter axiom deals with the continuity of the bargaining solution if the underlying bargaining problem slightly changes. The following axiom is the continuity axiom.

Axiom 5.12. *Let* $\langle S(k)\rangle_{k=1}^{\infty}$ *be a convergent sequence of strongly comprehensive sets (w.r.t. d) in the Hausdorf topology with limit the strongly comprehensive set S (w.r.t. d). Then for any sequence* $\langle s(k)\rangle_{k=1}^{\infty}$, $s(k) \in B(S(k), d)$, *such that* $\lim_{k\to\infty} s(k) = s$ *it must hold that* $s \in B(S, d)$.

The following theorem is a straightforward extension of the NBS characterized in theorem 4.5 to the class of strongly comprehensive bargaining problems and we refer to Kaneko (1980) for its proof.

Theorem 5.13. *The unique correspondence* $B(S,d)$ *that satisfies axioms 4.1-4.4 and 5.12 is the correspondence* $N(S,d)$ *given by*

$$N(S,d) = \left\{ \arg\max_{s \in S^{IR}} (s_1 - d_1)(s_2 - d_2) \right\}.$$

Note that (S, d) is strongly comprehensive implies that $s^* > d$ for each $s^* \in N(S, d)$ and that this solution also satisfies the individual rationality axiom, i.e. 4.6. The reason is that the Nash product curve through $(d_1, s_2^{\max})$ is a vertical line that cannot be tangent to the Pareto frontier of S^{IR} since vertical parts of this frontier are ruled out by the strong comprehensiveness. Hence, $(d_1, s_2^{\max}) \notin N(S, d)$. Similar arguments imply that $(s_1^{\max}, d_2) \notin N(S, d)$ and, therefore, $s^* \in N(S, d) > d$. As in theorem 4.7 the efficiency axiom can be replaced by the individual rationality axiom.

If $s^* \in S^{IR}$ is a generalized NBS, then s^* satisfies both geometrical properties discussed in section 4.2.2. There are some differences though. Define the set $T(S, d) \subseteq S^{IR}$ as the set of utility pairs $\hat{s}$ on the Pareto frontier of S^{IR} for which the Nash product curve $(s_1 - d_1)(s_2 - d_2)$ through $\hat{s}$ is tangent to the Pareto frontier in $\hat{s}$. Then obviously, each $\hat{s} \in T(S, d)$ also satisfies the second geometrical property and $\hat{s} > d$ for each $\hat{s} \in T(S, d)$. Moreover, $N(S, d) \subseteq T(S, d)$ and $N(S, d) = T(S, d)$ consists of a single element if S is a convex set. However, there is a generic subclass of strongly comprehensive bargaining problems (S, d) for which $N(S, d) \subset T(S, d)$ meaning $N(S, d) \neq T(S, d)$, as example 5.16 illustrates.

The correspondence $T(S, d)$ can also be obtained as a generalized NBS simply by relaxing axiom 4.4 that imposes the independence of irrelevant alternatives. In Herrero (1989) the latter axiom in theorem 5.13 is replaced by a local independence of irrelevant alternatives, which is stated as follows. (Recall that $f_2'^-$ and $f_2'^+$ are the left and right derivative of f_2.)

Axiom 5.14. *Let* $s^* \in B(S, d)$, $f_2'^-(s_1^*) = f_2'^+(s_1^*)$ *and for* $(\hat{S}, d)$ *either* $i)$ $\hat{S} \subset S$ *and* $s^* \in \hat{S}$, *or* $ii)$ $\hat{S} \supset S$ *and* $s^* \in \hat{S}^{PE}$.[7] *Then* $s^* \in B(\hat{S}, d)$.

The following theorem weakens the predictive power of theorem 5.13 to obtain $B(S, d) = T(S, d)$ as the generalized NBS. Beside the less restrictive axiom 5.14 the individual rationality axiom has to be added to the list of axioms. For a proof we refer to Herrero (1989).

Theorem 5.15. *The unique correspondence* $B(S, d)$ *that satisfies axioms 4.1-4.3, 5.14, 4.6 and 5.12 is the correspondence* $T(S, d)$.

It goes without saying that these results can be translated in terms of contracts and we omit doing so. Then each $c \in T\left(C, \{u_i\}_{i=1,2}, c^d\right)$ corresponds to one of the stationary points of the Kuhn-Tucker first-order conditions in section 4.2.4, whereas each $c^* \in N\left(C, \{u_i\}_{i=1,2}, c^d\right)$ corresponds to one of the global maximizers. Finally, the discussion of section 4.2.5-4.2.8 also applies to both generalizations of the NBS for convex bargaining problems and it will not be repeated here.

We conclude with an example that illustrates the difference between the two extensions of the NBS. This example also shows that the multiplicity of elements in the generalized NBS of theorem 5.13 is nongeneric.

Example 5.16. *Consider the bargaining problem of example 5.3 and 5.5 once more and additionally assume $\alpha_1, \alpha_2 < \frac{1}{2}$. Then*

$$\max_{s \in S^{IR}} s_1 s_2, \text{ s.t. } s_2 = \begin{cases} 1 - \frac{1-\alpha_2}{\alpha_1} s_1, & \text{if } s_1 \in [0, \alpha_1], \\ \frac{\alpha_2}{1-\alpha_1} - \frac{\alpha_2}{1-\alpha_1} s_1, & \text{if } s_1 \in (\alpha_1, 1]. \end{cases}$$

yields

$$T(S,d) = \left\{ \left(\tfrac{\alpha_1}{2(1-\alpha_2)}, \tfrac{1}{2} \right), (\alpha_1, \alpha_2), \left(\tfrac{1}{2}, \tfrac{\alpha_2}{2(1-\alpha_1)} \right) \right\}.$$

The corresponding Nash products are given (in the same order) by

$$\tfrac{1}{4}\alpha_1 (1-\alpha_2)^{-1}, \quad \alpha_1\alpha_2, \quad \tfrac{1}{4}\alpha_2 (1-\alpha_1)^{-1}.$$

Then $\alpha_i < \frac{1}{2}$, $i = 1, 2$, implies that $\frac{1}{4}(1-\alpha_i)^{-1} \geq \alpha_i$ and, thus, $\alpha_1\alpha_2$ is the smallest of the three Nash products. Furthermore,

$$\tfrac{\alpha_1}{4(1-\alpha_2)} \geq \tfrac{\alpha_2}{4(1-\alpha_1)} \Leftrightarrow \alpha_2 \in [\alpha_1, 1-\alpha_1].$$

So, we obtain that

$$N(S,d) = \begin{cases} \left(\frac{\alpha_1}{2(1-\alpha_2)}, \frac{1}{2}\right) & \alpha_2 \in (0, \alpha_1), \\ \left(\frac{\alpha}{2(1-\alpha)}, \frac{1}{2}\right), \left(\frac{1}{2}, \frac{\alpha}{2(1-\alpha)}\right) & \alpha_2 = \alpha_1 = \alpha, \\ \left(\frac{1}{2}, \frac{\alpha_2}{2(1-\alpha_1)}\right), & \alpha_2 \in (\alpha_1, \frac{1}{2}), \end{cases}$$

The multiplicity of $N(S,d)$ only occurs when $\alpha_1 = \alpha_2$ and is, therefore, nongeneric, because it has measure zero for $(\alpha_1, \alpha_2) \in (0, \frac{1}{2})^2$. Note that $N(S,d) \subset T(S,d)$ for all $\alpha_1, \alpha_2 < \frac{1}{2}$ and that this result is generic.

5.5.2 Limit set of SPE utility pairs

In this section the limit set of SPE utility pairs is characterized as the time between proposals vanishes or, equivalently, the risk of breakdown vanishes. Since the set of MPE utility pairs plays such an important role in characterizing the set of SPE utility pairs we first investigate MPE utility pairs and relate the limit results to the two axiomatic solutions of the previous section. This section is concluded by investigating the maximum SPE delay measured in continuous time. We postpone a discussion how to interpret these limit results to the next section.

Consider the set of MPE utilities or, equivalently, the set of fixed points of $p \times q$. We write $(x^*(\delta), y^*(\delta))$ for a typical pair of MPE proposals in order to indicate that these depend upon δ. The multiplicity of fixed points means that we no longer have a continuous vector function $(x^*(\delta), y^*(\delta))$ of δ but rather a upper-semi continuous correspondence (which is left to the proof of the next theorem) with domain $[0, 1]$ and subsets of $S \times S$ as its image. In Herrero (1989) it is shown that for every $s^* \in T(S, d)$ there exists a $\underline{\delta}(s^*) < 1$ and a corresponding continuous function $(x^*(\delta), y^*(\delta))$ for $\delta \in (\underline{\delta}(s^*), 1)$ in the correspondence of fixed points of $p \times q$, such that $\lim_{\delta \to 1} x^*(\delta) = \lim_{\delta \to 1} y^*(\delta) = s^*$.

Before we state this result we first reduce the image of the correspondence of fixed points of $p \times q$ to subsets of S by defining the correspondence $F : [0, 1] \to S$ as

$$F(\delta) = \{x^*(\delta), y^*(\delta) \mid (x^*(\delta), y^*(\delta)) \text{ is a fixed point of } p \times q\}. \qquad (5.6)$$

Then the convergence result reads as follows.

Theorem 5.17. *If S^{IR} is strongly comprehensive, then*

$$\lim_{\delta \to 1} F(\delta) = T(S, d) \supseteq N(S, d).$$

This theorem implies that the limit set of MPE proposals of the strategic alternating offers model corresponds to the generalized NBS of theorem 5.15. Only for bargaining problems that admit a unique fixed point of $p \times q$, or equivalently, $T(S, d)$ is a singleton, we have that the MPE corresponds to the maximization of the Nash product, i.e. $N(S, d)$. The class of convex bargaining problems is therefor special in this respect.

Combining proposition 5.6 and 5.9 together with theorem 5.17 immediately implies that the limit set of SPE utility pairs is closely related to the set $T(S, d)$. The following theorem is trivial.

Theorem 5.18. *Let $E_j(\delta)$, $j = 1, 2$, denote the history independent set of SPE utilities for $\delta \in (0, 1)$ at the beginning of the bargaining round where player j proposes. Then*

$$\lim_{\delta \to 1} E_j(\delta) = \left\{s \in \mathbb{R}^2 \mid s_1 \geq \underline{s}_1,\ s_2 \geq \underline{s}_2,\ s_2 \leq f_2(s_1)\right\}, \quad j = 1, 2,$$

where $\underline{s}_i = \min\{s_i | s \in T(S, d)\}$, $i = 1, 2$.

In section 5.4.4 the maximum SPE delay $\hat{t}(\Delta)$, where $\hat{t}(\Delta)$ expresses that $\hat{t}$ also depends upon Δ through $\delta = e^{-r\Delta}$, is measured in the number of bargaining rounds. Measured in real time the maximum SPE delay is given by $\hat{t}(\Delta) \cdot \Delta$. In order to derive this limit consider the following. For all $\Delta > 0$, the SPE strategies that feature delay are similar to the ones in table 5.2.

Now, denote the length of delay measured in real time as $\ell > 0$. The (discrete) number of rounds $t(\Delta; \ell) = [\ell / \Delta]$, where $[z]$ is the Entier function of z, ensures that $t(\Delta;\ell)\cdot\Delta \approx \ell > 0$, but always $t(\Delta;\ell)\cdot\Delta \leq \ell$. Can we adjust $t(\Delta;\ell)$ such that the SPE delay is equal to ℓ as Δ goes to 0? The answer is affirmative, which can be seen as follows. First, define the set of pairs of delay $\ell \in \mathbb{R}_+$ and agreements $s \in S$ that are strictly better for both players to the limit pair of worst SPE utilities $(\underline{s}_1, \underline{s}_2)$ as

$$L = \left\{ (\ell, s) \in \mathbb{R}_+ \times S \,\middle|\, (1-e^{-r\ell})d + e^{-r\ell}s > (\underline{s}_1, \underline{s}_2) \right\}.$$

Since for each $i = 1, 2$ player i's worst SPE payoff converges to $\underline{s}_i$ from above as Δ goes to 0 it follows that for every $(\ell, s) \in L$ there exists a $\overline{\Delta} > 0$ such that for all $\Delta < \overline{\Delta}$ we have that[8]

$$(1-e^{-r\ell})d + e^{-r\ell}s > \left(\underline{s}_1^1(\Delta), \underline{s}_2^1(\Delta)\right) > (\underline{s}_1, \underline{s}_2).$$

Furthermore, for all $\Delta < \overline{\Delta}$ we also have that

$$\begin{aligned}(1-e^{-rt(\Delta;\ell)\Delta})d + e^{-rt(\Delta;\ell)\Delta}s &\geq (1-e^{-r\ell})d + e^{-r\ell}s \\ &> \left(\underline{s}_1^1(\Delta), \underline{s}_2^1(\Delta)\right).\end{aligned}$$

This implies that for all $\Delta < \overline{\Delta}$ the strategies of table 5.2 are SPE if the negotiations are concluded at $t(\sigma) = t(\Delta;\ell)$ with agreement upon $s(\sigma) = s$.

We now return to characterizing the maximum delay measured in real time. As Δ goes to 0 all combinations (ℓ, s) that qualify in the limit as SPE strategies of table 5.2 belong to the set L. Since the set L is an open set, we can only take the supremum. Reformulating this maximization problem in terms of (5.5) yields that $(\hat{\ell}, \hat{s})$ solves

$$\sup_{\ell \in \mathbb{R}_+, s \in S} \ell, \quad \text{s.t.} \quad -(1-e^{-r\ell})d_i - e^{-r\ell}s_i < -\underline{s}_i,\ i = 1, 2, \quad s_2 \leq f_2(s_1).$$

However, if $T(S, d)$ consists of multiple elements, then $(\underline{s}_1, \underline{s}_2)$ is Pareto inefficient and there exist (infinitely) many $(\ell, s) \in L$ such that $\ell > 0$ exist. Then $\hat{\ell} > 0$ and the SPE delay does not vanish in the limit as Δ goes to 0. Finally, note that if $T(S, d)$ consists of a single element, then $(\underline{s}_1, \underline{s}_2)$ is Pareto efficient and there simply is no SPE delay in any SPE let alone in the limit, i.e., the supremum is equal to 0.

The following example illustrates theorem 5.18 and the maximum SPE delay measured in continuous time.

Example 5.19. *Consider the bargaining problem of example 5.3 and 5.16 once more and additionally assume* $\alpha_1, \alpha_2 < \frac{1}{2}$. *Then for all* $\delta > \max\left\{\frac{\alpha_1}{1-\alpha_2}, \frac{\alpha_2}{1-\alpha_1}\right\}$ *we have that* $p \times q$ *admits three fixed points and each fixed point in isolation is a continuous function in* δ *and* $x^*(\delta)$ *and* $y^*(\delta)$ *converge to a common limit point in* $T(S, d)$. *Moreover, for every* $s^* \in T(S, d)$ *there exists a pair* $x^*(\delta)$

and $y^*(\delta)$ that converges to it as δ goes to 1. Thus, the limit of $F(\delta)$ is the set $T(S,d)$ derived in example 5.16 and $N(S,d) \neq T(S,d)$.
Finally, application of proposition 5.6 implies that

$$\lim_{\delta\to1} \underline{s}_1^1(\delta) = \lim_{\delta\to1} \underline{s}_1^2(\delta) = \tfrac{\alpha_1}{2(1-\alpha_2)}, \quad \lim_{\delta\to1} \overline{s}_2^1(\delta) = \lim_{\delta\to1} \overline{s}_2^2(\delta) = \tfrac{1}{2},$$
$$\lim_{\delta\to1} \overline{s}_1^1(\delta) = \lim_{\delta\to1} \overline{s}_1^2(\delta) = \tfrac{1}{2}, \quad \lim_{\delta\to1} \underline{s}_2^1(\delta) = \lim_{\delta\to1} \underline{s}_2^2(\delta) = \tfrac{\alpha_2}{2(1-\alpha_1)}.$$

It is easy to verify that $\lim_{\delta\to1} \underline{s}_1^j(\delta) = \min\{s_1 | s \in T(S,d)\}$, $j = 1,2$. Next, with respect to the set $E_j(\delta)$, $j = 1,2$, we have that

$$\lim_{\delta\to1} E_j(\delta) = \left\{ s \in \mathbb{R}^2 \,\middle|\, s_1 \geq \tfrac{\alpha_1}{2(1-\alpha_2)},\ s_2 \geq \tfrac{\alpha_2}{2(1-\alpha_1)},\ s_2 \leq f_2(s_1) \right\}$$

and this is in accordance with theorem 5.18. Finally, directly taking the limit of the maximum SPE delay $\hat{t}(\Delta)\cdot\Delta > 0$ measured in real time means

$$\hat{\ell} = \lim_{\Delta\to0} -\left(\ln e^{-r\Delta}\right)^{-1} \Delta \ln\left(1+e^{-r\Delta}\right)(1-\alpha) = \tfrac{\ln 2(1-\alpha)}{r} > 0,$$

because $\alpha < \frac{1}{2}$. So, the maximum SPE delay measured in real time is bounded away from 0 and even in the limit many SPE featuring delay can be constructed.

5.5.3 Convergence or nonconvergence, that's the question

The idea of the strategic and axiomatic approach is that each complements each other in interpreting the results. For $T(S,d)$ it can be said that this solution concept envisions a bargaining procedure that is represented by the alternating offer model and that the players strategies are Markov. The drawback of game theory that it does not necessarily produce a unique equilibrium is also reflected in the indeterminacy of the axiomatic solution. Each $s^* \in T(S,d)$ requires the same type of equilibrium coordination between the players as for instance playing one of the Nash equilibrium in the battle-of-the-sexes game. Finally, in Nash's demand game and S is a convex set the H-essential equilibrium, which corresponds to $N(S,d)$, is also the unique risk-dominant Nash equilibrium, e.g., Young (1998). For the alternating offers model this result is also true if the set S is a convex set, but for the class of strongly comprehensive bargaining problems it fails to do so.

The axioms underlying the two generalizations of the NBS can also be used to interpret the strategic results. The difference between these two axiomatic solutions can be traced back to the differences in the independence axiom. The local variant describes the MPE behaviour better. The reason is that the players' threshold levels $p_2(y^*(\delta))$ and $q_1(x^*(\delta))$ restrict the area of proposals $s \in S$ that are acceptable for both players and, therefore, equilibrium coordination on these levels discard distant proposals $s \in S$ as relevant alternatives. For sufficiently large $\delta < 1$ the remaining area of relevant alternatives is simply a neighbourhood of the associated limit solution $s^* \in T(S, d)$. Note that one of the discarded alternatives is the generalized NBS $N(S, d)$, which is the risk-dominant Nash equilibrium in Nash's demand game. This explains why $T(S, d)$ is the limit set and not the risk-dominant $N(S, d)$.

Theorem 5.18 implies that SPE strategies that are not Markov strategies do not correspond to any of the two generalized NBS. With respect to Pareto inefficient outcomes the players' fear of being punished prevents each of them to make the following speech: 'Listen, we are both rational players and we both realize the equilibrium outcome is inefficient and that there are better outcomes for both of us. Therefore, let's work it out'. The efficiency axiom is more or less defended by arguing that rational players work it out. Nevertheless, also a curve of Pareto efficient SPE agreements exists that do not correspond to MPE proposals nor to $T(S, d)$ and that are based upon the same type of strategies.

To summarize, the analysis of strongly comprehensive bargaining problems gave us more insight in the relation between MPE strategies, SPE strategies and the two generalizations of the NBS. For convex bargaining problems all four reinforce each other and this extends as long as the strongly comprehensive bargaining problem admits a unique fixed point of $p \times q$ or, equivalently, $T(S, d)$ consists of a single element. If the latter does not hold, we obtain that MPE strategies and elements of $T(S, d)$ are one-to-one related. Thus, $N(S, d)$ and its associated MPE strategies reduce to one particular case. SPE strategies other than MPE are inconsistent with $T(S, d)$. So, there always exists an axiomatic solution and a subclass of equilibrium strategies in the alternating offer model that are mutually reinforcing. However, for a generic class of strongly comprehensive bargaining problems there exists an even larger set of SPE strategies that do not reinforce this axiomatic solution. This begs the question: do we have convergence or not? The answer is not clear cut.

5.6. Contract space

In this section the results obtained thus far are simply translated into the contract space. We then discuss sufficient conditions upon the players' utility functions for uniqueness of the pair of MPE proposals, where we restrict attention to the formulation of section 3.3.6. From this discussion a research agenda for deriving general conditions for (non)uniqueness emerges.

5.6.1 SPE contracts

Similar as in section 3.3.6 we translate the results obtained for the bargaining problem in utility representation into the associated SPE contracts, which is relatively easy.

Consider MPE strategies first. Then the possibly multiple pairs of MPE contracts (c^1, c^2) are found by solving the two maximization problems stated in proposition 3.14, i.e., (c^1, c^2) simultaneously solves

$$c^1 = \arg\max_c u_1(c), \text{s.t.} c \in \{c \in C | \ u_2(c) \geq (1-\delta) d_2 + \delta u_2(c^2)\}, \tag{5.7}$$

$$c^2 = \arg\max_c u_2(c), \text{s.t.} c \in \{c \in C | \ u_1(c) \geq (1-\delta) d_1 + \delta u_1(c^1)\}. \tag{5.8}$$

If the latter two maximization problems admit a unique pair of MPE contracts (c^1, c^2), then this unique pair of MPE contracts is also the unique pair of SPE contracts. Recall that concavity of the utility functions and C is a convex set is a sufficient condition for a generically unique pair of MPE contracts.

In case there are multiple pairs of contracts that solve (5.7) and (5.8), then for obvious reasons we have for $i, j = 1, 2$ that

$$\underline{s}_i^j = \min\{u_i(c^j) | \ (c^1, c^2) \text{ solves (5.7) and (5.8)}\}$$

and the set of SPE contracts associated with immediate agreement is given by

$$\{c \in C | \ u_i(c) \geq \underline{s}_i^1, i = 1, 2\}.$$

We leave the derivation of equilibrium conditions for SPE strategies featuring delay to the reader.

5.6.2 Sufficient conditions for uniqueness

Conditions upon the utility functions that guarantee the uniqueness of a MPE (or SPE) utility pair mean that one has to find sufficient conditions that guarantee the generic uniqueness of a pair of MPE contracts in (5.7) and (5.8). Before we continue, consider maximization problem (5.7) and (5.8) in isolation. Then u_1 and u_2 are quasiconcave is a sufficient condition for a generically unique contract in each of these two problems in isolation.[9] Furthermore, quasiconcave utility functions ensure that the associated bargaining problem in utility representation (S, d) is strongly comprehensive and we saw that the latter is not sufficient to guarantee uniqueness of a pair of MPE utilities $(x^*(\delta), y^*(\delta))$. Therefore, sufficient conditions upon the utility functions that guarantee generic uniqueness of a pair of MPE proposals $(c^1(\delta), c^2(\delta))$ must be stronger than quasiconcavity but weaker than concavity.

Sufficient conditions for uniqueness in the context of divide a dollar problem, e.g., see section 3.5.1, have been first derived in the seminal article Rubinstein (1982). In this reference six assumptions, labelled A1 up to A6, are imposed upon the utility functions. A1 simply states that the disagreement point is the worst outcome in the negotiations. The assumptions A2-A5 impose utility functions of the form $\delta^t \mu_i(c_i)$, $i = 1,2$, associated with agreement upon the contract $c \in C$ at period $t \in \mathbb{N}$, e.g., see Fishburn and Rubinstein (1982), Rubinstein (1982) and Osborne and Rubinstein (1990). In section 3.5.1 the assumptions A1-A5 are satisfied and the only difference concerns A6, called 'increasing loss of delay'. The latter simply means that the responding player's monetary costs of delay, i.e., rejecting, are low if the next round's agreement would give this player a relatively small share of the dollar, whereas these costs are high if the next round's agreement would bring a large share. The costs of delay for the responding player if the next round's agreement would be $c \in C$ are defined as the (maximum) amount the responding player is willing to give up to prevent waiting. The next round's share is worth its certainty equivalent in the current round, denoted as c_i^e, $i = 1,2$, and the difference between next round's share minus the certainty equivalent share is player i's loss of delay. Since c_i^e, $i = 1,2$, equates $\mu_i(c_i^e)$ with $\delta\mu_i(c_i)$ and, after taking the inverse on both sides, it is equal to $\mu_i^{-1}(\delta\mu_i(c_i))$. Increasing loss of delay, i.e. A6, imposes that the function $c_i - \mu_i^{-1}(\delta\mu_i(c_i))$, $i = 1,2$, is increasing in player i's share c_i. Every concave function μ_i, $i = 1,2$, satisfies increasing loss of delay and, therefore, the latter assumption is less restrictive than concavity, see e.g., Osborne and Rubinstein (1990). In Rubinstein (1982) it was first shown that the assumptions A1-A6 are sufficient conditions in order to obtain the unique SPE pair of proposals characterized in (3.20). In example 3.23 increasing loss of delay simply means $\alpha_i > 0$, $i = 1,2$,[10] instead of the more restrictive $\alpha_i \in (0,1]$ for concavity and the derived SPE shares are thus valid for all positive α_i even though $\alpha_i > 1$ means that μ_i is a convex function, i.e., player i is said to be risk loving.

In Hoel (1986) the uniqueness of the pair of SPE shares in divide a dollar is also investigated and it is shown that strictly logconcave utility functions μ_i, $i = 1,2$, are a sufficient condition for uniqueness, where μ_i is logconcave means that the function $\ln(\mu_i)$, $i = 1,2$, is concave. In example 3.23 logconcavity also means that $\alpha_i > 0$, $i = 1,2$, and it imposes the same restriction as increasing loss of delay. However, the latter does not generally hold.[11]

The above discussion for the divide a dollar problem has never been generalized and it is still an open issue. In our view the route to take is to concentrate on the Kuhn-Tucker first-order conditions for the NBS derived in section 4.2.4 and to derive sufficient conditions such that these conditions yield a unique stationary point, which is then equal to $N(C, \{u_i\}_{i=1,2}, c^d)$. As a bonus, theorem 5.17 then implies that the alternating offers model yields a unique MPE for sufficiently large $\delta < 1$. A very simple condition, which we are sure is known to many economists but for which we could not find a reference (with the exception of Hoel (1986) for the divide a dollar), is given in the next theorem.

Theorem 5.20. *If u_i, $i = 1, 2$, is strict logconcave, then (5.7) and (5.8) admit a unique pair of MPE contracts* $\left(c^1(\delta), c^2(\delta)\right)$.

In the barter economy of section 3.5.2 the utility functions are strict log-concave if the parameters $\alpha, \beta > 0$. It is easy to check that the pair of SPE proposals $\left(c^1(\delta), c^2(\delta)\right)$ is also unique for all $\alpha, \beta > 0$.

We make two final remarks on this subject matter.

Remark 5.21. *In Hoel (1986) a second sufficient condition for uniqueness of the pair of SPE shares in divide a dollar is given that states that the Nash product* $\mu_1(c_1)\mu_2(1 - c_1)$ *(recall* $d = 0$*) is single peaked on its domain.*[12] *Finding sufficient conditions for the Nash product has the methodological disadvantage that these do not have an economic interpretation, whereas (log)concavity and increasing loss of delay have. However, from an applied point of view, any condition that can easily be checked is of use.*

Remark 5.22. *In Hoel (1986) an example of divide a dollar with piecewise linear utility functions is presented that yields multiple pairs of MPE proposals. Such utility functions can be easily translated into a bargaining problem* (S, d) *with a piecewise linear Pareto frontier such as in example 5.3. This example and the discussion above hints at that nondifferentiable utility functions represent the tough case.*

5.7. Related Literature

The results for the strategic approach are based upon Herrero (1989), whereas the results for the axiomatic approach were conducted in Herrero (1989) and Kaneko (1980). With respect to the axiomatic approach we want to make one additional remark. Theorem 6.6 of the next chapter states an asymmetric version of the NBS for convex bargaining problems that was first axiomatized in Kalai (1977). In Roth (1979) it is shown that the efficiency axiom in theorem 6.6 can be replaced by the individual rationality axiom and in Zhou (1996) this result is extended to strongly comprehensive bargaining problems. The latter implies that the maximization of (what in chapter 6 is called) the asymmetric Nash product is the only axiomatic solution that satisfies 4.3-4.6 and 5.12. If additionally the symmetry axiom is added, then the $N(S, d)$ of theorem 5.13 is obtained. Therefore, similar as in section 4.2.1, the efficiency axiom can be replaced by the individual rationality axiom in theorem 5.13 also (and this is its informal proof).

Notes

1 A set $X \in \mathbb{R}^n$ is said to be connected if for each pair of points x and $\hat{x}$ in X there exists a finite sequence of $k+1$, $k \in \mathbb{N}$, distinct points $x^i \in X$, $i = 0, \ldots, k$, such that $x^0 = x$, $x^k = \hat{x}$ and each line piece between the points x^i and x^{i+1} also belongs to X. Note that X is a convex set if additionally $k = 1$ for every pair x and $\hat{x}$ in X.

2 To see this, consider example 5.3 for $\alpha_1 = \alpha_2 = \alpha < \frac{1}{2}$. Then the fixed points for the second and third case, respectively, are given by

$$x^* = \left(\tfrac{\alpha}{(1+\delta)(1-\alpha)}, \tfrac{\delta}{1+\delta}\right), \qquad y^* = \left(\tfrac{\delta\alpha}{(1+\delta)(1-\alpha)}, \tfrac{1}{1+\delta}\right),$$
$$x^* = \left(\tfrac{1}{1+\delta}, \tfrac{\delta\alpha}{(1+\delta)(1-\alpha)}\right), \qquad y^* = \left(\tfrac{\delta}{1+\delta}, \tfrac{\alpha}{(1+\delta)(1-\alpha)}\right),$$

and all *four* points lie on the same Nash product curve. This means that the latter Nash product curve intersects the Pareto frontier four times. Note that $\alpha_1 = \alpha_2$ is nongeneric, because it has measure zero in the (α_1, α_2)-space.

3 The function $g : [0,1] \to [0,1]$ given by

$$g(\alpha_1) = \delta^2 (1-\alpha_1) / \left(\alpha_1 + \delta^2 (1-\alpha_1)\right)$$

is decreasing and convex in α_1. Its curve goes through the points $(0,1)$, $\left(\frac{\delta}{1+\delta}, \frac{\delta}{1+\delta}\right)$ and $(1,0)$. The first two points also lie on the line $\alpha_2 = \delta^{-1}(\delta - \alpha_1)$ and, by convexity of g, for $\alpha_1 \in [0, \delta/(1+\delta)]$ this line lies above g. Similar, convexity of g implies that g lies above the line $\alpha_2 = \delta(1-\alpha_1)$ for $\alpha_1 \in [\delta/(1+\delta), 1]$. Thus, for all $\alpha_1 \in [0,1]$ we have that

$$\tfrac{\delta^2(1-\alpha_1)}{\alpha_1+\delta^2(1-\alpha_1)} \leq \max\left\{\delta(1-\alpha_1), \delta^{-1}(\delta-\alpha_1)\right\}.$$

4 Consider the decreasing and convex function g defined in the previous note. For $\alpha_1 \in [0, \delta/(1+\delta)]$ convexity of g implies that g lies above the line $\alpha_2 = \delta(1-\alpha_1)$ and for $\alpha_1 \in [\delta/(1+\delta), 1]$ it lies above the line $\alpha_2 = \delta^{-1}(\delta - \alpha_1)$. Thus, for all $\alpha_1 \in [0,1]$ we have that

$$\min\left\{\delta(1-\alpha_1), \delta^{-1}(\delta-\alpha_1)\right\} \leq \tfrac{\delta^2(1-\alpha_1)}{\alpha_1+\delta^2(1-\alpha_1)}.$$

5 Note that player 2 is indifferent between accepting and rejecting the proposal s^0 if one of the equal signs holds, but this simply implies that both *yes* and *no* are optimal. Formally, arg max $\cdot$ is a correspondence and we should have written *yes* $\in \{\arg\max \cdot\}$.

6 For completeness, we mention that $(s_1^{\max}, d_2)$ and $(d_1, s_2^{\max})$ could be replaced by arbitrary $s^t \in S$ for t is even and odd, respectively, provided that the responding player obtains at most his worst SPE utility, i.e., $s_2^t \leq \underline{s}_2^1$ for even t and $s_1^t \leq \underline{s}_1^2$ for odd t (these are additional equilibrium requirements). Formulated in this manner, these modified strategies fully characterize the set of SPE *strategies*. Verification is left to the reader.

7 The set $\hat{S}^{PE}$ denotes the Pareto frontier of the set $\hat{S}$.

8 Consider the case S is a convex set first. Then the Nash product curve through $N(S,d)$ lies north-east of the Nash product curve through x^* and y^*. Furthermore, x^* lies to the left of y^*. Then, obviously $y_1^* < N_1(S,d) < x_1^*$ and x_1^* converges from above to $N_1(S,d)$ as δ goes to 1. The same holds for every $\hat{s} \in T(S,d)$ and its corresponding pair (x^*, y^*) such that $\lim_{\delta \to 1} x^*(\delta) = \lim_{\delta \to 1} y^*(\delta) = \hat{s}$.

9 First, from the mathematical appendix M.K in Mas-Colell, Whinston, and Green (1995) it follows that strict quasiconcave utility functions are sufficient to guarantee uniqueness. The discussion of generic uniqueness in section 2.2.7 is also valid for (strict) quasiconcave utility functions.

10 For all $i = 1, 2$, $\delta \in (0, 1]$, $\alpha_i > 0$ it holds that $c_i^e = \delta^{1/\alpha_i} c_i$ and

$$\tfrac{\partial}{\partial c_i}\left(c_i - \mu_i^{-1}(\delta \mu_i(c_i))\right) = 1 - \delta \frac{\mu_i'(c_i)}{\mu_i'(c_i^e)} = 1 - \delta^{1/\alpha_i} > 0.$$

11 For twice differentiable functions μ_i, $i = 1, 2$, the condition for increasing loss of delay follows from the previous note and this expression differs from the condition of strict logconcavity given by

$$\tfrac{\partial^2}{\partial c_i^2} \ln \mu_i(c_i) = \frac{\mu_i''(c_i)\mu_i(c_i) - (\mu_i'(c_i))^2}{(\mu_i(c_i))^2} < 0.$$

12 It is easily checked that the utility functions μ_i, $i = 1, 2$, in example 3.23 also satisfy this condition for all $\alpha_i > 0$.

Chapter 6

COMPARATIVE STATICS

6.1. Introduction

This is the second chapter in which the robustness of the results obtained in chapters 3 and 4 is investigated. Contrary to the previous chapter where just one alternative assumption was addressed there is more diversity of alternative assumptions in this chapter. Having already established the consequences of nonconcave utility functions we return to our initial assumption of *concave* utility functions throughout this chapter. We also still assume a *fixed* disagreement point.

Recall that assumption 2.2, 2.7 and 2.9 impose restrictions upon the contract space and the utility functions (other than continuity and concavity) and in the first section one by one each of these assumptions is relaxed. Next, the constant risk of breakdown (or discount factor) is replaced by a nonstationary, time-dependent risk that alternates between even and odd rounds. We continue by doing the same with respect to the constant disagreement point and show that there is an equivalence between the model with alternating disagreement points and strongly comprehensive bargaining problems of the previous chapter. The latter result is also important because it is needed in the analysis conducted in chapter 9.

The final section deals with the impact of the rules of the game upon the equilibrium strategies. After all, there are many bargaining procedures one can think of and from game theory it is known that sometimes seemingly nonessential changes in the game tree can qualitatively change the equilibrium outcomes, e.g., Fudenberg and Tirole (1991), Mas-Colell, Whinston, and Green (1995), Meyerson (1991), Osborne and Rubinstein (1994) and Van Damme (1991). Two modifications are analyzed. The first version extends the bargaining procedure in such a way that a class of such procedures is considered and that each procedure in this class is uniquely determined by two parameters. Several of the well-known models in the literature fit into this class of bargaining procedures, including alternating offers, one-sided offers and random proposing players. We show that, similar as before, each bargaining procedure within this class admits a unique pair of SPE proposals and that this pair changes in a continuous manner with the two parameters that specify each procedure in the class of bargaining procedures. The second version of the bargaining procedure does not impose the rigid structure of moves that characterizes the alternating offers model, but instead a continuous time version is analyzed in which each player is free to make a proposal or refrain from doing so at any time. This continuous time model is analyzed under some mild restrictions concerning the minimum amount of time a proposal stands open, the minimum time needed to respond and the exclusion of simultaneous moves. It is then shown that this continuous time model generates unique SPE behaviour that mimics the bargaining behaviour in the alternating offer model. Therefore, there is also convergence of the SPE proposals to the axiomatic Nash bargaining solution. We regard the test of robustness with respect to the bargaining procedure as the most important one and we find that the alternating offers model gloriously passes this test.

6.2. Utility functions and the contract space

The alternative assumptions made in this section should be regarded as alternative assumptions with respect to the utility functions or the contract space. The consequences are investigated as briefly as is possible.

6.2.1 Mutual and conflicting interests

Assumption 2.9 excluded each player's best contract in the contract space from the set of strict individual-rational contracts. In this section we no longer make this assumption and we argue that the results do not change qualitatively. Bargaining problems with common interests are briefly discussed as a simple special case. We conclude this section with bargaining problems without mutual interests.

For convenience we consider the class of bargaining problems in utility representation (S, d) and assume that S is a convex set and $d \in S$. Recall from section 2.2.8 that the endpoints of the Pareto frontier of S^{IR} are given by $(d_1, s_2^{\max})$ and $(s_1^{\max}, d_2)$ and if we drop this assumption these endpoints can be thought of as shifting in the north-east direction. We denote the 'new' endpoints of the Pareto frontier as $(s_1^+, s_2^{\max})$ and $(s_1^{\max}, s_2^+)$, $s_i^+ \geq d_i$, $i = 1, 2$. The special case $s_i^+ = d_i$, $i = 1, 2$, corresponds to the class bargaining problems that are already extensively analyzed. See figure 6.1 for an illustration for the case $s_i^+ > d_i$, $i = 1, 2$. The subset $S^+ \subseteq S^{IR}$ is defined

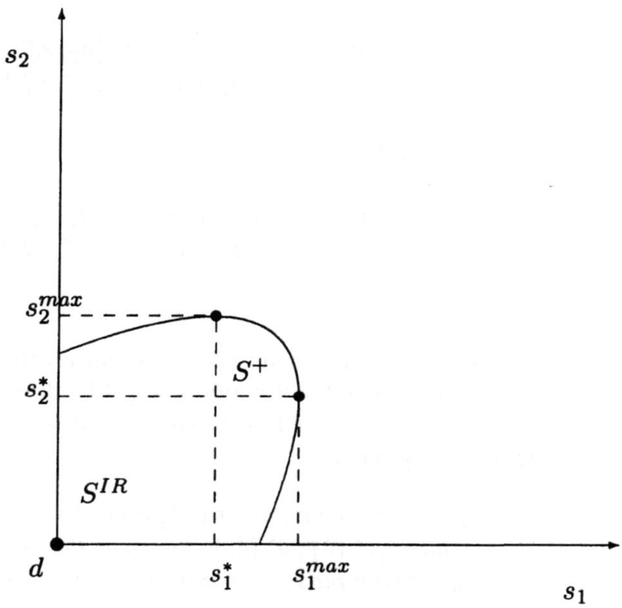

Figure 6.1. The set $S^+ \subset S^{IR}$ of the bargaining problem (S, d).

as $\{s \in S^{IR} | \, s \geq (s_1^+, s_2^+)\}$. Similar as before, the Pareto frontier of S^+ is described by the function $f_j : [s_i^+, s_i^{\max}] \to [s_j^+, s_j^{\max}]$, $i, j = 1, 2$ and $i \neq j$, and $(s_1, f_2(s_1))$ or $(f_1(s_2), s_2)$ is a typical element of this frontier.

Before we characterize the unique SPE for this case we first derive a result that states that the limit set of SPE utility pairs as δ goes to 1 is always a subset of S^+, i.e., $\lim_{\delta \to 1} E(\delta) \subseteq S^+$. To put it differently, every $s \in S^{IR} \setminus S^+$ is excluded as an element of the limit set of SPE utility pairs as δ goes to 1.

Proposition 6.1. $\lim_{\delta \to 1} E(h_1(t), t) \subseteq S^+$, $t \in \mathbb{N}$ *and* $h_1(t) \in H_1(t)$.

This proposition means that the set S^+ can be regarded as a (very) rough bound upon the limit set E. It also implies that assumption 2.9 is innocuous if limit sets of SPE utility pairs are studied.

The full analysis of SPE utilities for this special case goes roughly as follows. Similar as in section 3.3.4 the optimal response is to reject the proposal s^t, $t \in \mathbb{N}$, if it yields a utility less than some threshold level. New is that player 1's threshold level can drop below s_1^+ and player 2's utility maximizing proposal is $(s_1^+, s_2^{\max})$, which will be accepted by player 1. Then all that is needed is the following redefinition of the function $p \times q$. Similar as in section 3.3.1 we define the vector function $p : S^{IR} \to S^+$ as

$$p_1(y) = f_1(p_2(y)),$$
$$p_2(y) = \begin{cases} s_2^+, & \text{if } (1-\delta)\,d_2 + \delta y_2 \in \left[d_2, s_2^+\right], \\ (1-\delta)\,d_2 + \delta y_2, & \text{if } (1-\delta)\,d_2 + \delta y_2 \in \left(s_2^+, s_2^{\max}\right], \end{cases}$$

and the vector function $q : S^{IR} \to S^+$ as

$$q_1(x) = \begin{cases} s_1^+, & \text{if } (1-\delta)\,d_1 + \delta x_1 \in \left[d_1, s_1^+\right], \\ (1-\delta)\,d_1 + \delta x_1, & \text{if } (1-\delta)\,d_1 + \delta x_1 \in \left(s_1^+, s_1^{\max}\right], \end{cases}$$
$$q_2(x) = f_2(q_1(y)).$$

These redefined functions take the new situation of threshold levels below s_1^+ or s_2^+ into account. It can be shown that the function $p \times q$ still admits a unique fixed point and that proposition 3.18 still holds, e.g., Houba (1994). Combining these results implies the following theorem.

Theorem 6.2. *Let $(x^*(\delta), y^*(\delta))$ be the unique fixed point of $p \times q$. Then the unique SPE is the MPE in which $(x^*(\delta), y^*(\delta))$ represent the pair of MPE proposals and player 1's respectively player 2's threshold level are given by $p_2(y^*(\delta))$ and $p_1(x^*(\delta))$. Moreover, as δ goes to 1 both $x^*(\delta)$ and $y^*(\delta)$ converge to $N(S,d)$.*

Note that the class of bargaining problems (S,d) with *common interests* is a special case that corresponds to $s_i^+ = s_i^{\max}$, $i = 1,2$. Therefore, $S^+ = \{(s_1^{\max}, s_2^{\max})\}$ consists of a single point and the Pareto frontier collapses to a single point. Then also $x^*(\delta) = y^*(\delta) = N(S,d) = (s_1^{\max}, s_2^{\max})$. These results support the argument raised in section 2.2.5 that bargaining problems with common interests are trivial bargaining problems.

Finally, proposition 2.12 incorporates the assumption of mutual interests meaning that there exists an $s \in S$ such that $s > d$. In case the players fail to have such interests, then for every $s \in S$ either $s_1 < d_1$ or $s_2 < d_2$ or both. This case is easily captured, simply by taking $\hat{S} = S \cup \{d\}$ as the set of alternatives which means that $\hat{S}^{IR} = \{d\}$. Obviously, rational players never agree upon any payoff $s \in S$, because there does not exist an agreement $s \in S$ that is mutually beneficial for both players when compared to d. We could say that the players 'agree to disagree'. Note that $d \in S^{PE}$ is a special case in which agreement upon $s^* = d$ after a delay of t rounds is payoff equivalent to both immediate agreement upon $s^* = d$ and perpetual disagreement. So, every outcome (s^*, t), with $s^* = d$ and $t \geq 0$ (including $t = \infty$) can occur in equilibrium if $d \in S^{PE}$.

6.2.2 Imperfectly divisible goods

Thus far we treated all the variables of the contract as continuous variables. Doing so means that we implicitly assume that these variables are perfectly divisible. For example, in the divide a dollar problem of section 3.5.1 money is assumed to be perfectly divisible meaning that a split into $1/3$ and $2/3$ can be implemented. However, in reality the dollar cent is the smallest monetary unit and $1/3$ of a dollar is an amount in between 33 and 34 dollar cents that is impossible to obtain. Also in negotiations for a settlement in case of a divorce or an inheritance many items are simply not divisible at all such as cars, furniture, antique and houses. In this section it is shown that perfect divisibility rules out a large set of Nash equilibria in the standard alternating offers model, as analyzed in chapter 3, that are not subgame perfect. Imperfect divisibility may then lead to a multiplicity of SPE contracts and we derive sufficient conditions under which this is the case.

We start the discussion by identifying a large set of Nash equilibria in the standard alternating offers. These Nash equilibria have been implicitly identified in section 4.3.2, where we discussed how Nash's demand game can be obtained by imposing restrictions upon the players' strategies in the alternating-offers model. There we already mentioned that these strategies are not only Nash equilibrium (NE) strategies in Nash's demand game, but also NE strategies in the alternating-offers model, e.g., Rubinstein (1982) and Osborne and Rubinstein (1990). Formulated in the contract space these fully history-independent strategies specify that both players propose the same proposal $c^* \in C$ over and over again and that each player only accepts proposed contracts $c^t \in C$ that yield at least as much (in terms of utility) as c^*, i.e., player i, $i = 1, 2$, only accepts c^t if $u_i(c^t) \geq u_i(c^*)$. These NE strategies result in immediate agreement upon $c^* \in C^{PE} \cap C^{IR}$.

These NE strategies fail to be subgame perfect if all the variables in the contract are continuous variables, which can be seen as follows. For each $c^* \in C^{PE}$ there exists a $c' \in C^{PE} \cap C^{IR}$ such that $u_1(c') > u_1(c^*)$ and $u_2(c') > (1-\delta) d_2 + \delta u_2(c^*)$. Consider the subgame, i.e., history, $h_2(t) = (h_1(t), c')$, t even, which means that the contract c' is currently on the table. Then following the NE strategies (by both players) means that player 2 rejects c' and that the players agree upon c^* in the next round. This yields player 2 a utility of $(1-\delta) d_2 + \delta u_2(c^*)$. However, consider the one-stage deviation by player 2 where he accepts c' in the current subgame. Then this one-stage deviation does strictly better and, hence, this ruins the subgame perfectness of the NE strategies in case all variables are continuous.

In the opposite case all the contract variables are imperfectly divisible, which translates into a contract space that can be represented by some kind of grid over C meaning that the contract space is a discrete set instead of a connected set. If this grid is very coarse, then the contract c' described above may fail to exist and the NE strategies become SPE strategies. Thus, imperfect divisibility allows multiplicity of SPE strategies.

We proceed by making this insight formal in such a way that limits with respect to the grid size can be taken. We do not pursue the most general model and rather have an evenly spaced grid as the contract space. Therefore, we keep the set C as a *convex* set in $\mathbb{R}^n$ and denote the *grid size* of the contract space on C as $g \in \mathbb{R}_+$. The discrete contract space with grid size g is defined as

$$C(g) = \{c \in C \mid \exists k_1, \ldots, k_n \in \mathbb{Z} : \; c = (gk_1, \ldots, gk_n)\}.$$

The advantage of defining the set $C(g)$ is that it covers the set C as the grid size g vanishes, i.e., g goes to 0. The set of Pareto efficient contracts in $C(g)$ is denoted as $C^{PE}(g)$. As mentioned earlier, immediate agreement upon the contract $c^* \in C^{PE}(g)$ can always be supported as NE outcome. The following proposition states necessary and sufficient conditions such that this NE outcome is also an SPE outcome.

Proposition 6.3. *Immediate agreement upon the contract $c^* \in C^{PE}(g)$ is an SPE outcome if and only if the set $C(c^*; g, \delta) = \emptyset$, where $C(c^*; g, \delta) \subseteq C(g)$ is defined as the set of contracts $c \in C(g)$ such that either $i)$ $u_1(c) > u_1(c^*)$ and $u_2(c) > (1-\delta)d_2 + \delta u_2(c^*)$ or $ii)$ $u_2(c) > u_2(c^*)$ and $u_1(c) > (1-\delta)d_1 + \delta u_1(c^*)$.*

The set $C(c^*; g, \delta)$ represents all contracts in $C(g)$ such that $i)$ the responding player accepts each (deviating) proposal $c \in C(g)$ that yields a utility that is larger than the present value of agreement upon c^* in the next round and $ii)$ the proposing player is better off by proposing such deviating proposal. Since each SPE excludes profitable one-stage deviations it is obvious that immediate agreement upon c^* can only be supported in a SPE if and only if the set of such profitable deviating proposals is empty. Note that the set $C(c^*; g, \delta)$ depends upon the two important parameters of the model, namely g and δ.

What is the exact relation between the grid size g, the discount factor δ and the (non)emptiness of the set $C(c^*; g, \delta)$ of proposition 6.3? The answer is as follows for fixed $c^* \in C^{PE}$ (in order to avoid having c^* depend upon g too) and fixed $\delta < 1$. Then for $C(0)$, i.e., a grid size of 0, we have that $C(0) = C$ and $C(c^*; 0, \delta) \neq \emptyset$. If we gradually increase g in a continuous manner, then for sufficiently small $g > 0$ the grid is still sufficiently fine and there still exists contracts $c \in C(g)$ sufficiently close to c^* such that one (or both) of the players in the role of proposing player is able to profitably deviate meaning that $C(c^*; g, \delta) \neq \emptyset$. However, there is some threshold level for g such that all of the neighbours of c^* are so far away that for each neighbouring contract c the required $>$-signs fail. Then $C(c^*; g, \delta)$ has become empty and it remains empty if we would make the grid more coarse by increasing the grid size, i.e., pushing the neighbours even further away. So, for each $\delta < 1$ and $c^* \in C^{PE}$ there exists $\underline{g}(\delta) > 0$ such that $C(c^*; g, \delta) \neq \emptyset$ for all $g < \underline{g}(\delta)$ and $C(c^*; g, \delta) = \emptyset$ for all $g \geq \underline{g}(\delta)$. Thus, we can support immediate agreement upon $c^* \in C^{PE}$ as a SPE outcome provided the grid $C(g)$ on C is sufficiently coarse.

As an alternative exercise, we could fix g and let δ approach 1. Doing so means that, in the limit as δ goes to 1, the contract c has to satisfy $u_i(c) > \lim_{\delta \to 1} (1 - \delta) d_i + \delta u_i(c^*) = u_i(c^*)$, $i = 1, 2$. Since $c^* \in C^{PE}$ no such contract $c \in C$ exists and neither in $C(g)$. So, no matter how fine the gridsize of $C(g)$ is, there exists a threshold $\overline{\delta} < 1$ such that the set $C(c^*; g, \delta)$ is empty for all $\delta > \overline{\delta}$, meaning that immediate agreement upon $c^* \in C^{PE}$ is a SPE outcome.

The mathematical answer to the above informally derived results is that the order in which limits with respect to g and δ are taken matters. If first g goes to 0 and then δ goes to 1 then we obtain the standard results. However, reversing the order means that in the limit multiple SPE outcomes can be supported. This proves the following proposition.

Proposition 6.4. *Let* $c^* \in C^{PE}$. *Then*

$$\lim_{\delta \to 1} \lim_{g \to 0} C(c^*; g, \delta) \neq \emptyset, \quad \lim_{g \to 0} \lim_{\delta \to 1} C(c^*; g, \delta) = \emptyset.$$

In Van Damme, Selten, and Winter (1990) it is shown that for the divide a dollar problem with arbitrary $g > 0$ and sufficiently large $\delta < 1$ every possible outcome path can be supported as a SPE outcome path. This includes SPE strategies featuring delay and such strategies apply the idea of equilibrium switching similar as in section 5.4.4. In Van Damme, Selten, and Winter (1990) also the method in Shaked and Sutton (1984) is modified in order to study sufficient conditions for uniqueness, which we omit.

To conclude, these results confirm our intuition. Recall that the NBS should be seen as an approximation of the negotiations' outcome and we already saw that the NBS corresponds to the unique SPE strategies in case the variables are perfectly divisible. The results in this section imply that the NBS still captures the essence of the negotiations in case the grid size is sufficiently small. Then the grid size does not matter and can be discarded by assuming continuous variables. However, if the grid size becomes too large, then the grid size hampers the players in their negotiations and it can be said that imperfect divisibility introduces a sort of coordination problem with respect to the multiple equilibria that may prevent the players from reaching an Pareto efficient agreement. In such circumstances we cannot automatically rely on the NBS as a predictive theory for the negotiations under consideration and we have to take into account the possibility of SPE behaviour not corresponding to the NBS, which is similar as for strongly comprehensive games. Although it is merely speculation, perhaps in many divorces the custody of the children can be regarded as an 'all or nothing' variable with a large emotional value that the (almost) continuous variable money cannot buy and then the application of the NBS simply is not appropriate. Nevertheless, the results obtained in this section provide a yet crude framework to take these considerations into account.

6.3. Nonstationary bargaining problems

In this section nonstationary bargaining problems are considered in which the players' intertemporal preferences become time dependent. The general nonstationary and time-dependent case can be analyzed similar as in sections 3.3.2-3.3.5, but if the preferences are roughly speaking too nonstationary we no longer have that the set of SPE utility pairs at a particular bargaining round coincides with the set of SPE utility pairs two rounds later. As a consequence, the fixed point property that drove the earlier results is lost and we refer to e.g., Binmore (1987c) how to tackle nonstationary problems in general. In this section the latter is avoided by studying two nonstationary bargaining problems that differ only between odd and even rounds and yet these problems are still rich enough to obtain some insights in nonstationary bargaining problems. The results derived for these two nonstationary bargaining problems also return as building blocks in later chapters.

6.3.1 Alternating probabilities of breakdown

In this section the risk of breakdown is assumed to alternate between even and odd periods and it is shown that the qualitative results do not change. The limit of the unique SPE utility pair is related to an asymmetric version of the Nash bargaining solution, which is also introduced in this section.

Consider the bargaining problem (S, d) with S is a *convex* set. Denote $\delta(t) \in (0,1), t \in \mathbb{N}$, as the time dependent discount factor or probability. Here we assume that there exist $\delta_1, \delta_2 \in (0,1)$ such that $\delta(t) = \delta_2$ for all even t and $\delta(t) = \delta_1$ for all odd t. Note that we thus far imposed $\delta_1 = \delta_2 = \delta$ and that the nonstationary risk simply alternates between even and odd rounds. The alternating risks introduce an asymmetry in the bargaining process, because player 1 and 2 are no longer symmetric with respect to the risk of breakdown each of them faces as the responding player. The model can also be reinterpreted as a model in which the players have different discount factors.

The analysis in chapter 3 only needs a minor modification with respect to the risk faced by the responding player in the specification of the function $p \times q$ in order to directly apply here. To this end, we define the functions $\hat{p} : S^{IR} \to S^{IR}$ and $\hat{q} : S^{IR} \to S^{IR}$ as

$$\hat{p}_2(y) = (1-\delta_2)\, d_2 + \delta_2 y_2, \qquad \hat{p}_1(y) = f_1(\hat{p}_2(y)), \tag{6.1}$$

$$\hat{q}_1(x) = (1-\delta_1)\, d_1 + \delta_1 x_1, \qquad \hat{q}_2(x) = f_2(\hat{q}_1(x)),$$

By now it should be clear that the fixed points of $\hat{p} \times \hat{q}$ play a crucial role in the analysis. Furthermore, recall that if (x^*, y^*) is such a fixed point then x^* and y^* have the same Nash product in case $\delta_1 = \delta_2 = \delta$. Something similar holds here, but the asymmetry demands the following asymmetric version of this product which we define first. For $\alpha \in (0, 1)$ the *asymmetric Nash product* is defined as $(s_1 - d_1)^\alpha (s_2 - d_2)^{1-\alpha}$. This asymmetric product has the same properties as the Nash product and for $\alpha = \frac{1}{2}$ the asymmetric Nash product curve coincides with the Nash product curve introduced in section 3.3.1. Then it follows that for each fixed point (x^*, y^*) of $\hat{p} \times \hat{q}$ the points x^* and y^* have the same asymmetric Nash product corresponding to $\hat{\alpha} = \ln \delta_2 (\ln \delta_1 + \ln \delta_2)^{-1}$. To see this, substitute $x_2^* = p_2(y^*)$ and $y_1^* = q_1(x^*)$ in the asymmetric Nash products for x^* and y^* in order to obtain[1]

$$(x_1^* - d_1)^{\hat{\alpha}} (x_2^* - d_2)^{1-\hat{\alpha}} = \delta_1^{\hat{\alpha}} (x_1^* - d_1)^{\hat{\alpha}} (y_2^* - d_2)^{1-\hat{\alpha}},$$
$$(y_1^* - d_1)^{\hat{\alpha}} (y_2^* - d_2)^{1-\hat{\alpha}} = \delta_1^{\hat{\alpha}} (x_1^* - d_1)^{\hat{\alpha}} (y_2^* - d_2)^{1-\hat{\alpha}}.$$

Obviously, symmetry $\hat{\alpha} = \frac{1}{2}$ prevails for $\delta_1 = \delta_2 = \delta$. We leave it to the reader to verify that the arguments underlying lemma 3.2 and 3.3 still apply. Therefore, the function $\hat{p} \times \hat{q}$ admits a fixed point (x^*, y^*). Moreover, all the other derivations in chapter 3 are still valid and we immediately obtain the following result.

Theorem 6.5. *Let (x^*, y^*) be the unique fixed point of $\hat{p} \times \hat{q}$. Then the unique SPE is the single MPE given by the pair of proposals (x^*, y^*) and the sets of acceptable proposals given by*

$$T_1 = \{s \in S |\ s_1 \geq y_1^* = q_1(x^*)\}, \quad T_2 = \{s \in S |\ s_2 \geq x_2^* = p_2(y^*)\}.$$

The SPE proposals x^* and y^* depend upon δ_1 and δ_2. For instance, x_1^* and y_1^* are increasing in δ_1 and decreasing in δ_2. This can be seen as follows. If (x^*, y^*) is the unique fixed point of $\hat{p} \times \hat{q}$ and δ_1 changes to $\delta_1' > \delta_1$, then the old y_1^* lies below player 1's increased threshold level and player 2 has to offer player 1 more, meaning that the new y_1^* increases and player 2 obtains less than the old y_2^*. So, player 2's new threshold level decreases, meaning the new x_2^* decreases, and player 1's new x_1^* increases.

As before, the limit outcome as the time between bargaining rounds vanishes is also of interest. Similar as in section 2.3.7, where we introduced $\delta = e^{-\theta\Delta}$ for some $\theta > 0$, we assume that there is some $\theta_i > 0$, $i = 1, 2$, such that $\delta_i = e^{-\theta_i\Delta}$. As a consequence, $\hat{\alpha} = \theta_2 (\theta_1 + \theta_2)^{-1}$ is independent of Δ. It is obvious that the asymmetry in the Nash products of x^* and y^*, which is reflected in $\hat{\alpha}$, prevents these two proposals to converge to the symmetric NBS as the time between proposals vanishes. However, intuitively one might expect that x^* and y^* converge to the maximization of some asymmetric Nash product and indeed this is true. Moreover, the latter limit solution is also an axiomatic solution that satisfies three axioms and this result is stated in the following theorem. For a prove we refer to Kalai (1977).

Theorem 6.6. *The unique class of functions $B(S,d)$ that satisfies axioms 4.1, 4.3 and 4.4 is the class of asymmetric Nash bargaining solutions $N_\alpha(S,d)$, $\alpha \in (0,1)$, given by*

$$N_\alpha(S,d) = \arg\max_{s\in S^{IR}} (s_1 - d_1)^\alpha (s_2 - d_2)^{1-\alpha}.$$

Moreover, if $B(S,d)$ additionally satisfies axiom 4.2, then $\alpha = \frac{1}{2}$ meaning that $N_{\frac{1}{2}}(S,d) = N(S,d)$.

Every asymmetric Nash bargaining solution also satisfies the two geometrical properties. The first states that the asymmetric Nash product curve is tangent to the Pareto frontier. The second property follows from the first-order conditions, assuming that f_2 is differentiable, and now reads

$$\frac{s_2^* - d_2}{s_1^* - d_1} = -\frac{1-\alpha}{\alpha} \cdot f_2'(s_1^*).$$

Then an increase in α reduces the right hand side of this equality and, taking into account that $s_2^* = f_2(s_1^*)$ is decreasing in s_1^*, the left hand side can only become level again by increasing s_1^*. So, s_1^* is increasing in α. The weight α is often interpreted as a measure of player 1's *bargaining power*.

The intuition behind the following convergence result is similar to the arguments of section 4.3.4: The distinct points $x^*(\Delta)$ and $y^*(\Delta)$ lie on the same asymmetric Nash product curve corresponding to $\alpha = \hat{\alpha}$ and, as the time Δ between rounds goes to 0, these two points converge to a common limit point meaning their associated asymmetric Nash product curve shifts in the north-east direction. In the limit this curve has shifted so far that it touches the Pareto frontier in the common limit point, i.e., the curve is tangent to the Pareto frontier. This is stated in the following theorem.

Theorem 6.7. *Let $\delta_i = e^{-\theta_i\Delta}$, $i = 1,2$. Then*

$$\lim_{\Delta\to 0} x^*(\Delta) = \lim_{\Delta\to 0} y^*(\Delta) = N_{\hat{\alpha}}(S,d).$$

Impatience deteriorates each player's bargaining position. To see this, $\hat{\alpha}$ is decreasing in θ_1 and increasing in θ_2. The parameter θ_i, $i = 1,2$, is a measure for player i's impatience and larger values of θ_i correspond to a more impatient player.

The following example illustrates the results of this section for the divide a dollar problem, which is of course formulated in terms of the contract space.

Example 6.8. *Consider the divide a dollar problem of section 3.5.1 once more and suppose $\mu_i(c_i) = c_i$, $i = 1,2$, and $\delta_1 \neq \delta_2$. Then (S,d) is given by $S = \left\{ s \in \mathbb{R}^2_+ \mid s_1 + s_2 \leq 1 \right\}$ and $d = 0$. The unique fixed point of $\hat{p} \times \hat{q}$ is given by*

$$x^* = \left(\tfrac{1-\delta_2}{1-\delta_1\delta_2}, \tfrac{\delta_2(1-\delta_1)}{1-\delta_1\delta_2} \right), \quad y^* = \left(\tfrac{\delta_1(1-\delta_2)}{1-\delta_1\delta_2}, \tfrac{1-\delta_1}{1-\delta_1\delta_2} \right)$$

and the asymmetric Nash products corresponding to $\alpha = \hat{\alpha}$ are

$$\left(x_1^*\right)^{\hat{\alpha}} \left(x_2^*\right)^{1-\hat{\alpha}} = \left(y_1^*\right)^{\hat{\alpha}} \left(y_2^*\right)^{1-\hat{\alpha}} = \tfrac{\delta_1^{\hat{\alpha}}(1-\delta_2)^{\hat{\alpha}}(1-\delta_1)^{1-\hat{\alpha}}}{1-\delta_1\delta_2}.$$

Furthermore, application of L'Hôpital's rule implies that

$$\begin{aligned} \lim_{\Delta\to 0} \tfrac{1-e^{-\theta_i\Delta}}{1-e^{-(\theta_1+\theta_2)\Delta}} &= \lim_{\Delta\to 0} \tfrac{\frac{\partial}{\partial\Delta}\left(1-e^{-\theta_i\Delta}\right)}{\frac{\partial}{\partial\Delta}\left(1-e^{-(\theta_1+\theta_2)\Delta}\right)} \\ &= \lim_{\Delta\to 0} \tfrac{\theta_i e^{-\theta_i\Delta}}{(\theta_1+\theta_2)e^{-(\theta_1+\theta_2)\Delta}} = \tfrac{\theta_i}{\theta_1+\theta_2}. \end{aligned}$$

Therefore,

$$\lim_{\Delta\to 0} x^*(\Delta) = \lim_{\Delta\to 0} y^*(\Delta) = N_{\hat{\alpha}}(S,d) = \left(\tfrac{\theta_1}{\theta_1+\theta_2}, \tfrac{\theta_2}{\theta_1+\theta_2} \right).$$

In terms of contracts, direct application of $s_i = \mu_i(c) = c_i$, $i = 1,2$, yields the unique pair of SPE proposals $\left(c^1, c^2\right)$ given by

$$c^1 = \left(\tfrac{1-\delta_2}{1-\delta_1\delta_2}, \tfrac{\delta_2(1-\delta_1)}{1-\delta_1\delta_2} \right), \quad c^2 = \left(\tfrac{1-\delta_2}{1-\delta_1\delta_2}, \tfrac{\delta_2(1-\delta_1)}{1-\delta_1\delta_2} \right)$$

and for $\alpha = \hat{\alpha}$ the common limit point of c^1 and c^2 coincides with the asymmetric NBS $N_\alpha(C, \{u_i\}_{i=1,2}, 0) = (\alpha, 1-\alpha)$, where

$$N_\alpha(C, \{u_i\}_{i=1,2}, 0) = \arg\max_c c_1^\alpha c_2^{1-\alpha}, \quad c_1 + c_2 \leq 1, c_1 \geq 0, c_2 \geq 0.$$

6.3.2 Alternating disagreement points

In this section the disagreement point is assumed to alternate between even and odd periods and this does dramatically change the qualitative results. The reason is that for a generic class of bargaining problems the presence of alternating disagreement point turns out to be equivalent to strongly comprehensive bargaining problems that fail to be convex. So, even though we assume a convex bargaining problem the presence of nonstationary disagreement points is likely to destroy the uniqueness of the SPE. The results of this section become relevant again in the policy bargaining model of chapter 9.

Consider the bargaining problem $(S, \{d(t)\}_{t=0}^\infty)$ with S is a *convex* set and $d(t) \in S$ as the disagreement point at round t, $t \in \mathbb{N}$. Here we assume that there exist $d^{even}, d^{odd} \in S$ such that $d(t) = d^{even}$ for all even t and $d(t) = d^{odd}$ for all odd t. For convenience we sometimes write $\left(S, \left\{d^{even}, d^{odd}\right\}\right)$. Note that we thus far imposed $d^{even} = d^{odd} = d$ and that in this section the nonstationary disagreement point simply alternates between even and odd rounds. This preserves the stationary nature of the alternating offers model, because cycles last two rounds. The expected disagreement point in even numbered rounds in case of perpetual disagreement is equal to

$$m^{even} = (1-\delta)\sum_{t=0}^{\infty}\delta^t d(t) = \tfrac{1}{1+\delta}d^{even} + \tfrac{\delta}{1+\delta}d^{odd} \in S.$$

Define $S^{IR}(t) = \{s \in S | s \geq m^{even}\}$ for even t. Similar, for odd numbered rounds t, define $m^{odd} = \frac{\delta}{1+\delta}d^{even} + \frac{1}{1+\delta}d^{odd} \in S$ and $S^{IR}(t) = \left\{s \in S | \, s \geq m^{odd}\right\}$.

As usual, the characterization of SPE utility pairs starts by deriving MPE strategies. Doing so means the straightforward application of the arguments in sections 4.2.3-4.2.5. This yields (formulated in terms of value functions):

1 Player 1 accepts every $s^t \in S$ if $s_1^t \geq (1-\delta)\, d_1^{odd} + \delta V_1^1$.

2 If $f_2\left((1-\delta)\, d_1^{odd} + \delta V_1^1\right) \geq (1-\delta)\, d_2^{odd} + \delta V_2^1$, the player 2 proposes $y = \left((1-\delta)\, d_1^{odd} + \delta V_1^1, f_2\left((1-\delta)\, d_1^{odd} + \delta V_1^1\right)\right)$. Otherwise, player 2 makes an unacceptable proposal at t is odd.

3 Player 2 accepts every $s^t \in S$ if $s_2^t \geq (1-\delta)\, d_2^{even} + \delta V_2^2$.

4 If $f_1\left((1-\delta)\, d_2^{even} + \delta V_2^2\right) \geq (1-\delta)\, d_1^{even} + \delta V_1^2$, then player 1 proposes $x = \left(f_1\left((1-\delta)\, d_2^{even} + \delta V_2^2\right), (1-\delta)\, d_2^{even} + \delta V_2^2\right)$. Otherwise, player 1 makes an unacceptable proposal at t is even.

5 If x and y are accepted in SPE, then $V^1 = x$ and $V^2 = y$.

Inspection of these results reveals two things. First, we must not forget to check whether or not the proposing player is better of by making an acceptable proposal.[2] Second, if in a MPE both players make acceptable proposals then the only component of $d(t)$ that is relevant at round $t \in \mathbb{N}$ is the responding player's component of $d(t)$. So, only d_2^{even} and d_1^{odd} matter for any MPE with immediate agreement in every subgame.

Before tackling the whole problem let us first concentrate on MPE strategies that feature immediate agreement at every subgame. At first glance, the only thing we have to do is to take $d = \left(d_1^{odd}, d_2^{even}\right)$ in the standard bargaining problem (S, d) of chapter 3 and substitute this d into the function $p \times q$, i.e.

$$p_2(y) = (1-\delta)d_2^{even} + \delta y_2, \quad p_1(y) = f_1(p_2(y))$$

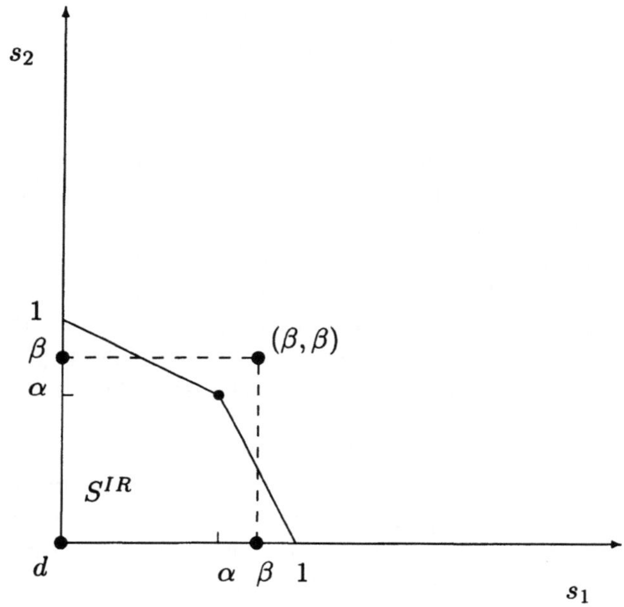

Figure 6.2. The constructed $d = (d_1^{odd}, d_2^{even}) = (\beta, \beta)$ of example 6.9 lies above the Pareto frontier of the convex set S.

and

$$q_1(x) = (1-\delta)d_1^{odd} + \delta x_1, \quad q_2(x) = f_2(q_1(x)).$$

However, to conclude that the unique fixed point of $p \times q$ specifies the unique pair of MPE proposals (x, y) in the model would be jumping to the wrong conclusion. The reason is that this insight is only true if $d = \left(d_1^{odd}, d_2^{even}\right) \in S$ and, unfortunately, this is not generally true as the following example shows.

Example 6.9. *Consider the bargaining problem* $\left(S, \left\{d^{even}, d^{odd}\right\}\right)$, *where for* $\alpha \in \left(\frac{1}{2}, 1\right)$ *the set* S *is equal to the convex hull of the four points* $(0,0)$, $(0,1)$, (α,α), $(1,0)$ *and for* $\beta \in [0,1]$ *the disagreement points* $d^{even} = (0,\beta)$ *and* $d^{odd} = (\beta, 0)$. *Then even though* d^{even} *and* d^{odd} *both belong to* S *for all* β *the constructed point* $d = \left(d_1^{odd}, d_2^{even}\right) = (\beta,\beta)$ *lies* above *the Pareto frontier of* S *for all* $\beta > \alpha$. *Unfortunately, the uniqueness of the fixed point of* $p \times q$ *is no longer guaranteed, because for* d *above the Pareto frontier the bargaining problem* (S, d) *is equivalent to a strongly comprehensive bargaining problem and in this example it is not convex. To see the equivalence, observe* *that the strongly comprehensive bargaining problem* $(S', 0)$ *of example 5.3 with* $\alpha_1 = \alpha_2 = \beta - \alpha < \frac{1}{2}$ *can be obtained after applying the linear transformation* $-(s_i - \beta)$, $i = 1, 2$ *to the bargaining problem* (S, d). *The transformed problem* $(S', 0)$ *admits three fixed points for sufficiently large* $\delta < 1$ *and, after applying the inverse transformation to obtain* (S, d) *back, so does* $(x, y) = (p(y), q(x))$ *for each* $\beta > \alpha$.

Example 6.9 belongs to a generic class of bargaining problems for which $i)$ $d^{even}, d^{odd} \in S$ admit the constructed point $d = \left(d_1^{odd}, d_2^{even}\right)$ to lie *above* the Pareto frontier of S and $ii)$ the function $p \times q$ admits a multiplicity of fixed points. The interpretation of this constructed point d should not be mistaken for the points m^{even} and m^{odd}, which represent the expected disagreement points in case of perpetual disagreement. The true meaning is that the constructed d represents the per-period disagreement utilities of the responding players in the threshold levels and these utilities determine each players bargaining position.

In case the constructed disagreement point d lies above the Pareto frontier, then in general the uniqueness of the fixed point is no longer guaranteed. The reason is that for d above the Pareto frontier the bargaining problem (S, d) is equivalent to a strongly comprehensive bargaining problem which may admit multiple fixed points. To see the equivalence, observe that applying the transformation $-(s_i - d_i)$, $i = 1, 2$ to the bargaining problem (S, d) yields a bargaining problem $(S', 0)$ in which the Pareto frontier can be described by a strictly-decreasing *convex* function instead of the usual concave function (if S is convex as we assumed). Note however, that a linear Pareto frontier is a special case since the associated bargaining problem $(S', 0)$ remains convex.

So, consider the fixed point problem $(x, y) = (p(x), q(y))$. Then proposition 5.4 implies that there is a one-to-one relation between the fixed points (x, y) and the MPE strategies. We have to distinguish two cases. The first case is $d = \left(d_1^{odd}, d_2^{even}\right) \in S$. For this case all the results of chapter 3 and 4 immediately apply. The second case is $d = \left(d_1^{odd}, d_2^{even}\right) \notin S$. Then the function $p \times q$ may admit multiple fixed points and the analysis for strongly comprehensive bargaining problems in chapter 5 applies. This means that for each player there is a MPE utility pair that is this player's worst SPE and that this worst SPE can then be used in equilibrium switching to derive similar sets of SPE utilities $E_1 \subseteq S^{IR}(t)$, t even, and $E_2 \subseteq S^{IR}(t)$, t odd as in proposition 5.10. The set of MPE utility pairs converges to the set $T(S, d)$ as δ goes to 1, where the definition of $T(S, d)$ is modified in the obvious way to allow for $d = \left(d_1^{odd}, d_2^{even}\right) \notin S$. To summarize, the following theorem states the convergence result for all $d = \left(d_1^{odd}, d_2^{even}\right)$ (so for both cases) and this result will be applied in the analysis of chapter 9.

Theorem 6.10. *Let* $\left(S, \left\{d^{even}, d^{odd}\right\}\right)$ *be a bargaining problem and* $F(\delta)$ *in (5.6) be modified for fixed points of* $p \times q$ *corresponding to* $d = \left(d_1^{odd}, d_2^{even}\right)$. *Then* $\lim_{\delta \to 1} F(\delta) = T(S, d)$.

The analysis thus far was conducted under the hypothesis that each player would be better of by making an acceptable proposal, but we did not yet check whether this hypothesis is correct. The following proposition states that making an acceptable proposal is indeed better for all $d^{odd}, d^{even} \in S$.

Proposition 6.11. *Let* $\left(S, \left\{d^{even}, d^{odd}\right\}\right)$ *be a bargaining problem. If* (x^*, y^*) *is a fixed point of* $p \times q$ *for* $d = \left(d_1^{odd}, d_2^{even}\right)$, *then player 1 (2) prefers to propose* x^* (y^*) *compared to agreement upon* y^* (x^*) *in the next round. Moreover,* $x^* \geq m^{even}$ *and* $y^* \geq m^{odd}$.

Corollary 6.12. *For each* $V^1 \in S$ *and* $V^2 \in S$:

$$
\begin{aligned}
f_1\left((1-\delta)\, d_2^{even} + \delta V_2^2\right) &\geq (1-\delta)\, d_1^{even} + \delta V_1^2, \\
f_2\left((1-\delta)\, d_1^{odd} + \delta V_1^1\right) &\geq (1-\delta)\, d_2^{odd} + \delta V_2^1.
\end{aligned}
$$

These results imply that each player in each MPE has an incentive to propose an agreement that is acceptable for his opponent. Hence, strategies must feature immediate agreement in every subgame. Moreover, the MPE agreements x^* and y^* are individual rational with respect to perpetual disagreement.

To summarize, alternating disagreement points yield the standard results as long as $d = \left(d_1^{odd}, d_2^{even}\right)$ belongs to S. Moreover, the standard uniqueness result is robust for alternating disagreement points d^{even} and d^{odd} that are relatively close. However, it yields results similar to those for strongly comprehensive bargaining problems (with possibly multiple SPE's) if the constructed point $d = \left(d_1^{odd}, d_2^{even}\right) \notin S$ lies above the Pareto frontier.

We conclude this section by showing that for $d = \left(d_1^{odd}, d_2^{even}\right) \notin S$ the property of the first-mover advantage reverses into a first-mover disadvantage, i.e., player 1 prefers $\hat{y}$ to $\hat{x}$. The underlying intuition is that the responding player's current period disagreement payoff is higher than this player's payoff in the agreement that will be reached one period later. Therefore, this player does not incur waiting costs if this player would reject any proposal. It is the proposing player who is harmed if no agreement is reached in the current period and this player can only avoid this by giving in.

Proposition 6.13. *Let* (x^*, y^*) *be a fixed point of the function* $p \times q$ *corresponding to* $d = \left(d_1^{odd}, d_2^{even}\right)$. *Then player* 1 *strictly prefers* y^* *to* x^*, *respectively, player* 2 *strictly prefers* x^* *to* y^* *if and only if d lies above the Pareto frontier of S.*

Proposition 6.13 is also important in the computation of MPE proposals. For instance, if the Pareto frontier is piecewise linear and x^* and y^* do not lie on the same line piece, then d lies above the Pareto frontier implies that x^* lies to the left of y^* and not to the right as in the standard alternating offers model.

The following example shows a first-mover disadvantage for bargaining problems with a linear Pareto frontier and alternating disagreement point

Example 6.14. *Consider the bargaining problem* $\left(S, \left\{d^{even}, d^{odd}\right\}\right)$*, with* $S = \left\{s \in \mathbb{R}^2_+ | s_1 + s_2 \leq 1\right\}$ *and* $d^{even}, d^{odd} \in S$*. Substitution of* $d = \left(d_1^{odd}, d_2^{even}\right)$ *into the function* $p \times q$ *of example 3.4 yields the unique fixed point given by*

$$x^* = \left(d_1^{odd} + \tfrac{1}{1+\delta}\left(1 - d_1^{odd} - d_2^{even}\right), d_2^{even} + \tfrac{\delta}{1+\delta}\left(1 - d_1^{odd} - d_2^{even}\right)\right),$$
$$y^* = \left(d_1^{odd} + \tfrac{\delta}{1+\delta}\left(1 - d_1^{odd} - d_2^{even}\right), d_2^{even} + \tfrac{1}{1+\delta}\left(1 - d_1^{odd} - d_2^{even}\right)\right).$$

Then $d = \left(d_1^{odd}, d_2^{even}\right) \notin S$ *means that* $1 - d_1^{odd} - d_2^{even} < 0$ *and, thus, we obtain the first-mover disadvantage* $x_1^* < y_1^*$ *and* $y_2^* > x_2^*$*. It is as if the constructed* $d = \left(d_1^{odd}, d_2^{even}\right)$ *is some kind of utopia point and that the players have to compromise with respect to this point in order to reach an agreement. Furthermore,* $x_1^* \geq m_1^{even}$ *follows after a simple substitution of* $d_2^{even} \leq 1 - d_1^{even}$ *into the expression for* x_1^* *and straightforward rewriting. Similar arguments show that* $x_2^* \geq m_2^{even}$ *and* $y_i^* \geq m_i^{odd}$*,* $i = 1, 2$*. Since the fixed point is unique we also have that the MPE is also the unique SPE.*

6.4. Alternative bargaining procedure

The alternating offer model assumes that the order in which the players propose is fixed. In this section two models are discussed that abandon this assumption. The first model still is a discrete time model in which the sequence of who proposes when is randomly determined by a Markov process. The advantage of this approach is that the alternating offer model and several alternative models in the literature can be considered as special cases of this generalized model. The second model we discuss assumes continuous time and 'no' bargaining procedure at all.

6.4.1 Markov process

In this section we consider a generalization of the alternating offers model in which the fixed alternating order of proposing players is a random sequence that is governed by a simple Markov process. The reason for doing so is that this Markov process captures several of the extensively studied bargaining procedures in the literature.

The subject of the negotiations is the bargaining problem (S, d), with S is a *convex* set and $d \in S$. The Markov process that governs who proposes has two states, numbered 1 and 2. State i, $i = 1, 2$, at round t means that player i proposes at this round. Denote $p_{ij} \in [0, 1]$, $i, j = 1, 2$, as the probability that the Markov process switches from state i at round t to state j at round $t+1$ meaning that p_{ij} expresses the probability that player j becomes the next proposing player if player i is currently the proposing player. Obviously, $p_{11} = 1 - p_{12}$ and $p_{22} = 1 - p_{21}$ are the probabilities that the state remains in the current state for another round.

Bargaining Procedure 6.15. *The proposing player at round* $t \in \mathbb{N}$ *is determined by a two-state Markov process with states* 1 *and* 2. *If the state at round* t *is* i, *then player* i *proposes* $s^t \in S$ *and his opponent either accepts* s^t *or not. If a proposal is accepted the negotiations end and, otherwise, the negotiations proceed to the next round* $t+1$ *governed by the Markov process.*

Markov strategies in this bargaining procedure are strategies in which player 1 always proposes $x \in S$ whenever the state is 1 at round t and this player always accepts any $s^t \in T_1$ in state 2 at round t and, similar player 2 proposes $y \in S$ in state 2 at t and accepts $s^t \in T_2$ in state 1 at t.

The advantage of introducing this particular Markov process is that several models that are well known in the economic literature fit nicely into this framework as special cases. First, the alternating offers model corresponds to $p_{12} = p_{21} = 1$, which means that state changes every round with probability one. But also the model with randomly proposing players in Hoel (1987) can be captured by imposing the state independent transition probabilities $p_{12} = p_{21} = \frac{1}{2}$. The latter model is extended to state independent $p_{12} = p_{22} = q$, $q \in (0,1)$, and $p_{21} = p_{11} = 1 - q$ in Muthoo (1999). The final class of bargaining procedures captured is the one-sided offer model in which one of the players has the 'monopoly' to make proposals. If player 1 has this privilege, then this model corresponds to $p_{12} = 0$ and $p_{21} = 1$. Note that the ultimatum game, which is very popular in many textbooks and experimental studies, is the one round version of one-sided offers model and this game yields a unique SPE outcome that can be regarded as being similar to the unique SPE outcome of the one-sided offers model.

The analysis in chapter 3 can be easily adapted to characterize the unique SPE. Suppose player 1 is the proposing player at the current round, i.e., the state is 1. If player 2 rejects player 1's proposal, then with probability $(1-\delta)$ the negotiations end in a permanent breakdown, with probability $\delta(1-p_{12})$ player 1 remains the proposing player for one more round and with probability δp_{12} player 2 becomes the next proposing player. So, Markov strategies such that $y \in T_1$ and $x \in T_2$ mean that rejecting the proposal s^t at even numbered rounds t imply an expected payoff of $(1-\delta)\, d_2 + \delta\,(p_{11}x_2 + p_{12}y_2)$ for player 2. So, accepting is a best response for player 2 if

$$s_2 \geq (1-\delta)\, d_2 + \delta\left((1-p_{12})\, x_2 + p_{12}y_2\right).$$

As before, player 1 in the role of the proposing player solves the maximization problem given by

$$\max_{x\in S} x_1, \text{ s.t. } x_1 \le f_1(x_2), x_2 \ge (1-\delta) d_2 + \delta((1-p_{12}) x_2 + p_{12} y_2),$$

which is equivalent to

$$\max_{x\in S} x_1, \quad \text{s.t.} \quad x_1 \le f_1(x_2), \quad x_2 \ge \tfrac{1-\delta}{1-\delta+\delta p_{12}} d_2 + \tfrac{\delta p_{12}}{1-\delta+\delta p_{12}} y_2.$$

So, it is as if player 2 faces $\delta_2 = \frac{\delta p_{12}}{1-\delta+\delta p_{12}}$ in the model of section 6.3.1 whenever this player is the responding player. As before, both constraints are binding in player 1's utility maximizing proposal meaning

$$x_1 = f_1(x_2), \quad x_2 = \tfrac{1-\delta}{1-\delta+\delta p_{12}} d_2 + \tfrac{\delta p_{12}}{1-\delta+\delta p_{12}} y_2 = \hat{p}_2(y).$$

Similar, player 1 accepts every proposal $y \in S$ proposed by player 2 if

$$y_1 \ge (1-\delta) d_1 + \delta((1-p_{21}) y_1 + p_{21} x_1)$$

and player 2's utility maximizing proposal y in state 2 is given by

$$y_1 = \tfrac{1-\delta}{1-\delta+\delta p_{21}} d_1 + \tfrac{\delta p_{21}}{1-\delta+\delta p_{21}} x_1 = \hat{q}_1(x), \quad y_2 = f_2(y_1).$$

So, effectively it is as if player 1 has $\delta_1 = \frac{\delta p_{21}}{1-\delta+\delta p_{21}}$ in the model of section 6.3.1. Having reformulated the model in terms of different probabilities δ_1 and δ_2 implies that we can apply the results for the model of section of 6.3.1 in order to obtain the following proposition.

Proposition 6.16. *Let $\delta_1 = \frac{\delta p_{21}}{1-\delta+\delta p_{21}}$ and $\delta_2 = \frac{\delta p_{12}}{1-\delta+\delta p_{12}}$ in the function $\hat{p} \times \hat{q}$ of (6.1). The unique SPE of bargaining procedure 6.15 is the MPE corresponding to the unique fixed point $(x^*, y^*) = (\hat{p}(y^*), \hat{q}(x^*))$ and*

$$\begin{aligned} T_1 &= \{s \in S \,|\, s_1 \ge (1-\delta) d_1 + \delta(p_{21} x_1^* + p_{22} y_1^*)\}, \\ T_2 &= \{s \in S \,|\, s_2 \ge (1-\delta) d_2 + \delta(p_{11} x_2^* + p_{12} y_2^*)\}. \end{aligned}$$

The probability δ_i, $i = 1, 2$, is increasing in p_{ij}, $j = 1, 2$ and $j \neq i$. Being a proposing player yields a first-mover advantage and the higher the probability that one stays the proposing player the longer the expected wait for the opponent before he can become the proposing player.

For the three special cases mentioned earlier we obtain the following:

1 Alternating offers. Then $p_{12} = p_{21} = 1$ imply that $\delta_1 = \delta_2 = \delta$.

2 Random proposing players. Then $p_{12} = p_{22} = q$, $q \in [0, 1]$, implies that $\delta_1 = \frac{\delta(1-q)}{1-\delta q}$ and $\delta_2 = \frac{\delta q}{1-\delta(1-q)}$. Thus, $\delta_1 \ge \delta_2$ if and only if $q \le \frac{1}{2}$. In particular, $q = \frac{1}{2}$ in Hoel (1987) implies that $\delta_1 = \delta_2 = \frac{\delta}{2-\delta}$.

3 One-sided offers. If player 1 has the privilege to propose, then $p_{12} = 0$ and $p_{21} = 1$ implies that $\delta_1 = \delta$ and $\delta_2 = 0$.

The unique pair of SPE proposals (x^*, y^*) is continuous in the switching probabilities of the underlying Markov process. In the bargaining and game theoretic literature it is popular to think that equilibrium outcomes are sensitive to 'small' changes in the game tree and the change from alternating to one-sided offers is regarded as such change. However, the Markov bargaining process shows that the results of going from one model to another gradually change in a continuous manner in the probabilities p_{12} and p_{21} and, therefore, the equilibrium outcomes of going from one model to another gradually change in a continuous manner and that there is no abrupt change at all.

We conclude this section by investigating the limit of the MPE proposals as the time between bargaining rounds vanishes, i.e., Δ goes to 0. Similar as for $\delta = e^{-r\Delta}$ we assume that the switching probability of the Markov process also depend upon Δ and we write $p_{ij}(\Delta)$, $i, j = 1, 2$ and $j \neq i$. We assume that $p_{ij} : \mathbb{R}_+ \to [0, 1]$ is continuously differentiable in Δ and $\lim_{\Delta \to 0} p_{ij}(\Delta) = \rho_{ij} > 0$. For example, if the Markov process is a continuous time Markov process, then $p_{ij}(\Delta) = \Pr\{x(\Delta) = j | x(0) = i\} = \rho_{ij}\Delta + O(\Delta) \to \rho_{ij}$ as Δ goes to 0 satisfies these conditions, e.g., Karlin and Taylor (1975). Letting the time between proposals vanish yields the following theorem.

Theorem 6.17. *As Δ goes to 0 the SPE proposals $x^*(\Delta)$ and $y^*(\Delta)$ converge to $N_{\hat{\alpha}}(S, d)$ with weight $\hat{\alpha} = \frac{\rho_{12}}{\rho_{12} + \rho_{12}}$.*

With respect to the special cases of alternative bargaining procedures we obtain $\hat{\alpha} = \frac{1}{2}$ for the alternating offers model, $\hat{\alpha} = q$ in Muthoo (1999) and, provided player 1 has the privilege to propose, $\hat{\alpha} = 1$ in the one-sided offers model. The weights of the asymmetric Nash's bargaining solution solely depend upon the switching probabilities in the Markov process. Moreover, the limit vector of weights $(\hat{\alpha}, 1 - \hat{\alpha})$ corresponds to the invariant distribution over the states of the continuous time Markov process introduced above and represents the fraction of time the state is in state i, $i = 1, 2$.

Note that the symmetric NBS corresponds to symmetry in the limit switching probabilities, i.e., $\rho_{12} = \rho_{21}$. These results show that the qualitative results obtained in chapter 3 for the alternating offers model are robust with respect to alternative bargaining procedures.

6.4.2 Strategic timing of proposals

The alternating offer model has been criticized, because it assumes a fixed order in which the players propose and respond to each other. The question where this order of moves comes from cannot be answered in the alternating offer model (nor in the bargaining procedure of the previous section). In this section a model with continuous time is introduced in which the players endogenously determine when to make a proposal including that they refrain from proposing. Also the timing of accepting a proposal is endogenous. The full model is a technically complicated model but its results are too important to be dismissed.

Consider the bargaining problem (S, d) with S is a *convex* set and $d \in S$. The continuous time model with endogenously timed proposals is specified as follows. At time $t \in \mathbb{R}_+$ player i, $i = 1, 2$, either makes the proposal $s^i(t) \in S$ or 'waits'. Each player cannot instantaneously adjust his last proposal but instead each proposal stands open for some minimum time $\Delta > 0$. During this time this player is committed to his standing proposal meaning that he has to wait until the expiration date of his standing proposal before he is allowed to make his next proposal. Formally, if player i, $i = 1, 2$, proposes $s^i(t) \in S$ at time t, then this player is committed to his proposal for each $t' \in [t, t + \Delta)$ and his opponent has the right to accept this proposal at any time during this time interval, provided his opponent is not committed to the opponent's standing proposal. However, if player i, $i = 1, 2$, proposes $s^i(t)$ at time t then his opponent requires some minimum time $\rho\Delta$, $\rho \in \left(0, \frac{1}{2}\right]$, to evaluate the proposal made, called the *reaction* and *response* time. Player i's proposal $s^i(t)$ is accepted at $t' \in [t + \rho\Delta, t + \Delta)$ by his opponent, called player j for $j = 1, 2$ and $j \neq i$, if the latter proposes $s^j(t') = s^i(t)$, provided player j is not committed at t', i.e., player j did not propose after $t' - \Delta$.

Following Sákovics (1993) we rule out simultaneous proposals at any moment of time. This is accomplished as follows. If both players simultaneously propose at time t, i.e., $s^1(t)$ and $s^2(t)$, then nature decides at random with equal probabilities which of these two proposals is to be validated as the standing proposal meaning that the other proposal is discarded and considered as not being made at all. The validated proposal at t stands at each $t' \in [t, t + \Delta)$ and the player who made it is not allowed to propose during this time. The player whose proposal failed to be validated at t is free to propose, except that we have to take into account this players reaction time meaning that he has to wait until $t + \rho\Delta$ before he can make a proposal.

Technically speaking, these assumptions ensure that each player's number of proposals on every bounded time interval is finite. Finally, waiting involves a risk of breakdown and $1 - e^{-rt}$ is the risk the negotiations break down before time t. The model can be summarized as follows.

Bargaining Procedure 6.18. *At time $t \in \mathbb{R}_+$ player i, $i = 1, 2$, either makes the proposal $s^i(t) \in S$ or 'waits'. If player i, $= 1, 2$, proposes at time t his proposal stands open at $t' \in [t, t + \Delta)$, player i has to wait at t' and his opponent has to wait at $t' \in (t, t + \rho\Delta)$. If both players simultaneously propose at time $t \in \mathbb{R}_+$, then each proposal has a probability of a half to be validated.*

A proposal made at $t \in \mathbb{R}_+$ is accepted at time $t' \in \mathbb{R}_+$, $t < t' < t + \Delta$, if either $s^1(t') = s^2(t)$ or $s^2(t') = s^1(t)$.

The model admits a unique SPE that can be regarded as some kind of Markov strategies, because the strategies are history independent up to the 'last proposal' on the table. In order to make this formal we introduce some additional notation. At time $t \in \mathbb{R}_+$ the history up to time t is denoted as h_t and belongs to the set of all histories at time t, denoted as H_t. For explanatory simplicity we neglect all histories in which both players have to wait according to the rules of the game. Then each history has one of the following two properties: Either both players are not committed and free to propose at time t, or player i, $i = 1, 2$, is committed to his proposal $s^i(t_i) \in S$ made at $t_i \in (t - \Delta, t)$ and his opponent is not. From now on we refer to the opponent of player i, $i = 1, 2$, as player j, $j = 1, 2$ and $j \neq i$. Formally, each history $h_t \in H_t$ at time t can be summarized by a pair of proposals $\left(s^1(t_1), s^2(t_2)\right)$, where $s^i(t_i)$, $i = 1, 2$, denotes player i's last proposal made at $t_i \in \mathbb{R}_+$, with the convention that i) $t_i \notin (t - \Delta, t)$ imposes $s^i(t_i) \in \emptyset$ meaning that player i is not committed to a standing proposal and ii) player j's response time has passed at t, i.e., $t \geq t_i + \rho\Delta$. Obviously, for every $h_t \in H_t$ either $s^1(t_1) \in \emptyset$, or $s^2(t_2) \in \emptyset$, or both. The initial history h_0 is trivially summarized by $(\emptyset, \emptyset)$.

This additional notation allows us to introduce the 'history independent' Markov strategies. Let $s^i \in S$, $i = 1, 2$, be player i's history independent proposal and the history independent set $T_i(t, t_j) \subseteq S$, $t \in [t_j + \rho\Delta, t_j + \Delta)$, denotes player i's set of acceptable proposals at time t if player j has committed himself at time $t_j \in (t - \Delta, t - \rho\Delta]$ to his opponent's proposal $s^j(t_j) \in S$. The Markov strategies are as follows. If both players are free to propose at time t, then player i, $i = 1, 2$, proposes s^i at t and nature validates one of these simultaneously made proposals. This means that almost always one of the players is committed except for a countable number of points in time. Suppose player j, $j = 1, 2$, is committed at t to his proposal $s^j(t_j)$ made at $t_j \in (t - \Delta, t - \rho\Delta]$. Then the strategies of player i prescribe the following. If $s^j(t_j)$ belongs to player i's set of acceptable proposals, then immediately match (i.e., accept) player j's proposal. However, if $s^j(t_j)$ is not acceptable, then player i waits if $t < t_j + (1 - \rho)\Delta$ and, otherwise, player i proposes $s^i(t) = s^i$. Note that $t < t_j + (1 - \rho)\Delta$ means that, at time t, the expiration date at $t_j + \Delta$ of $s^j(t_j)$ is not within $\rho\Delta$ of distance measured in time.

The Markov strategies σ are as follows. Consider a pair of proposals $\left(s^1, s^2\right)$ and a pair of sets $(T_1(t, t_2), T_2(t, t_1))$ and strategies σ given by

1 For any h_t such that $\left(s^1(t_1), s^2(t_2)\right) = (\emptyset, \emptyset)$:

$$\sigma_i(h_t) = s^i, \quad i = 1, 2.$$

2 For any h_t such that $s^i(t_i) \in \emptyset$, $s^j(t_j) \in S$ and $t_j \in (t-\Delta, t-\rho\Delta]$:

$$\sigma_i(h_t) = \begin{cases} s^j(t_j), & \text{if } s^j(t_j) \in T_i(t,t_j), \\ \text{wait}, & \text{if } s^j(t_j) \notin T_i(t,t_j) \text{ and } t < t_j + (1-\rho)\Delta, \\ s^i, & \text{if } s^j(t_j) \notin T_i(t,t_j) \text{ and } t \geq t_j + (1-\rho)\Delta. \end{cases}$$

The following proposition relates these Markov strategies to the unique fixed point of the function $p \times q$ as already defined in section 6.3.1, provided the risk of breakdown is properly chosen. So, once more this unique fixed point plays a major role in the characterization of MPE proposals.

Proposition 6.19. *The Markov strategies σ are SPE strategies if and only if (s^1, s^2) is the unique fixed point of the function $p \times q$ for $\delta = e^{-r(1-\rho)\Delta}$ and for $i,j = 1,2$ and $j \neq i$ the set $T_i(t,t_j)$ is given by*

$$\left\{ s \in S \middle| s_i \geq \left(1 - e^{-r\max\{\rho\Delta, t_j+\Delta-t\}}\right) d_i + e^{-r\max\{\rho\Delta, t_j+\Delta-t\}} s_i^i \right\}.$$

Moreover, with probability $1 - e^{-r\rho\Delta}$ the negotiations break down and with $\frac{1}{2}e^{-r\rho\Delta}$ the players agree upon s^i, $i = 1,2$, at $t = \rho\Delta$.

The rational of the strategic timing is the following. Suppose player 1 is currently not committed while his opponent has a standing proposal until $t_2+\Delta$. Then player 1 could delay his own counter proposal from t_1 to $t_1' = t_1 + \varepsilon$, $\varepsilon > 0$. By doing so, player 1 also delays his commitment period from $t_1 + \Delta$ to $t_1 + \Delta + \varepsilon$ meaning that he postpones his first opportunity to accept his opponent's *next* counter proposal to the proposal player 1 is currently delaying (in case his opponent prefers to counter propose at all). If player 2's next counter proposal is accepted by player 1, then a delay from $t_1+\Delta$ to $t_1+\Delta+\varepsilon$ inflicts some extra costs of waiting upon player 2 which is reflected in the decrease of player 2's present value of an agreement at $t_1+\Delta+\varepsilon$. Therefore, such a delay makes player 2 more eager to give in. So, if a utility of s_2^1 is acceptable for player 2 if player 1 proposes at t_1 then $e^{-r\varepsilon}s_2^1$ is acceptable if player 1 delays his proposal an ε of time. However, player 1 should not delay his proposal for ever, because player 2 does not remain committed for ever. So, if player 2 is committed until $t_2 + \Delta$ and $t < t_2 + (1-\rho)\Delta$, then player 1's optimal time to counter propose is $t_2 + (1-\rho)\Delta$ meaning that player 2 is able to respond to player 1's counter proposal immediately after player 2's expiration date of his current standing proposal. Any further delay in proposing is costly to the proposing player and is avoided in equilibrium. So, effectively each player's maximum of time that he can refrain himself from proposing is equal to $(1-\rho)\Delta$. In equilibrium each player's proposal is such that it is acceptable for his opponent just after the reaction time has elapsed and each player inflicts the maximum costs of delay upon his opponent. So,

$$s_1^2 = \left(1 - e^{-r(1-\rho)\Delta}\right) d_1 + e^{-r(1-\rho)\Delta} s_1^1, \quad s_2^2 = f_2\left(s_1^2\right),$$
$$s_2^1 = \left(1 - e^{-r(1-\rho)\Delta}\right) d_2 + e^{-r(1-\rho)\Delta} s_2^2, \quad s_1^1 = f_1\left(s_2^1\right).$$

Being the first to commit has a strategic advantage, because by committing himself each player can subtract the inflicted costs of waiting from his opponent. This is reflected in the 'first-mover-advantage' property of each pair of fixed points, i.e. $s_1^1 > s_1^2$ and $s_2^2 > s_2^1$. Since no player has a strategic advantage at the beginning of the negotiations and both players simultaneously propose in order to be the first to commit himself. Then nature has to decide which proposal is validated and the other player accepts the validated equilibrium proposal just after his reaction time elapses at $t = \rho\Delta$.

Finally, note that in deriving limit outcomes in the alternating offer model we assumed that $\delta = e^{-r\Delta}$ and not $\delta = e^{-r(1-\rho)\Delta}$. This means that the unique SPE outcome in the alternating offer model can be thought of as resulting from these strategies in the continuous time model where players are allowed to react (almost) immediately, i.e., $\rho \approx 0$.

The analysis is rather straightforward from here. In Sákovics (1993) the method in Shaked and Sutton (1984) is slightly modified for the continuous time framework and, then, application of this method yields that the MPE strategies of proposition 6.19 are the unique SPE strategies.

Theorem 6.20. *The MPE strategies σ of proposition 6.19 are the unique SPE strategies. Moreover, as Δ goes to 0, both s^1 and s^2 converge to $N(S,d)$ and both players immediately agree upon $N(S,d)$ in this limit.*

This theorem states a very strong result. The by now familiar convergence of SPE proposals to the NBS also occurs in the continuous time model with endogenous timing of the proposals and vanishing times of commitment Δ. The continuous time model also gives an answer to the question where the fixed order of moves in the alternating model comes from. The strategies are such that once one of the players is committed his opponent tries to avoid the situation where both players are free to propose and, therefore, each player proposes before the expiration date of the current commitment period taking into account the opponent's response time. So, even though in equilibrium the first proposal to be validated is accepted, an endogenous order determined by the players' equilibrium strategies emerges.

6.5. Related Literature

The results in section 6.2.1 were derived in Houba (1997). Imperfect divisibility was first considered in Van Damme, Selten, and Winter (1990), where the presence of a smallest monetary unit in the divide a dollar problem is studied. A similar analysis can also be found in e.g., Muthoo (1991) and Osborne and Rubinstein (1990). The discussion in section 6.2.2 is a direct translation of these results for an evenly spaced grid over the contract space in the general model. We refer to Weinberger (2000) for an interesting analysis of two-dimensional contract spaces, where players have the opportunity to selectively accept proposals. The latter means that if the proposed contract is (c_1, c_2), then the responding player may either accept $i)$ the entire contract, $ii)$ reject the entire contract or $iii)$ accept only the k-th component, $k = 1, 2$, and the negotiations continue over the remaining component. The case of two perfectly divisible variables is compared to one perfectly divisible issue and one (indivisible) yes/no decision. The latter situation typically yields (some) inefficient MPE agreements and SPE featuring delay. Selective acceptance of proposals is also closely related to multi-issue bargaining problems and the importance of the agenda, e.g., see Busch and Horstman (1997), Fershtman (1990) and Fershtman and Seidmann (1993).

Alternating risks of breakdown, or alternatively different discount factors, in the general divide a dollar problem was first analyzed in Rubinstein (1982), and this reference also contains the unique pair of SPE utilities of example 6.8. The asymmetric NBS of theorem 6.6 was first axiomatized in Kalai (1977) for convex bargaining problems and in Roth (1979) it is shown that the efficiency axiom in theorem 6.6 can be replaced by the individual rationality axiom. The convergence of the unique SPE to the asymmetric Nash bargaining solution was first conducted in Binmore, Rubinstein, and Wolinsky (1986) and surveys can be found in e.g., Hoel (1986), Muthoo (1999), Osborne and Rubinstein (1990) and Van Damme (1991).

Alternating disagreement points, as studied in section 6.3.2, were first analyzed in Houba (1997) and later in Corominas-Bosch (2000). Proposition 1 in Busch and Wen (1995) characterizes the unique SPE for the alternating offer model with a *linear* Pareto frontier and an arbitrary sequence $\{d(t)\}_{t=0}^{\infty}$ of fixed disagreement points. The results of section 6.3.2 imply that this uniqueness result does not extend to bargaining problems with a strictly concave Pareto frontier. The proof of proposition 6.11 relies on f_i, $i = 1, 2$, is a concave function and it is an open problem whether this result (and the other results in section 6.3.2) extends to strongly comprehensive bargaining problems. For an analysis of alternating disagreement points in combination with alternating risk of breakdown we refer to Bolt (1995). In this reference a case is studied in which the result of proposition 6.11 breaks down and, then, the SPE features that one of the players always makes an unacceptable proposal.

Nonstationary bargaining problems also arise if nonexpected utility functions are assumed. We refer to Shalev (1997) for an analysis of nonexpected utility functions featuring loss aversion in the alternating offers model. However, also in this reference convergence of the unique SPE outcome to the asymmetric NBS is obtained.

The bargaining procedure in section 6.4.1 is first proposed in Merlow and Wilson (1995) and is more general than the models with random proposing players in Hoel (1987) and Muthoo (1999). In Merlow and Wilson (1995) the set S is also stochastic and governed by some Markov process and uniqueness and efficiency of the equilibrium outcomes is investigated. However, the limit Δ goes to 0 is not investigated in this reference. The continuous time model in section 6.4.2 is based upon the second continuous-time bargaining procedure in Sákovics (1993). In this reference as well as in Perry and Reny (1993) also a closely related model is analyzed in which simultaneous proposals are allowed. For the latter model it is shown that a large set of SPE utility pairs can be supported and that each player's worst SPE utility corresponds to the unique fixed point of the function $\hat{p} \times \hat{q}$ provided δ_1 and δ_2 are suitably chosen. In the limit, as both Δ and ρ go to 0, the set of SPE utilities collapses into the Nash bargaining solution $N(S, d)$. In these two references, convergence to the asymmetric Nash bargaining solution is derived for the general case with response and commitment times that differ across players. So, the results of the alternating offer model are quite robust. We refer to Sákovics (1993) for a valuable discussion on the difference between the two versions of the continuous time model. We also note that the restriction $\rho \in \left[0, \frac{1}{2}\right]$ imposed in Sákovics (1993) ensures that the interval $\left[t + \rho\Delta, t + (1 - \rho)\Delta\right)$ is nonempty and that this cast some doubts upon the correctness of the results in Perry and Reny (1993) where this restriction is not imposed.

Articles that can be considered to deal with different bargaining procedures can also be found in Muthoo (1990), Muthoo (1992) and Andrelini and Felli (2001). In the latter reference imperfect recall is modelled as part of the information sets in the game tree and its consequences are studied. In Muthoo (1990) it is assumed that the proposing player can withdraw the proposal after the responding player has accepted it and a large set of SPE utility pairs is derived, including SPE strategies featuring delay exist. This hints at that a juridical system that forbids the withdrawal of an accepted proposal is essential as an institution that promotes Pareto efficient agreements in negotiations. Many countries, including the Netherlands, forbid such withdrawal for a large number of juridical situations. A discussion of Muthoo (1992) is postponed to the next chapter.

For a recent and complete survey of incomplete information and its consequences we refer to Muthoo (1999). Other references concerning incomplete information are e.g., Osborne and Rubinstein (1990) and Watson (1998). Incomplete information destroys Pareto efficiency, because it is no longer certain that the players immediately agree. The reason is that delay of an agreement should be regarded as the only credible 'signal' that can be used to learn or to transmit information depending upon whether a player lacks information or wants to reveal information. In a complete information setting players know all the relevant information and there is no need to transmit information by delaying an agreement. This is an alternative way to look at many of the no-delay results in negotiations models with complete information.

A topic of research that attracted a lot of attention is the Coase conjecture. In Andrelini and Felli (2001), among other things, the role of transaction costs is shown to be devastating for the Coase conjecture. In Wang (1998) it is shown that simultaneous negotiations on multiple issues with incomplete information admits a unique sequential equilibrium that is not only separating but also results in immediate agreement. This result is derived under the assumption that menus of contracts can be proposed. However, incomplete information on behalf of the uninformed party can be regarded as weakening this parties' bargaining position, which also violates the Coase conjecture.

Notes

1 To see that $\delta_2^{1-\hat{\alpha}} = \delta_1^{\hat{\alpha}}$ first note that $\delta_1^{\ln \delta_2} = e^{\ln \delta_1 \ln \delta_2} = \delta_2^{\ln \delta_1}$ and that

$$\delta_1^{\frac{\ln \delta_2}{\ln \delta_1 + \ln \delta_2}} = \delta_2^{\frac{\ln \delta_1}{\ln \delta_1 + \ln \delta_2}} = \delta_2^{1-\frac{\ln \delta_2}{\ln \delta_1 + \ln \delta_2}}.$$

2 In the context of wage bargaining this mistake was made in Fernandez and Glazer (1991), as was pointed out in Bolt (1995).

II

ENDOGENOUS THREATS

Chapter 7

COMMITMENT AND ENDOGENOUS THREATS

7.1. Introduction

The bargaining problems with fixed disagreement actions studied in the previous chapters can be understood as representing the last 'phase' of some more complex negotiation model in which the parties choose the disagreement actions prior to the negotiations. This point of view requires that the 'fixed' disagreement actions will be carried out if the parties are called upon to do so. This means that either players are committed upon these disagreement actions and cannot alter these actions if actually called upon to act, or the disagreement actions can be changed at will but these actions are anticipated as being credible, i.e., equilibrium actions, in case of a breakdown.

The ability to commit can be regarded as a powerful ingredient in negotiations. Intuitively, the underlying idea is that once a negotiator has convinced his opponents that he will not retreat from a specific course of action during the process, the opponents may decide to concede to his demands. Hence, deliberately reducing one's flexibility in the bargaining process may induce the other side to behave in a way that is favorable to the committed player. In this sense, we may say that in bargaining 'weakness may also represent a strength'. This was already understood by Thomas Schelling who was one of the first to informally explore the role, credibility and feasibility of commitment mechanisms in bargaining situations, e.g. Schelling (1960). In his view a bargaining process represents a struggle between negotiators to commit themselves to favorable bargaining positions. Obviously, the credibility of the mechanism which enforces these commitments is of great importance. A commitment which lacks credibility looses its strategic effect against rival opponents and they will merely regard this commitment action as a way of 'tactical bluffing' and act accordingly. Schelling further suggested that mutual incompatible commitments may cause bargaining impasses.

As lucidly described in Dixit and Nalebuff (1991), 'burning one's bridges' provides a good example of credible commitment. In the sixteenth century, in his conquest of Mexico, upon arrival in Cempoalla, Cortès commanded that all but one of his ships had to be destroyed, thus making an unconditional commitment to fight rather than to retreat. For his soldiers, although vastly outnumbered, there was no choice: to fight and win, or to loose and die. In fact, destroying the ships gave Cortès two advantages. First, his camp was unified, each soldier knowing that they would all fight till the end since desertion or even retreat was impossible. Second, and more important, was the effect this irreversible commitment to fight had on the enemy. The enemy knew that Cortès could only win or die, while they had the option to retreat into the hinterland. In the end, they chose to retreat rather than to fight such a determined opponent. The burning down of ships and Cortès' soldiers knowing it created a credible commitment to fight. If Cortès had failed it might well have seemed as an act of suicide. Yet, because he was victorious, it was the fruit of good strategic contemplation.[1] Still, these strategic commitment devices are at play in everyday bargaining situations. Consider for example a labour union engaged in collective wage bargaining which takes a firm position at the bargaining table by announcing a massive strike. It cannot back out from this announcement without 'losing its face'. If employers believe this to be true, then they may ultimately be better off by giving in to the union's demands.

In this chapter endogenous disagreement actions and the ability (not) to commit are studied in the simplest setting as is possible in order to understand the difference between the two notions. The natural starting point is Nash's variable-threat game, which is a simple two-stage model in which the parties commit themselves on their disagreement actions prior to the negotiations. This model can be easily modified to study negotiations without commitment, which we intend to do. Moreover, the negotiations in Nash's variable-threat game are modelled as Nash's demand game and, as extensively argued in section 4.3, the parties' demands can be regarded as representing the behavioural strategies of some underlying, unspecified dynamic bargaining process. The alternating offers procedure could represent this unspecified bargaining model given that it yields similar results. Therefore, it is obvious to replace Nash's demand game by the alternating offers model and study the issue of commitment versus no commitment in greater detail. Nevertheless, doing so does not yield duplicated results as will be seen below. Recall from chapter 6 that despite the rigid order of moves in the alternating offers model this procedure produces results that are robust if endogenous timing or some randomized sequence of proposing parties is taken into account.

This chapter is organized as follows. The starting point of the analysis is the introduction of Nash's variable threat game in the next section. After a brief analysis the unspecified negotiation process is explicitly modelled as the alternating offers procedure and the consequences are studied. Then, in the subsequent section, both models are studied under the assumption that players cannot commit themselves meaning that disagreement actions are chosen only in case of a breakdown. We note that the models without commitment can be regarded as simple in terms of mathematical complexity, because it is assumed that disagreement actions played in case of a terminal breakdown have no feedback on future behaviour in the negotiations (which are over by then).

7.2. Optimal threats with commitment

The simplest modification in terms of complexity is the extension in which the threats are chosen prior to the negotiations. As mentioned earlier, this requires that the players are committed to their disagreement actions if called upon to carry them out and players cannot back out from these actions. In the first subsection, Nash's variable-threat game is introduced and analyzed. Then, in the subsequent subsection, Nash's demand game that constitutes the second stage of the variable-threat game is replaced by the alternating offers procedure.

7.2.1 Nash's original variable-threat game

One of the first strategic bargaining games where players can influence the disagreement outcome was analyzed, not surprisingly, by Nash himself. In a now classical paper, e.g., Nash (1953), an elegant two-stage negotiation game is introduced, which is usually dubbed as *Nash's variable-threat game.* In the first stage, the players commit themselves to threats prior to the negotiations which they have to carry out in case they fail to reach an agreement in the second stage. In this second stage the players simultaneously play Nash's demand game of section 4.3.1. The multiplicity of equilibria in the latter model is resolved by resorting to the H-essential equilibrium of the demand game. The latter means that the parties agree upon the axiomatic Nash bargaining solution and the choice of threats in the first stage actually determines the equilibrium agreement through the disagreement outcome in the second stage. As a consequence, every player tries to choose its threat such as to increase its final payoff when agreement is struck.

For notational convenience, we briefly mention some of the notation already introduced in chapter 2. This means that the triple $\langle S, \{A_i\}_{i=1,2}, \{d_i\}_{i=1,2}\rangle$ specifies the non-empty and convex set S of attainable utility pairs, player i's set A_i of 'threats' and player i's utility function d_i over the set of disagreement outcomes A. Assumption 2.7 implies that for each $a \in A$ there exists an agreement $s(a) \in S$ such that $d(a) \leq s(a)$, i.e. mutual interests. This means that the Pareto frontier in terms of utility pairs associated with disagreement outcomes in the set A lies to the south-west of the Pareto frontier of the set S. Alternatively, we may regard the last two ingredients as representing a two-player game in normal form characterized by $\langle\{1,2\},\{A_i\}_{i=1,2},\{d_i\}_{i=1,2}\rangle$, where $\{1,2\}$ denotes the set of players, A_i denotes player i's set of 'threats' and d_i denotes player i's utility function.

Nash's variable-threat game is a two-stage game with almost perfect information, with two players and both players simultaneously and independently of each other choose their actions in each stage. In the first stage of the game, each player i, $i = 1, 2$, chooses his threat $a_i^T \in A$. Each player's threat determines this player's disagreement action in case the negotiations of the second stage break down and it is assumed that each player is committed to his threat in case of such breakdown. Once the threats have been chosen by the players, each player observes the threat chosen by his opponent. Then the game moves to the second stage which consists of Nash's demand game, as already described in section 4.3.1. Briefly recalling, the players simultaneously and independently state their demands $x_i \in \mathbb{R}$ in terms of utility and if the players' demands are compatible, i.e. $x \in S$, then both demands can be fulfilled, each player obtaining x_i, $i = 1, 2$, and the game ends. If both demands are incompatible, i.e. $x \notin S$, then players fail to reach agreement and have to carry out their threats a^T, each player securing $d_i(a^T)$, $i = 1, 2$, and the game ends. The formal definition of Nash's variable-threat game is as follows.

Bargaining Procedure 7.1. *Nash's variable-threat game is the two-stage game with almost perfect information characterized by the set of players $N = \{1,2\}$, player i's threat $a_i^T \in A_i$, $i = 1, 2$, in the first stage, player i's demand $x_i \in \mathbb{R}$ in the second stage, and player i's utility function $d_i(a^T) + i(x)\left(x_i - d_i(a^T)\right)$, where the indicator function $i(x)$ is defined in bargaining procedure 4.29.*

As argued in chapter 2 the equilibrium concept for multi-stage games with almost-perfect information is the SPE concept. Unfortunately, this concept is rather weak in Nash's variable-threat game, because it only imposes that a Nash equilibrium of the demand game is played in the second stage. As already discussed in section 4.3.1, under quite general conditions the Nash demand game admits a continuum of equilibria: any possible agreement which is both Pareto efficient and individually rational corresponds to a particular Nash equilibrium. In the variable-threat game this multiplicity problem aggravates, because the SPE strategies allow for the Nash equilibrium to depend upon the threats $a^T \in A$ chosen in the first stage. Formally, every vector function $x^* : A \to \mathbb{R}^2$ describes the players' SPE strategies in the second stage of the threat game if and only if $x^*(a^T)$, $a^T \in A$, is a Nash equilibrium of Nash's demand game. In order to deal with this indeterminacy, John Nash proposed the equilibrium selection described in section 4.3.1. In modern terminology, the

H-essential equilibrium concept is applied to the second stage and according to theorem 4.31 the unique H-essential equilibrium of the second stage is given by $x^*(a^T) = N(S, d)$ for $d = d(a^T)$. So, the players' demands coincide with the axiomatic Nash bargaining solution of chapter 4. Recall that this solution is Pareto efficient, which is important in the sequel.

Having 'solved' the SPE strategies for the second stage means that the analysis continues with deriving the optimal threats in the first stage of the variable-threat game. Imposing the H-essential equilibrium outcome of the demand game as the SPE continuation in the variable-threat game means that both players anticipate agreement upon $x^*(a^T) = N(S, d(a^T))$ in the second stage. Then the SPE actions in the first stage of the threat game correspond to the Nash equilibrium actions of the game in normal form in which A_i, $i = 1, 2$, represents player i's set of feasible actions and $N_i(S, d(a^T))$ represents player i's utility. In Nash (1953) it is argued that the latter game is strategically equivalent to a zero-sum game, because Pareto efficiency of the agreement implies that an improvement enforced by one player is at the expense of his opponent. This equivalence has several important implications. First, zero-sum games are known to have a unique *value* meaning that there exists a unique number that expresses player 1's Nash equilibrium utility and, due to the antagonistic character of zero-sum games, it also expresses player 2's utility. Translated to the threat game, this means that there is uniqueness in terms of a unique utility pair on the Pareto frontier. So, the threat game resolves in a unique Pareto efficient agreement. With respect to the equilibrium threats, (generic) uniqueness cannot be established, because zero-sum games like all other games often admit multiple Nash equilibria for a generic class of problems.[2]

Second, all Nash equilibrium actions in zero-sum games are *minmax* strategies and vice versa. A minmax strategy for player 1 maximizes this player's worst outcome his opponent can impose upon him and, due to the antagonistic setting, player 2 has every incentive to keep player 1 down at his lowest utility level. Translated to the threat game this means that $(a_1^T, a_2^T) \in A$ are SPE disagreement actions in the first stage if and only if

$$a_i^T = \arg \max_{a_i^T \in A_i} \min_{a_j^T \in A_j} N_i\left(S, d(a^T)\right), \quad i, j = 1, 2 \quad j \neq i. \tag{7.1}$$

So, player 1 wants to pull the bargaining solution along the Pareto frontier in the south-east direction, whereas his opponent wants to pull it in the opposite direction. Finally, minmax strategies are said to be interchangeable meaning that if a_i^* and a_i^{**} are both minmax strategies for player i, then both actions do equally well against all of his opponents minmax actions and player i can choose any of these two actions regardless of what his opponent will choose. This implies that the players in a zero-sum game that admits multiple Nash equilibria do not have the coordination problem of playing the same Nash equilibrium actions as other games in normal form have, as for example the battle of the sexes.

We state the following proposition without further proof.

Theorem 7.2. *Consider the class of SPE strategies in which the players play the H-essential equilibrium $x^*\left(a^T\right) = N\left(S, d\left(a^T\right)\right)$ in the second stage.*

i Then $a^T \in A$ is a pair of SPE threats in the first stage of Nash's variable-threat game if and only if for both $i = 1, 2$ the threat a_i^T is a minmax action associated with (7.1).

ii All SPE strategies result in the same Pareto efficient agreement in S, i.e., there exist a unique $s^* \in S$ such that $x^*\left(a^T\right) = s^*$ for all SPE threats a^T.

This proposition is illustrated in the following proposition.

Example 7.3. *Consider the class of variable-threat games such that the set S is given by*

$$S = \left\{ s \in \mathbb{R}_+^2 \mid s_1 + s_2 \leq 1 \right\}.$$

The H-essential equilibrium concept singles out the equilibrium demands $x^*\left(a^T\right) = N\left(S, d(a^T)\right)$ in the second stage that are given by

$$x^*(a^T) = \tfrac{1}{2}\left(1 + d_1(a^T) - d_2(a^T), 1 - d_1(a^T) + d_2(a^T)\right), \; a^T \in A.$$

According to theorem 7.2, in choosing optimal threats, player 1 tries to maximize the quantity $d_1(a^T) - d_2(a^T)$, whereas player 2 tries to minimize this same quantity. Effectively, this means that the players play a zero-sum game in which $d_1(a^T) - d_2(a^T)$ represents player 1's utility function. The optimal threats a^T need to constitute a Nash equilibrium of this zero-sum game.

7.2.2 The variable-threat game with alternating offers

As argued in section 4.3.2, John Nash thought of the demands as representing the players' strategies of some unspecified bargaining procedure. In this section we replace the demand game by one specific bargaining procedure, being the procedure with alternating offers. Since this procedure showed quite some robustness with respect to alternative bargaining procedures, as shown in chapter 6, there is not much loss in generality by choosing this specific bargaining procedure. At first glance the reader might consider this exercise as being trivial, because the unique SPE of the alternating offers also corresponds to the Nash bargaining solution. However, we will show that the results of theorem 7.2 need to be modified unless some additional assumption is imposed.

Like in the previous section we consider bargaining problems in utility representation $\langle S, \{A_i\}_{i=1,2}, \{d_i\}_{i=1,2} \rangle$. The choice of the threats takes place prior to the negotiations and, similar as before, we adopt the convention that the first bargaining round takes place at $t = 0$ meaning that the choice of disagreement actions takes place at let's say $t = -1$. It is again assumed that each player perfectly observes the other player's choice of disagreement actions before the negotiations start at round $t = 0$ and this guarantees that the players have perfect information in the bargaining process. For explanatory reasons we think of $\delta \in (0, 1)$ as representing the probability of a next bargaining round. By doing so, a breakdown is an irreversible event once it occurs. A novel feature is that we assume that the players may condition their disagreement actions upon the bargaining round t in which breakdown occurs. Therefore, the players choose an infinite sequence of actions $\langle a^t \rangle_{t=0}^{\infty}$, $a^t \in A$, before the negotiations start. Let us therefore state the informally described bargaining procedure.

Bargaining Procedure 7.4.
Round -1*:*
Player i*,* $i = 1, 2$*, chooses the infinite sequence of disagreement actions* $\langle a_i^t \rangle_{t=0}^{\infty}$*,* $a_i^t \in A_i$*.*

Round t $(t$ *even,* $t \geq 0)$*:*
Player 1 *proposes an offer* $s^t \in S$ *and, then, player 2 either accepts ('yes') or rejects ('no') the offer* s^t*. If player 2 accepts, then the negotiations end and the players implement the offer* s^t *yielding a final utility* s_i^t*,* $i = 1, 2$*, in round* t*. However, if player 2 rejects, then with probability* $\delta \in [0, 1)$ *the bargaining process continues to the next round* $t + 1$*, and with probability* $1 - \delta$ *the bargaining process breaks down forever and the players are committed to carry out the pair of disagreement actions chosen at round* -1 *with associated pair of utilities* $d(a^t)$*,* $a^t \in A$*.*

Round t $(t$ *odd,* $t \geq 1)$*:*
The procedure is similar as for t *is even except that now player* 2 *has the initiative to propose an offer* $s^t \in S$ *and, afterwards, player 1 responds.*

As in Nash's variable-threat game the bargaining procedure is a multi-stage game with almost-perfect information and the SPE concept is appropriate. This implies that the alternating offers has to be solved for every possibly infinite sequence of actions $\langle a^t \rangle_{t=0}^{\infty}$, $a^t \in A$. Since these actions change over time, the disagreement point $d\left(a^t\right)$ will also vary over time, which include the special case of alternating disagreement points studied in section 6.3.2. Recall that alternating disagreement points cause the technical problem that the standard analysis is only applicable if some 'constructed' disagreement point is taken into account. The latter point may lie above the Pareto frontier of the set S and generic multiplicity cannot be ruled out. In this section we want to avoid these technical problems by restricting the analysis to the class of normalized bargaining problems with a linear Pareto frontier, that is, we assume,

$$S = \left\{ s \in \mathbb{R}^2 \middle| \; s_1 + s_2 \leq 1 \right\}. \tag{7.2}$$

This implicitly means that we also impose $d_1(a) + d_2(a) \leq 1$ for all $a \in A$.

Consider an arbitrary subgame starting at the first bargaining round after the players have chosen their disagreement actions. This means that the infinite sequence of actions $\langle a^t \rangle_{t=0}^{\infty}$, $a^t \in A$, is given and observed by the two players. The corresponding subgame is equivalent to a non-stationary version of the alternating offers procedure, because the disagreement actions may vary over time. The first lemma states the unique SPE payoff of the subgame starting at the first bargaining round. A proof can be found in Busch and Wen (1995) and is therefore omitted.

Lemma 7.5. *For every infinite sequence of disagreement actions $\langle a^t \rangle_{t=0}^{\infty}$ the alternating offers process admits a unique SPE. Moreover, at $t = 0$ the players immediately agree upon the SPE proposal $x(\delta, \langle a^t \rangle) \in S$ given by*

$$x_1(\delta, \langle a^t \rangle) = (1+\delta)^{-1} + (1-\delta) \sum_{k=0}^{\infty} \delta^{2k} \left[-d_2(a^{2k}) + \delta d_1(a^{2k+1}) \right]$$

and

$$x_2(\delta, \langle a^t \rangle) = 1 - x_1(\delta, \langle a^t \rangle).$$

Having solved for the SPE utilities at $t = 0$ means that the choice of the equilibrium disagreement actions at $t = -1$ simply boils down to playing a normal-form game in which player i, $i = 1, 2$, chooses an infinite sequence of actions $\langle a_i^t \rangle_{t=0}^{\infty}$ with payoff function $x_i\left(\delta, \langle a^t \rangle\right)$. Moreover, this normal-form game is equivalent to a zero-sum game, which is similar as in Nash's variable-threat game.

This zero-sum game is easy to solve. First, the utility function $x_1(\delta, \langle a^t \rangle)$ is linear in the disagreement payoffs $d_2(a^{2k})$ and $d_1(a^{2k+1})$, $k \in \mathbb{N}$, and independent of the disagreement payoffs $d_1(a^{2k+1})$ and $d_2(a^{2k+2})$. Therefore, we can divide the 'large' zero-sum game into infinitely many smaller subproblems, each of them also being a zero-sum game. So, a^{2k}, $k \in \mathbb{N}$, is found by solving the smaller zero-sum game with player 1's utility function equal to $-d_2(a^{2k})$. The unique value of this smaller zero-sum game is simply the minmax value, as explained in section 7.2.1. Now, von Neumann's famous minmax theorem for two-player zero-sum games states that the minmax value is equal to the minmax value, e.g., von Neumann (1928) or Mas-Colell, Whinston, and Green (1995). Hence, player 2's value for this zero-sum game is equal to his minmax value m_2 as already defined in chapter 2. Formally, the pair of SPE disagreement actions $a^{2k} \in A$, $k \in \mathbb{N}$, is given by

$$a_1^{2k} = \arg \max_{a_1 \in A_1} \left[\min_{a_2 \in A_2} -d_2(a) \right] \quad \text{and} \quad a_2^{2k} = \arg \max_{a_2 \in A_2} \left[\min_{a_1 \in A_1} d_2(a) \right], \tag{7.3}$$

where the argument, i.e., arg, only refers to the maximization. For a good understanding: The action a_1^{2k} corresponds to player 1's minmax strategy that keeps player 2 at the utility level m_2. Recall from chapter 2 that the set of actions that minmax player 2 is denoted as A^2. Similar, the SPE disagreement actions $a^{2k+1} \in A$, $k \in \mathbb{N}$, solve the smaller zero-sum game in which player 1's utility function is equal to $d_1(a^{2k+1})$ and the corresponding minmax actions are

$$a_1^{2k+1} = \arg \max_{a_1 \in A_1} \left[\min_{a_2 \in A_2} d_1(a) \right] \quad \text{and} \quad a_2^{2k+1} = \arg \max_{a_2 \in A_2} \left[\min_{a_1 \in A_1} -d_1(a) \right] \tag{7.4}$$

This implies that the (mixed) actions in A^i, $i = 1, 2$, are the Nash equilibrium actions in a two-player zero-sum game with payoff functions $d_i(a)$, $a \in A$, for player i and $-d_i(a)$ for the other player. As in chapter 2, the set of actions that minmax player 1 is denoted as A^1. A final substitution of $d_2(a^{2k}) = m_2$ and $d_1(a^{2k+1}) = m_1$ into the expressions of lemma 7.5 yields the elegant expression

$$x(\delta, \langle \hat{a}^t \rangle) = \tfrac{1}{1+\delta}(1 + \delta m_1 - m_2, \delta + m_2 - \delta m_1).$$

These arguments imply the following proposition.

Proposition 7.6. *The sequence $\langle \hat{a}^t \rangle_{t=0}^{\infty}$, $\hat{a}^t \in A$, is a sequence of SPE threats at $t = -1$ if and only if*

$$\hat{a}^t \in \begin{cases} A^2, & \text{if } t \text{ is even,} \\ A^1, & \text{if } t \text{ is odd.} \end{cases}$$

Moreover, $x(\delta, \langle \hat{a}^t \rangle) = (1+\delta)^{-1}(1+\delta m_1 - m_2, \delta + m_2 - \delta m_1)$ is the unique SPE agreement for all SPE sequences $\langle \hat{a}^t \rangle_{t=0}^{\infty}$ and $x(\delta, \langle \hat{a}^t \rangle)$ converges to the axiomatic Nash bargaining solution $N(S, (m_1, m_2))$ as δ goes to 1.

From proposition 7.6 it is clear that the equilibrium disagreement actions alternate over time. Moreover, the disagreement actions are independent of the probability δ. This result is quite different from the result obtained in theorem 7.2 or, given the linear Pareto frontier, example 7.3, where the disagreement actions are equilibrium actions of the zero-sum game with player 1's utility given by $d_1(a) - d_2(a)$. A similar result can only be obtained in the modified model of this section by imposing the restriction $a^t = a$ for all $t \in \mathbb{N}$ to the sequence $\langle \hat{a}^t \rangle_{t=0}^{\infty}$. Doing so, would mean that

$$x(\delta, \langle \hat{a}^t \rangle) = \tfrac{1}{1+\delta}(1 + \delta d_1(a) - d_2(a), \delta + d_2(a) - \delta d_1(a))$$

and the SPE threats $a \in A$ solve a zero-sum game in which player 1's utility is given by $\delta d_1(a) - d_2(a)$. Thus, the associated equilibrium threats now depend upon the probability δ. Then, as δ goes to 1, the player 1's utility function converges to this player's utility function as derived in example 7.3. Without going into the details, it can be shown that the set of Nash equilibria is upper semi-continuous in $\delta \in [0, 1]$ and this suffices for the set of SPE disagreement actions corresponding to $\delta d_1(a) - d_2(a)$ to converge to some subset of the set of optimal threats of example 7.3. Thus, we have arrived at the following theorem.

Theorem 7.7. *Let the set of sequences $\langle \hat{a}^t \rangle_{t=0}^{\infty}$ be restricted to sequences such that $a^t = a$, $t \in \mathbb{N}$. Then, as δ goes to 1, the limit set of SPE disagreement actions of bargaining procedure 7.4 is contained in the set of SPE disagreement utilities in Nash's variable threat game.*

Note that the limit set of Nash equilibrium of theorem 7.7 is always contained in the set of equilibrium actions corresponding to Nash's variable-threat game and that it could be a strict subset in which case this limit set can be regarded as a refinement.[3] In the previous section the H-essential equilibrium was proposed on an ad hoc basis, whereas the analysis in this section can be regarded as a more plausible underpinning. First, the demand game represents the reduced normal-form of the alternating offers model and, second, players are not allowed to condition their disagreement actions upon the exact bargaining round in case of a breakdown occurs. So, we may say that in Nash (1953) it is implicitly assumed that the disagreement actions a^t, $a^t \in A$, are constant, i.e. $a^t = a$ for all $t \in \mathbb{N}$, during the underlying unspecified dynamic bargaining process. If the players are able to condition the disagreement actions upon the exact round of breakdown, then the actions drift away from the equilibrium threats in Nash's variable-threat game.

7.3. Credible threats without commitment

The assumption of commitment is restrictive in that it requires an adequate enforcement mechanism for the players to stick to their threats in case of breakdown. Without commitment each player is free to choose his disagreement actions in case of a breakdown and each player will only carry out a previously announced threat if it is still in the best interest of this party to do so. The SPE threats derived in the previous section often induce a player to choose a threat which may indeed hurt himself if he is forced to carry it out. Thus, we can not expect that the SPE threats under commitment survive in any environment where an adequate enforcement is proven infeasible and this calls for a new analysis, which is conducted in this section.

In the first section we remove the enforcement of commitment in Nash's variable-threat game. This means that the players cannot commit themselves upon the disagreement actions before the negotiations start. Instead, the players choose their disagreement actions only in case of a breakdown. In this section also briefly touch upon the question whether or not the parties would choose to commit if adequate enforcement would be at their disposal. Then we proceed by replacing the unspecified negotiation process thought to underlie Nash's demand game by the alternating offers process. The latter model also serves as a first step to the more complex bargaining models that are postponed to subsequent chapters. The analysis is conducted in several subsections. In the first equilibria in Markov strategies are investigated. Then the set of SPE utility pairs is characterized and necessary and sufficient conditions for uniqueness are derived. Finally, the limit set of SPE utility pairs is investigated as the risk of breakdown vanishes.

One remark is in place. Here, the disagreement actions are chosen in case of a breakdown and, therefore, the players' past behaviour in the negotiations can influence the disagreement actions but not vice versa. The absence of this feedback will be restored in latter chapters.

7.3.1 Nash's variable-threat game: no commitment

In case the enforcement of commitment is removed from Nash's variable-threat game the players choose their disagreement actions only in case of a breakdown. So, disagreement actions are no longer chosen prior to the negotiations but after the negotiations ended, i.e., a permanent breakdown occurred, meaning the demand game is played first and only in case of incompatible demands the disagreement actions are chosen.

From here on we consider the triple $\langle S, \{A_i\}_{i=1,2}, \{d_i\}_{i=1,2}\rangle$ without additional assumptions other than the by now standard assumptions made in chapter 2. Recall that the assumption of mutual interests means that the Pareto frontier in terms of utility pairs associated with disagreement outcomes in the set A lies to the south-west of the Pareto frontier of the set S. The last two terms of the triple specify the disagreement game in case of a breakdown. The assumptions made in chapter 2 ensure that the set of Nash equilibria of this disagreement game, denoted as $A^N \subseteq A$, is a non-empty and compact set.

The formal definition of Nash's variable-threat game without commitment runs now as follows.

Bargaining Procedure 7.8. *Nash's variable-threat game without commitment is the two-stage game with almost perfect information characterized by the set of players* $N = \{1,2\}$ *and player* i*'s demand* $x_i \in \mathbb{R}$, $i = 1,2$, *in the first stage. In* *case $x \in S$, i.e., compatible demands, the negotiations end with the agreement x. Otherwise, the game proceeds to the second stage where player i chooses his threat $a_i^T \in A_i$. Player i's utility function $d_i(a^T) + i\,(x)\left(x_i - d_i(a^T)\right)$, where the indicator function $i\,(x)$ is defined in bargaining procedure 4.29.*

As before, the two-stage game has almost perfect information and the SPE concept is the appropriate solution concept. Note that strategies specify actions at the second stage conditional upon the history $x \notin S$ meaning that strategies prescribe some function $a_i^T : \mathbb{R}^2 \to A_i$, $i = 1, 2$, that specifies player i's action at the second stage. Now, application of the SPE concept implies that the actions played at the final stage of the game are Nash equilibrium actions. Thus, every SPE strategy satisfies $a^T(x) \in A^N$ for all $x \notin S$. The simplest functions specify the same Nash equilibrium for every subgame, i.e. there exists a $a^N \in A^N$ such that $a^T(x) = a^N$ for all $x \notin S$. The latter is the only possibility in case the disagreement game admits a single Nash equilibrium. However, with multiple Nash equilibria in the disagreement game numerous discontinuous functions can also be constructed, but we will not pursue this line of research. So, for explanatory simplicity we assume SPE functions of the first form only.

At the first stage, the players have to state their demands while both anticipate the same Nash equilibrium $a^N \in A^N$ to be played in case of incompatible demands. It will be clear that for every $a^N \in A^N$ the multiplicity of equilibrium demands as discussed in section 4.3.1 arises. Thus, the SPE concept has too little bite in the two-stage game and, again, we resort to the H-essential equilibrium of the demand game to obtain some results. Doing so means that we simply characterize SPE strategies as follows: For some $a^N \in A^N$ the SPE demands in the first stage are $x^* = N\left(S, d(a^N)\right)$ and the disagreement actions $a^T(x) = a^N$ for all $x \notin S$. These strategies induce agreement upon the compatible demands $x^* = N\left(S, d(a^N)\right)$ at the first stage and permanent breakdown never occurs, i.e., the second stage is never reached along any SPE path.

Proposition 7.9. *Consider the class of SPE strategies in which for some $a \in A$ the threats $a^T(x) = a$, $x \notin S$, and the players play the H-essential equilibrium $x = N(S, d(a))$ at the first stage. Then these strategies are SPE strategies if and only if $a \in A^N$. Moreover, the set of SPE outcomes consists of the set of compatible demands (or agreements)*

$$\left\{x \in S \mid\ x = N(S, d(a^N)),\ a^N \in A^N\right\}. \tag{7.5}$$

From this results it follows that the ability (not) to commit has a huge impact on the type of results that can be obtained. First, at the heart of Nash's variable-threat game lies the assumption of *irrevocable* commitment to the ex-ante specified threats $a^T \in A$. However, note that in general the SPE disagreement actions of theorem 7.2 do not constitute a Nash equilibrium of the disagreement game, i.e. $a^T \notin A^N$, as is clearly demonstrated by example 7.24. This simple observation alone requires the existence of some strong commitment device, because, by definition, in a non-cooperative environment players cannot be assumed to play one-shot non-credible actions. As Nash (1953), (p. 130), puts it:

> ... we must assume there is an adequate mechanism for forcing the players to stick to heir threats and demands once made; and one to enforce the bargain, once agreed. Thus, we need a sort of umpire, who will enforce contracts or commitments.

Proposition 7.9 states a one-to-one relationship between the Nash equilibrium actions in A^N and the set of SPE outcomes, provided the class of SPE strategies is restricted. Note that we no longer can guarantee uniqueness of the SPE agreement as in Nash's variable-threat game unless all Nash equilibria of the disagreement game result in the same agreement, i.e., there exists a $s^* \in S$ such that $x^*(a^N) = s^*$ for all $a^N \in A^N$. The latter condition can be reinterpreted in terms of the second geometrical property of the axiomatic Nash bargaining solution, as discussed in section 4.2.2. Consider one particular $a^N \in A^N$ and suppose the function f_1 (or f_2) that describes the Pareto frontier of S, i.e., $(f_1(s_2), s_2)$ lies on the Pareto frontier, is differentiable. Then uniqueness requires that for all $\hat{a}^N \in A^N$ the line through $d(\hat{a}^N)$ and $N(S, d(\hat{a}^N))$ coincides with the line through for $d(a^N)$ and $N(S, d(a^N))$. So, in order to have a unique SPE agreement all disagreement points corresponding to Nash equilibria of the disagreement game should lie on one line and all the associated SPE demands x^* should also lie on this line. This condition trivially holds for the class of disagreement games that admit a single Nash equilibrium. However, outside this class this condition is nongeneric. This implies the following proposition.[4]

Proposition 7.10. *Let f_1 be differentiable. Then Nash's variable-threat game without commitment admits a unique SPE agreement if A^N consists of a single element and it generically admits multiple SPE agreements otherwise.*

This result implies that the uniqueness of the SPE agreement obtained for the case with commitment is only partially preserved. Since disagreement points are linked in both models to the compatible SPE demands in the same manner and SPE threats differ, this implies that the SPE agreements are different as well.

To conclude, suppose the possibility of some costless but strong commitment device is available to each of the players and that each player has a choice whether or not to commit upon his threat prior to the negotiations, where a player who does not wish to commit remains flexible in choosing his disagreement point in case of a breakdown. What should each player do? Commit or remain flexible? In Mao (1993) this important question is addressed. The answer is that each player chooses to commit himself prior to the negotiations.

The underlying intuition is rather simple. Suppose that both players do not commit. Then a deviation by let us say player 1 to commit himself prior to the negotiations means that this player sets his threat before his opponent does and that his opponent has to react to player 1's choice, which cannot be undone due to the commitment. Then player 1 can be regarded as the leader in the famous Stackelberg equilibrium from industrial organization, e.g., Fudenberg and Tirole (1991), Mas-Colell, Whinston, and Green (1995) and Tirole (1988), and it is known that the leader can always obtain the Nash equilibrium profit of the symmetric Cournot or Bertrand model, whichever is considered. Manoeuvering oneself in the position of the 'leader' is not only profitable in the Stackelberg model, but also in the negotiations with endogenous threats for the same underlying reason. So, both players refraining from committing themselves cannot be an SPE outcome. In Mao (1993) it is further shown that the outcome in which one player refrains from committing himself cannot be an SPE outcome either, because this player can do better by choosing to commit himself upon a suitable threat. Hence, if costless commitment devices are available, *both* players will commit themselves.

7.3.2 Variable threats with alternating offers: no commitment

For similar reasons as in section 7.2.2 we replace the demand game by the alternating offers procedure and investigate SPE utility pairs. In this section the bargaining procedure is specified and SPE in Markov strategies are derived. In the subsequent sections we derive upper and lower bounds upon the players' SPE utilities and characterize the set of SPE utility pairs.

Consider the class of triples $\langle S, \{A_i\}_{i=1,2}, \{d_i\}_{i=1,2}\rangle$ that satisfy the assumptions made in chapter 2. Bargaining procedure 7.8 is modified to allow for alternating offers and becomes the following bargaining procedure.

Bargaining Procedure 7.11.
Round t (t even, $t \geq 0$):
Player 1 proposes an offer $s^t \in S$ and, then, player 2 either accepts ('yes') or rejects ('no') the offer s^t. If player 2 accepts, then the negotiations end and the players implement the offer s^t yielding a final utility s_i^t, $i = 1, 2$, in round t. However, if player 2 rejects, then with probability $\delta \in [0, 1)$ the bargaining process continues to the next round $t + 1$, and with probability $1 - \delta$ the bargaining process breaks down forever and players play the disagreement game once that determines the final disagreement outcome $d(a^t)$, $a^t \in A$.

Round t (t odd, $t \geq 1$):
The procedure is similar except that now player 2 has the initiative to propose an offer $s^t \in S$ and, afterwards, player 1 responds.

State		Markov	
Player 1	proposes	$x^*(\delta, d(a^N))$	always
	responds *yes*		if $s_1^t \geq y_1^*(\delta, d(a^N))$
	action	a_1^N	if breakdown at $t \in \mathbb{N}$
Player 2	proposes	$y^*(\delta, d(a^N))$	always
	responds *yes*		if $s_2^t \geq x_2^*(\delta, d(a^N))$
	action	a_2^N	if breakdown at $t \in \mathbb{N}$
Transitions		absorbing	

Table 7.1. The MPE strategies of proposition 7.13, where $(x^*(\delta, d(a^N)), y^*(\delta, d(a^N)))$ is the unique fixed point of the function $p \times q$ for $d = d(a^N)$, $a^N \in A^N$, where $p \times q$ is defined in (3.1) of section 3.3.1.

As before, this bargaining procedure is a multi-stage game with almost perfect information and we therefore apply the SPE concept. As in section 7.3.1 the strategies specify disagreement actions in round t, $t \in \mathbb{N}$, in case of a breakdown at this round conditional upon the history $h_2(t)$ meaning that strategies prescribes some function $a_i^t : H_2(t) \to A_i$, $i = 1, 2$, that specify player i's action at the second stage. Now, application of the SPE concept implies that the actions played at the final stage of the game are Nash equilibrium actions. Thus, every SPE strategy satisfies $a^t(h_2(t)) \in A^N$ for all $h_2(t) \in H_2(t)$ including a permanent breakdown in round t. The following lemma summarizes the main results in this (sub)section.

Lemma 7.12. *Consider a pair of strategies σ such that $a^t(h_2(t))$ are the disagreement actions at round t, $t \in \mathbb{N}$, for histories $h_2(t) \in H_2(t)$ that include a breakdown in round t. If the strategies σ are SPE strategies, then $a^t(h_2(t)) \in A^N$ for all $h_2(t) \in H_2(t)$ and all t.*

Recall from section 2.3.8 that Markov strategies are, loosely stated, restricted to the 'payoff-relevant' history of the play. For the bargaining model under consideration, Markov strategies ignore all past offers that have been rejected and do not condition the disagreement actions on the bargaining round in which breakdown occurs. Formally, this means that Markov strategies impose (among others) that there exists an $\hat{a} \in A$ such that $a^t(h_2(t)) = \hat{a}$ for all rounds t, $t \in \mathbb{N}$, and histories $h_2(t) \in H_2(t)$ that include a breakdown in round t. Thus, Markov strategies induce the simplest function possible upon the disagreement actions $a^t : H_2(t) \to A$. Combined with lemma 7.12 this immediately implies that in every MPE strategy $a^t(h_2(t)) = a^N$ for some $a^N \in A^N$.

The full derivation of MPE strategies then becomes almost trivial. Since the disagreement actions do not depend upon time nor on the history of proposals during the negotiations we can replace the infinite sequence of endogenous disagreement points by a time-independent disagreement point and analyze the alternating offers with this disagreement point. Doing so means that we analyze the standard alternating offers of chapter 3, where d is equal to $d(a^N)$ for some $a^N \in A^N$. The results obtained in the latter chapter immediately apply leading to the following proposition.

Proposition 7.13. *For $a^N \in A^N$ let $x^*\left(\delta, d(a^N)\right)$ and $y^*\left(\delta, d(a^N)\right)$ form the unique fixed point of the function $p \times q$ for $d = d(a^N)$, where $p \times q$ is defined in (3.1) of section 3.3.1. Then the Markov strategies of table 7.1 are SPE strategies in bargaining procedure 7.11. Moreover, in the limit, as δ goes to 1, both $x^*\left(\delta, d(a^N)\right)$ and $y^*\left(\delta, d(a^N)\right)$ converge to the axiomatic Nash bargaining solution $N\left(S, d(a^N)\right)$.*

Note that each Nash equilibrium in A^N corresponds to one particular pair of MPE strategies in the bargaining model. Since the set A^N is non-empty it immediately follows that the set of Markov perfect equilibria is non-empty as well. Hence, this implies existence of at least one such equilibrium.

7.3.3 Bounds for SPE utilities

Upper and lower bounds upon SPE utilities are derived by modifying the method of Shaked and Sutton, as extensively discussed in section 3.4.1. Lower and upper bounds upon SPE utilities require knowledge of these bounds in the 'final' subgame associated with permanent breakdown. Since SPE disagreement actions correspond to some pair of actions in A^N contingent upon the exact history of the negotiations each player's lowest and highest utilities in the final subgame coincide with this player's extreme Nash equilibria in A^N. Formally, for each player i, $i = 1, 2$, the minimum and maximum Nash equilibrium utilities in A^N are defined as

$$n_i = \min\{d_i(a^N) | \; a^N \in A^N\} \quad \text{and} \quad N_i = \max\{d_i(a^N) | \; a^N \in A^N\}$$

(the symbol N_i is used twice, namely for N_i and $N_i(S, d)$, and it will be clear from the context which interpretation N_i should have) and the corresponding Nash equilibrium actions which attain these payoffs are defined by

$$\underline{a}^i = \arg \min_{a^N \in A^N} d_i(a^N) \quad \text{and} \quad \bar{a}^i = \arg \max_{a^N \in A^N} d_i(a^N).$$

By definition $d_2(\bar{a}^2) = N_2$ and $n_1 \leq d_1(\bar{a}^2)$. But then it holds that $(m_1, m_2) \leq (n_1, N_2) \leq d(\bar{a}^2)$ and $d(\bar{a}^2) \in S^{IR}$ imply that $(n_1, N_2) \in S^{IR}$. Similar arguments apply to $(N_1, n_2) \in S^{IR}$. We arrived at the following crucial result in the analysis.

Lemma 7.14. *The utility pairs (n_1, N_2) and (N_1, n_2) belong to S^{IR}.*

State			1
Player 1	proposes	$x^*(\delta,(n_1,N_2))$	always
	responds *yes*		if $s_1^t \geq y_1^*(\delta,(n_1,N_2))$
	action	$\overline{a}_1^2$	if breakdown at t even
	action	$\underline{a}_1^1$	if breakdown at t odd
Player 2	proposes	$y^*(\delta,d(n_1,N_2))$	always
	responds *yes*		if $s_2^t \geq x_2^*(\delta,(n_1,N_2))$
	action	$\overline{a}_2^2$	if breakdown at t even
	action	$\underline{a}_2^1$	if breakdown at t odd
Transitions		absorbing	

Table 7.2. Player 1's worst pair of SPE strategies of proposition 7.15, where $x^*(\delta,(n_1,N_2))$ and $y^*(\delta,(n_1,N_2))$ form the unique fixed point of the function $p \times q$ for $d = (n_1,N_2)$, where $p \times q$ is defined in (3.1) of section 3.3.1. Note that reversing the roles of the players yields player 2's worst SPE strategy.

Before deriving the lower and upper bounds on the players' SPE payoffs, we postulate a pair of strategies with alternating disagreement actions and show that these strategies are SPE strategies. The rationale is simple. Consider a linear Pareto frontier and player 1's utility of lemma 7.5 once more, i.e.,

$$(1+\delta)^{-1} + (1-\delta)\sum_{k=0}^{\infty} \delta^{2k}\left[-d_2(a^{2k}) + \delta d_1(a^{2k+1})\right].$$

Observe that player 1's utility depends positively on his disagreement payoff $d_1(a^{2k+1})$ if breakdown occurs in a round in which player 1 responds, and negatively on player 2's disagreement payoff $d_2(a^{2k})$ if breakdown occurs in a round in which player 2 is the responding player. Therefore, in order to incur the greatest loss of delay on player 1's utility given that SPE disagreement actions always belong to A^N, player 1's disagreement utility $d_1(a^{2k+1})$, $k \in \mathbb{N}$, are minimized if the actions $\underline{a}^1$ are played in case of a breakdown in every even period. The opposite case applies to player 2's disagreement utilities $d_2\left(a^{2k}\right)$, $k \in \mathbb{N}$, that are maximized by the actions $\overline{a}^2$ in case of a breakdown in every odd period. The latter actions minimize the loss of delay on player 2's payoff and, thereby, strengthen his bargaining position. So, player 1's utility becomes $(1+\delta)^{-1}(1+\delta n_1 - N_2)$.

This idea also works in general, especially because lemma 7.14 ensures that the constructed disagreement point $d = (n_1, N_2)$ lies below the Pareto frontier of S^{IR}. This rules out the multiplicity encountered in section 6.3.2. The informally described strategies are formally presented by table 7.2 for general bargaining problems, where $x^*(\delta,d)$ and $y^*(\delta,d)$ refer to the unique fixed point of the function $p \times q$ for $d = (n_1, N_2)$. The next proposition states that the strategies of table 7.2 are SPE. As will be shown in theorem 7.17 below these SPE strategies constitute player 1's worst pair of SPE strategies.

Proposition 7.15. *Let $x^*(\delta, (n_1, N_2))$ and $y^*(\delta, (n_1, N_2))$ form the unique fixed point of the function $p \times q$ for $d = (n_1, N_2)$, where $p \times q$ is defined in (3.1). Then the strategies of table 7.2 are SPE strategies for all $\delta \in [0, 1)$. Moreover,*

$$\lim_{\delta\to 1} x^*(\delta, (n_1, N_2)) = \lim_{\delta\to 1} y^{**}(\delta, (n_1, N_2)) = N(S, (n_1, N_2)).$$

By reversing the roles of both players in the previous proposition we obtain player 2's worst pair of SPE strategies.

Proposition 7.16. *Let $x^*(\delta, (N_1, n_2))$ and $y^*(\delta, (N_1, n_2))$ form the unique fixed point of the function $p \times q$ for $d = (N_1, n_2)$. Then the strategies that are obtained by reversing the roles of the players in table 7.2 are SPE strategies for all $\delta \in [0, 1)$. Moreover,*

$$\lim_{\delta\to 1} x^*(\delta, (N_1, n_2)) = \lim_{\delta\to 1} y^*(\delta, (N_1, n_2)) = N(S, (N_1, n_2)).$$

Note that the strategies of table 7.2 are non-Markov, since the disagreement actions played in case of breakdown do depend on the specific bargaining round in which breakdown occurs.

Obviously the main question now concerns whether the strategies of proposition 7.15 correspond to player 1's worst SPE utility. The following theorem states an affirmative answer. Its proof is based upon the method of Shaked and Sutton, as introduced in section 3.4.1.

Theorem 7.17. *For all $\delta \in [0, 1)$ it holds that the pair of SPE strategies in proposition 7.15 constitute player 1's pair of worst SPE strategies. Similar, the pair of SPE strategies in corollary 7.16 constitute player 1's pair of worst SPE strategies for all $\delta \in [0, 1)$.*

From this theorem a similar conclusion as at the end of section 7.2.2 can be drawn with respect to the results for bargaining procedure 7.8. Namely, theorem 7.17 reveals the implicit assumption that the strategies in the bargaining process underlying the modified variable-threat game in bargaining procedure 7.8 need to be Markov strategies. This implicit restriction to Markov strategies implies that disagreement actions are constant over time and have to be Nash equilibrium actions, that is, $a^t = a^N$, for all t and $a^N \in A^N$. If the unspecified bargaining procedure is modelled as the alternating offers procedure, then richer SPE strategies emerge and a possibly larger set of SPE utility pairs arises.

7.3.4 The set of SPE utility pairs

The upper and lower bounds upon the players' SPE utilities derived in theorem 7.17 enables us to characterize the set of SPE utilities, which is based upon

State		*IA*	
	proposes	s^*	
Player 1	responds *yes*		if $s_1^t \geq s_1^*$
	action	$\underline{a}^2$	if breakdown at t even
	action	$\underline{a}^1$	if breakdown at t odd
	proposes	s^*	
Player 2	responds *yes*		if $s_2^t \geq s_2^*$
	action	$\underline{a}^2$	if breakdown at t even
	action	$\underline{a}^1$	if breakdown at t odd
Transitions		Go to the strategies of proposition 7.15 if either $s^t \neq s^*$, t even, or player 1 rejects $s_1^t = s^*$, t odd. Go to the strategies of proposition 7.16 if either $s^t \neq s^*$, t odd, or player 2 rejects $s^t = s^*$, t even.	

Table 7.3. A pair of SPE strategies that induce immediate agreement upon $s^* \in E_1(\delta)$, where $E_1(\delta)$ is characterized in proposition 7.18. If one of the players deviates equilibrium switching to either the strategies of proposition 7.15 or 7.16 takes place. The latter states are absorbing.

the idea of equilibrium switching, as described in section 5.4.2. The discussion then continues by deriving sufficient and necessary conditions for generic uniqueness.

As said, the idea of equilibrium switching is employed to characterize the set of SPE utility pairs. This is accomplished by considering all SPE strategies that induce immediate agreement upon some possibly inefficient agreement. Equilibrium switching is already extensively discussed in section 5.4.2 and 5.4.3. The strategies represented by table 7.3 can be considered as a straightforward translation of table 5.1 to the situation under consideration. The strategies start in state 'IA', where 'IA' stands for immediate agreement. In this initial state player 1 makes the possible inefficient proposal $s^* \in S$ that is accepted by his opponent. In case player 1 deviates by making a different proposal the strategies prescribe an immediate switch to the SPE strategies in proposition 7.15. The latter strategies constitute player 1's most severe equilibrium punishment and, supposing acceptance by his opponent, this player will not deviate if and only if $s_1^* \geq x_1^*(\delta, (n_1, N_2))$. Similar, player 2 is deterred from rejecting the proposal s^* if and only if $s_2^* \geq x_2^*(\delta, (N_1, n_2))$, because rejecting s^* induces an immediate switch to player 2's most severe equilibrium punishment, i.e., the SPE strategies in proposition 7.15. Obviously, we have arrived at the following proposition, where $E_1(\delta) \subseteq S$ denotes the set of SPE utility pairs at the start of every even bargaining round.

Proposition 7.18. *The strategies represented by table 7.3 are SPE strategies featuring immediate agreement upon s^* at round $t(\sigma) = 0$ if and only if $s_1^* \geq x_1^*(\delta, (n_1, N_2))$ and $s_2^* \geq x_2^*(\delta, (N_1, n_2))$ for all $\delta \in [0, 1)$. Moreover, the* *history independent set of SPE utilities $E_1(\delta)$ at the start of every t even is given by*

$$E_1(\delta) = \{s \in S \mid s_1^* \geq x_1^*(\delta, (n_1, N_2)),\ \ s_2^* \geq x_2^*(\delta, (N_1, n_2))\ \}.$$

Note that this proposition is also valid in case the disagreement game admits a unique Nash equilibrium, i.e., A^N consists of a single element, because then combining $(n_1, N_2) = (N_1, n_2) = d\left(a^N\right)$, $a^N \in A^N$, and the definition of the function $p \times q$ simply imply that

$$\begin{aligned} x_2^*(\delta, (N_1, n_2)) &= f_2\left(x_1^*(\delta, (N_1, n_2))\right) = f_2\left(x_1^*(\delta, (n_1, N_2))\right) \\ &= x_2^*(\delta, (n_1, N_2)). \end{aligned}$$

Thus, the upper and lower bounds coincide and the set $E_1(\delta)$ reduces to a single point on the Pareto frontier.

Without going through all the details, we mention that it is also straightforward to characterize the history independent set $E_2(\delta)$ of SPE utility pairs at the subgame that coincides with the start of every odd numbered bargaining round. This set is given by

$$E_2(\delta) = \{s \in S \mid s_1^* \geq y_1^*(\delta, (n_1, N_2)),\ \ s_2^* \geq y_2^*(\delta, (N_1, n_2))\ \}.$$

Proposition 7.18 also makes it easy to characterize the limit set of SPE utility pairs. Then we obtain the following results, simply by taking the limit δ goes to 1.

Theorem 7.19. *Let $E_j(\delta)$, $j = 1, 2$, denote the history independent set of SPE utilities for $\delta \in (0, 1)$ at the beginning of the bargaining round where player j proposes. Then for $j = 1, 2$ it holds that*

$$\lim_{\delta \to 1} E_j(\delta) = \{s \in S \mid s_1 \geq N_1(S, (n_1, N_2)),\ s_2 \geq N_2(S, (N_1, n_2))\ \},$$

where the function $N_i(.)$, $i = 1, 2$, denotes player i's utility according to the Nash bargaining solution.

We conclude this section with the derivation of conditions for uniqueness. For $\delta < 1$ this task is rather complicated, because conditions have to be specified under which a multiplicity of fixed points of the function $p \times q$ can be regarded as coinciding with each other. Since the Nash program is already identified with the limit δ goes to 1, we investigate the conditions for uniqueness in this limit, which is a case that is easy to handle.

Consider the limit set of theorem 7.19 once more. Obviously, uniqueness (in the limit) in bargaining procedure 7.11 simply requires that the two Nash bargaining solutions $N(S, (n_1, N_2))$ and $N(S, (N_1, n_2))$ coincide, i.e.,

State		Initial state
Player 1	proposes responds *yes* action	$(s_1^{\max}, d_2)$ if $s_1^t \geq q_1(s_1^{\max}, d_2)$ a_1^N if breakdown at $t \in N$
Player 2	proposes responds *yes* action	$(d_1, s_2^{\max})$ if $s_2^t \geq p_2(d_1, s_2^{\max})$ a_2^N if breakdown at $t \in N$
Transitions		Go to the strategies in table 7.3 if $t \geq t(\sigma)$. Go to the strategies in proposition 7.15 if $s_1^t < s_1^{\max}$, t even. Go to the strategies in proposition 7.16 if $s_2^t < s_2^{\max}$, t odd.

Table 7.4. These strategies induce agreement upon s^* at round $t(\sigma)$, where the functions $q_1(.)$ and $p_2(.)$ refer to $p \times q$. Equilibrium conditions are stated in proposition 7.6.

$N(S, (n_1, N_2)) = N(S, (N_1, n_2))$. In order to derive a condition for uniqueness we make use of the second geometrical property of the Nash bargaining solution, as discussed in section 4.2.2. For explanatory reasons we suppose that the Pareto frontier is smooth in the unique point, i.e., the function f_1 that describes the this frontier is differentiable in $s_2 = n_2 = N_2$. Then uniqueness and the second geometrical property imply that the point (N_1, n_2) should lie on the line through (n_1, N_2) and $N(S, (n_1, N_2))$.[5] Of course, the latter condition trivially holds for the class of bargaining problems that admit a single Nash equilibrium in the disagreement game. However, outside this class of bargaining problems the condition for uniqueness is nongeneric, because slight changes in the players' utility functions are sufficient to will shift (N_1, n_2) off this line. This implies the following proposition.

Proposition 7.20. *Consider the class of bargaining problems with a smooth Pareto frontier of S in $N(S, d(a^N))$ for every $a^N \in A^N$. Then bargaining procedure 7.11 admits a unique SPE agreement if A^N consists of a single element and it generically admits multiple SPE agreements otherwise.*

So, this proposition states the same sufficient and almost necessary condition for uniqueness as in proposition 7.10, namely the set A^N should consists of a unique Nash equilibrium. The latter means that the disagreement game rules out the coordination problem in case of multiple Nash equilibria. If a coordination problem over Nash equilibria arises, then this carries over to the negotiations.

Finally, for the case where the Pareto frontier of S is not smooth in the point $N(S, d(a^N))$, $a^N \in A^N$, we refer to section 9.2.5. The results in the latter section can be easily translated into necessary and sufficient conditions for the nonsmooth case here.

7.3.5 SPE with delay

Similar as in section 5.4.4 SPE strategies with delay can be supported in case the necessary and sufficient conditions for uniqueness, as stated in proposition 7.20 do not hold. Recall that the maximum equilibrium delay in section 5.4.4 is finite. Here a new feature may arise: Perpetual disagreement cannot be ruled out if some simple condition holds.

Since equilibrium strategies require that a Nash equilibrium is played in case of a breakdown we might as well start by considering some $a^N \in A^N$ and take this a^N as the players' threats prescribed by the strategies. Recall that strategies also induce the intended round $t(\sigma) \in \mathbb{N}$ of agreement (delay means $t(\sigma) > 0$), the final agreement $s^{t(\sigma)}(\sigma) \in S^{IR}$ at $t(\sigma)$ and the endogenous risk of breakdown $1 - \delta^{t(\sigma)}$. Then the associated utilities are given by $\left(1 - \delta^{t(\sigma)}\right) d(a^N) + \delta^{t(\sigma)} s^{t(\sigma)}(\sigma)$. The question is what are the equilibrium conditions upon $t(\sigma) \in \mathbb{N}$ and $s^{t(\sigma)}(\sigma) \in S^{IR}$ such that $\left(1 - \delta^{t(\sigma)}\right) d + \delta^{t(\sigma)} s^{t(\sigma)}(\sigma) \in E_1(\delta)$. Similar as in section 5.4.4 we extend the strategies represented by table 7.3 by imposing a new initial state. In this initial state players must first make proposals until round $t(\sigma)$ that are rejected and we regard $(s_1^{\max}, d_2)$ and $(d_1, s_2^{\max})$ as natural candidates for these proposals, i.e., the proposing player demands his utopia utility in S^{IR} and is not willing to settle for anything less. In case player j, $j = 1, 2$, would make an offer that leaves less to himself than $s_j^{\max}$, then equilibrium switching to this player's worst SPE is immediately triggered. At round $t(\sigma)$ the strategies move to state IA (immediate agreement) in table 7.3, including the same state transitions. These modified strategies implement the agreement $s^{t(\sigma)}(\sigma) = s^*$ at round $t(\sigma)$ and the players disagree until round $t(\sigma) \in \mathbb{N}$. Table 7.4 represents the initial state of the strategies that induce delay. Deriving equilibrium conditions upon s^* and $t(\sigma)$ is similar as in the previous section and section 5.4.4 and therefore omitted.

Proposition 7.21. *Let $a^N \in A^N$, $t(\sigma)$ be even and $s^* \in E_1(\delta)$. The strategies of table 7.4 are SPE if and only if*

$$\left(1 - \delta^{t(\sigma)}\right) d\left(a^N\right) + \delta^{t(\sigma)} s^* \geq \left(x_1^*(\delta, (n_1, N_2)),\ x_2^*(\delta, (N_1, n_2)) \right). \quad (7.6)$$

Note that in case of uniqueness we automatically have that $t(\sigma) = 0$ and s^* coincides with the unique SPE outcome.

From equilibrium condition (7.6) the maximum delay of any SPE can be derived and for the details we refer once more to section 5.4.4. However, there is an interesting difference. With a fixed disagreement point the disagreement point is always located south-west of the point where both players are at their minimum SPE utility and this guarantees that the maximum delay has to be finite. Here, $d\left(a^N\right)$ corresponds to a Nash equilibrium utility pair and multiplicity of SPE outcome is definitely associated with multiple Nash equilibria in the disagreement game. Now, the following interesting case cannot be ruled out: The disagreement game may admit a Nash equilibrium $a^N \in A^N$ such that $d\left(a^N\right) \in E_1(\delta)$. Then for every $s^* \in E_1(\delta)$ and $t(\sigma) > 0$ equilibrium condition (7.6) holds meaning that even $t(\sigma) = \infty$ can be supported by these SPE strategies. In this special case the negotiations break down with probability equal to 1 without the players reaching an agreement, i.e., perpetual disagreement.

Corollary 7.22. *If there exists a pair of actions $a^N \in A^N$ such that $d(a^N) \in E_1(\delta)$, then for this particular a^N and every $s^* \in E_1(\delta)$ equilibrium condition (7.6) imposes no maximum bound upon $t(\sigma)$. Moreover, for every $s^* \in E_1(\delta)$ there is one SPE strategy with perpetual disagreement.*

The results obtained can be related to Thomas Schelling's point of view mentioned in the introduction of this chapter. If commitment fails the players' threats have to be credible. In case of multiple Nash equilibria there is an implicit struggle for coordination on one of these equilibria to obtain favourable bargaining positions. If the Nash equilibrium utilities lie relatively scattered, this struggle results in an indeterminacy and the possibility that this indeterminacy is never resolved and may cause bargaining impasses.

7.3.6 A comparison between models

In this section the equilibrium outcomes of the two alternating offers procedures with and without commitment on threats are compared with each other. This is done by first investigating the conditions under which these two models produce the same equilibrium conditions. For a short analysis we assume that the Pareto frontier of the bargaining problem is smooth.

Recall from section 7.2.2 that the bargaining procedure with commitment yields the unique outcome $N(S,(m_1,m_2))$. So, the question becomes when does the alternating offers procedure without commitment yield the same SPE outcome. First, the latter outcome has to be unique and for a smooth Pareto frontier this means that, generically, the disagreement game may admit at most one Nash equilibrium. Second, the unique SPE outcome $N(S,d(a^N))$ has to coincide with $N(S,(m_1,m_2))$. Making use of the second geometrical property, this means that the unique Nash equilibrium utilities $d\left(a^N\right)$ have to lie on the line through (m_1,m_2) and $N(S,(m_1,m_2))$ and this condition is relatively easy to check. This yields the following proposition.

Proposition 7.23. *Let the Pareto frontier of S be smooth in the point $s = N(S,(m_1,m_2))$. Then, in the limit as δ goes to 1, the set of SPE utility pairs of procedure 7.4 coincides with the limit set of procedure 7.11 if and only if the disagreement game admits a unique Nash equilibrium and the associated pair of Nash equilibrium utilities lie on the line piece with endpoints (m_1,m_2) and $N(S,(m_1,m_2))$.*

The necessary and sufficient conditions specify a generic class of bargaining problems that meet these conditions. This is easy to see if one realizes that the set S need not be restricted at all for the class of (disagreement) games where the unique pair of Nash equilibrium utilities coincides with the players minmax levels. This is for example the case for the class of prisoners' dilemmas. However, there is also a large class of bargaining problems, probably larger, for which these conditions are not met at all.

7.4. Numerical examples

In this section two numerical examples are presented that illustrate the variety of results derived in this chapter. In the first example the players' threats correspond to the actions in the battle-of-the-sexes game. This example shows a multiplicity of SPE outcomes, including SPE outcomes featuring perpetual disagreement. The disagreement actions in the second example arise from the prisoners' dilemma. For this example all four bargaining procedures predict the same and unique agreement and the same threats

Example 7.24.
Consider the set $S = \{s \in \mathbb{R}^2_+ \,|\, s_1 + s_2 \leq 10\}$ *and the battle of the sexes given by*

	L_2	R_2
L_1	6,2	0,0
R_1	0,0	2,6

The game admits three Nash equilibria, namely two pure given by (L_1, L_2) *and* (R_1, R_2), *with corresponding utility pairs* $(6,2)$ *and* $(2,6)$, *and one in mixed strategies* $(a_1, a_2) = ((\frac{3}{4}, \frac{1}{4}), (\frac{1}{4}, \frac{3}{4}))$ *with expected utilities* $(\frac{3}{2}, \frac{3}{2})$. *Player* i*'s minmax value is given by* $m_i = \frac{3}{2}$, $i = 1, 2$, *and the corresponding minmax strategies coincide with the mixed Nash equilibrium.*
Application of theorem 7.2 and example 7.3 to Nash's variable threat game implies that the equilibrium threats are found by solving the zero-sum game given by

	L_2	R_2
L_1	4,-4	0,0
R_1	0,0	-4,4

and this game yields (L_1, R_2) *as the unique equilibrium threats. Note that these threats do not correspond to any of the three Nash equilibria of the battle of the sexes. The unique equilibrium agreement reached in the negotiations is given by* $N(S, (0,0)) = (5, 5)$. *Note that the actions* (L_1, R_2) *do not correspond to Nash equilibrium actions in the battle of the sexes and that commitment to these threats in case of a breakdown is necessary. Without commitment the players would have an incentive to deviate.*

In case the alternating offers procedure replaces the demand game, as in procedure 7.4, the equilibrium threats change to the minmax actions, e.g., proposition 7.6. Here, the actions that minmax player i, $i = 1, 2$, coincide with the mixed Nash equilibrium. Thus, the players threaten each other with the mixed Nash equilibrium in case of a breakdown in every round and final agreement is $N(S, (\frac{3}{2}, \frac{3}{2})) = (5, 5)$ *in the limit as* δ *goes to* 1.

Procedure 7.8 describes Nash's variable threat game in the absence of an enforcement mechanism, i.e., players lack the ability to commit themselves prior to the negotiations. In that case, the players play one of the Nash equilibria in case of a breakdown, which means that there are three cases to consider. First, consider the SPE threats (L_1, R_2). *Then the SPE agreement is given by, with some abuse of notation,* $N(S, (L_1, R_2)) = (7, 3)$. *Second, reversal of the players roles yields* $N(S, (R_1, L_2)) = (3, 7)$. *Third, the results for the mixed Nash equilibrium case coincide with the results for the previous procedure. Which agreement actually results depends upon the self-fulfilling prophecy of which threats are anticipated by both players. Since the first two procedures always predict a unique equilibrium agreement, these results hint at that commitment plays an important role as some kind of coordination device.*

Finally, consider procedure 7.11. Then the gap between a player's worst and best SPE agreement increases even further. Proposition 7.15 implies that in, for example, player 1's worst SPE the threats are conditioned upon even and odd rounds. Here it is the threat of the actions (R_1, L_2) *in even rounds and the mixed Nash equilibrium in odd rounds. These yield the constructed disagreement point* $(n_1, N_2) = (\frac{3}{2}, 6)$ *with associated agreement* $N(S, (n_1, N_2)) = (2\frac{3}{4}, 7\frac{1}{4})$. *By symmetry,* $(N_1, n_2) = (6, \frac{3}{2})$ *with associated agreement* $N(S, (N_1, n_2)) = (7\frac{1}{4}, 2\frac{3}{4})$. *Theorem 7.19 then yields that the limit set of SPE utility pairs is given by*

$$\left\{ s \in \mathbb{R}^2_+ \mid s_1, s_2 \geq 2\tfrac{3}{4},\ s_1 + s_2 \leq 10 \right\}.$$

This game is capable of generating perpetual disagreement, as described in corollary 7.22, provided we apply it to the correlated equilibrium in which (L_1, L_2) *and* (R_1, R_2) *are equally probable.*[6] *Then the associated expected utility pair* $(4, 4)$ *lies in the limit set of payoffs and corollary 7.22 states that perpetual disagreement resulting in a breakdown and threats played according to this correlated equilibrium is a SPE outcome.*

Example 7.25.

Consider $S = \{s \in \mathbb{R}^2_+ \,|\, s_1 + s_2 \leq 10\}$ *and the prisoners' dilemma given by*

	L_2	R_2
L_1	4,4	0,5
R_1	5,0	1,1

The unique Nash equilibrium is (R_1, R_2) and also constitutes the actions that minmax player i, $i = 1, 2$.
This bargaining problem satisfies the necessary and sufficient conditions of proposition 7.23. To see this, first note that by theorem 7.2 and example 7.3 the equilibrium threats are found by solving the zero-sum game given by

	L_2	R_2
L_1	0,0	-5,5
R_1	5,-5	0,0

and this game yields (R_1, R_2) as the unique equilibrium threats with utility pair $(1, 1)$ in the original prisoners' dilemma. The unique equilibrium agreement reached in Nash's variable threat game is given by $N(S, (1,1)) = (5, 5)$. Then it is straightforward to see that for the prisoners' dilemma both bargaining procedure 7.4 and 7.11 yield the same (and unique) SPE agreement $N(S, (1,1)) = (5, 5)$.

7.5. Related literature

Nash's variable threat game is first analyzed in Nash (1953). This chapter is based upon Bolt and Houba (1998), where it was first suggested to replace the demand game by the alternating offers procedure. An analysis of Nash's variable threat game and its modified version without commitment can also be found in Van Damme (1991). In Mao (1993) the question is addressed whether or not players commit themselves if they have the freedom to do so and, in this reference it is shown that both players do commit themselves. This references is also the only reference that applies the H-essential equilibrium concept to the entire two-stage negotiation procedure, which is more appropriate then applying it to only the stage where the players play the demand game.

In this chapter the demand game is interpreted as the normal-form representation of some unspecified bargaining procedure. In Muthoo (1992) and (1996) another interesting point of view is proposed. In these papers the demand game is interpreted as a dynamic bargaining procedure with irrevocable commitments to the players' demands. As an alternative procedure a bargaining model is presented where commitments are revocable and where the players incur costs of revoking their commitments. In this way, a unifying framework is created in which Nash's demand game and the alternating offers procedure both fit.

In this chapter the notion of (non)commitment is strictly reserved to whether or not the players can commit to disagreement actions. In Fershtman and Seidmann (1993), Muthoo (1992) and (1996), it is argued that the alternating offers procedure can be regarded as a bargaining procedure where the players can make no commitments to their respective offers and in these references the standard model is modified in order to allow for 'partially' revocable commitments. Here, in these references, the disagreement point in the bargaining process is assumed to be fixed and exogenously given. Therefore, they restrict the notion of commitment to proposals and not to disagreement actions. Thus, in these models commitment is looked upon from a different angle.

Notes

1 For an informal but entertaining treatment of the concept of commitment in situations of strategic interaction we refer to Dixit and Nalebuff (1991).

2 This is easy to see, because in case of a linear Pareto frontier, as in example 7.3, the equilibrium threats are the Nash equilibrium actions of a true zero-sum game and many of the early game-theoretic literature contains several examples that admit multiple Nash equilibria.

3 For example, consider the game Γ as the 2×2 bimatrix game with each player's payoff matrix the 'identity' matrix $\frac{1}{2}I$. According to example 7.3 the set of optimal threats is the complete set A, while theorem 7.7 predicts the unique equilibrium in symmetric mixed strategies.

4 The case for f_1 is not differentiable everywhere introduces a convex set of disagreement point that all yield the same Nash bargaining solution, as discussed in section 4.2.2. This case is left to the reader at this stage of the book. For the nondifferentiable case we also refer to section 9.2.5.

5 If $N(S,(n_1,N_2)) = N(S,(N_1,n_2))$ and the function f_1 is not differentiable in $s_2 = n_2 = N_2$ then the second geometrical property implies that the necessary and sufficient conditions for uniqueness are given by

$$(n_1,N_2),(N_1,n_2) \in D(s^*), \quad s^* = N(S,(N_1,n_2)).$$

where the set $D(s^*)$ is already defined in section 4.2.2. Rewriting the inequalities that characterize the set $D(s^*)$ yields less demanding conditions for uniqueness.

6 There is a notationally more demanding alternative approach: Condition the Nash equilibrium to be played upon even and odd rounds. For example, (L_1,L_2) at an even round and (R_1,R_2) at an odd rounds. Then for sufficiently large $\delta < 1$ the same result can be obtained.

Chapter 8

BARGAINING OVER WAGES

8.1. Introduction

Nowadays, unemployment and inflation are persistent real life economic phenomena in many industrialized countries and, as such, pose major research questions for economists. Since unemployment and inflation are affected by the system of wage setting, a good insight in the functioning of the labour market is required as well. Because in most industrialized countries workers have their wages set in collective bargaining, this process has received increased attention by economists over the last decade. Collective bargaining is concerned with the strategic process of decision-making between parties that represent employer and employee interests. Its central task is to reach a formal collective agreement on a set of rules which governs the terms and conditions of the employment relation. Negotiation is always the essence of collective wage bargaining, but conflict situations may pop up since the parties share opposing interests regarding wages. These conflicting interests may disrupt the bargaining process and possibly trigger industrial action, such as executing a strike by the union delaying the negotiations.

In this chapter we will focus on the wage bargaining process between a single union and a single firm. Here, we focus on what can be regarded as the asymmetric case where only the union is able to use threats during the bargaining process. The union's power in wage negotiations depends to a great extent on the ability to mobilize its members for industrial action. Industrial action often takes the form of a strike executed by the union, but may also induce a more 'peaceful' form of delay, during which the union's workers continue to work under the terms of the expired contract. In the literature, this type of delay is often called a *holdout*. Moreover, this type of delay is generally not without costs for the union or the firm, which will be motivated later on. These efficiency costs generally depend on the exact type of holdout chosen by the union. We will make a distinction between holdouts with and without 'work-to-rule'. Holdouts without work-to-rule are the least costly: the union just complies to the old wage contract and production continues as before. In holdouts with work-to-rule the union *deliberately* follows the work rules in an inflexible manner in order to reduce the firm's profits, but without actually violating the old contract. By distinguishing three types of industrial action, we will focus on the credibility of any of these threats during the negotiations and what impact they have on wages. An interesting question concerns whether inefficient delays such as lengthy strikes or work-to-rule can occur in equilibrium.

The aim of this chapter is threefold. First, within an extended version of the wage bargaining model with three possible types of industrial action, we investigate to which extent holdouts are substitutes for strikes. It is shown that the union's most effective action is the action that inflicts the highest costs upon the firm among the union's options that are *credible*. An option is credible for the union if its cost do not exceed its benefits. Second, we show that our wage bargaining model is able to capture the time-consuming wage negotiations with lengthy holdouts that are observed in the Netherlands. Since the union's actions may inflict costs upon both parties during these lengthy holdouts, it is not clear from the outset that such industrial action may be worthwhile for the union. Hence, one may question the strategic potential of industrial action. The third aim is to strategically explore the aspect of *backdating* of new wage contracts to the expiration date of the old wage contract, a practice commonly observed in the Netherlands. The length of the holdout period has an unambiguously negative but small effect upon the settlement wage when wage contracts are backdated. This finding confirms empirical evidence for the Netherlands. Furthermore, backdating does not affect the bargaining position of each party in terms of utilities, which is in line with common wisdom that backdating is a minor detail of wage negotiations.

This chapter is organized as follows. After introducing the model of wage negotiations and exploring the underlying assumptions, the starting point of the analysis is the characterization of a restricted wage bargaining game in which holdouts do not impose any costs upon the players. We proceed by showing that the assumption of a common discount factor is not minor and that it matters for deriving the set of perfect equilbria. Then, in a subsequent section, the theory is applied to the Dutch case of wage negotiations, which features lengthy delays and backdating of new wage contracts.

8.2. A model of wage negotiations

The situation in which a union and a firm bargain over the division of the firm's profits is modelled as an alternating offers bargaining procedure in which the union must decide on the type of industrial action in every period as long as agreement has not been reached. This is the asymmetric case in which only the union has threats at its disposal. The symmetric case where both players can threaten to disrupt the bargaining process is left for chapter 9.

In this section we first focus on some stylized facts of wage bargaining situations and explore some of the underlying assumptions and concepts. Then the wage bargaining model is introduced formally.

8.2.1 Wage bargaining: some facts and assumptions

Collective bargaining in the Netherlands has a large impact on wage formation. It covers the employment terms of 70-80% of the labour force in the private sector, while employer coverage is about 90% of all firms. Although strike incidence in the Netherlands is very low compared to other European countries, Dutch wage formation is a time-consuming process with lengthy negotiations between unions and employers' associations. As studied in Van de Wijngaert (1994), holdout periods, the time between expiration of an old contract and the signing of a new contract, take an average of 7-8 months in the Netherlands. In contrast, in the US the average holdout period only takes 2 months, see Cramton and Tracy (1992). During a holdout production continues and the terms of the old contract apply.

In another empirical study, Van Ours and Van de Wijngaert (1996) present an exploratory analysis of the relationship between holdouts and wage bargaining in the Netherlands. Their estimations show that holdouts have a significant *negative* effect of 0.1% per two months of holdout on the negotiated wage increase. A theoretical explanation for this finding is not offered, it is merely concluded that 'to some extent holdouts are substitutes for strikes'. This chapter makes a contribution by addressing the endogenous choice between various types of industrial action and the way this affects equilibrium behaviour and settlement wages.

An essential ingredient of any wage bargaining model is that the union may use different types of industrial action. We assume that there are three types of possible industrial action: *i)* strike, *ii)* holdout, and *iii)* work-to-rule. These three types of industrial action differ in terms of costs they inflict upon the players. The important question concerns which type of action is optimal in what circumstances and can be supported as part of equilibrium behaviour.[1]

In the economic literature a holdout is the period in between the expiration date of the old contract and the date a new contract is signed. During this period production continues under the terms of the old contract and meanwhile the parties negotiate. During holdouts the union may carry out strategic threats, such as 'work-to-rule' or 'go-slow'. Work-to-rule means that workers deliberately follow the work rules in an inflexible manner without breaking the expired contract in order to reduce profits. Crucial to work-to-rule is that there are no verifiable violations of the old contract and, therefore, workers are paid the full wage as specified by the old contract. However, it is also argued that the pay system may allow for some flexibility and could include for instance bonus payments which can be suspended under a holdout, see e.g., Holden (1997). In addition, costs of organizing work-to-rule may exist. Defined in this manner, the union bears some costs in adopting work-to-rule. Strike, on the other hand, disrupts production and implies a complete work stoppage.

In the wage bargaining model as specified below, the union has three options and these actions are ranked with respect to the costs the union has to bear. Strike has the highest cost, holdout with work-to-rule has 'intermediate' costs and holdout without work-to-rule has lowest costs, which will be normalized to zero. With three strategic options for the union competition among these options enters the analysis. There is no loss of generality, because the results for three options can be easily extended to allow for more options.

What about the costs the firm has to bear? From the discussion above the answer seems simple: Work-to-rule and strike reduce labour productivity and, therefore, reduce profitability. Indeed, the above mentioned empirical studies explore this possibility. Yet, another possibility is also discussed in Cramton and Tracy (1994a). They claim that due to continuous technological progress, production under the old contract might be inefficient and a new contract is needed in order to improve efficiency. Another explanation could be an efficiency wage argument: A wage increase boosts the workers' motivation and, therefore, a new contract increases labour productivity. So, even without a work-to-rule policy the firm may already suffer opportunity costs from not having reached a new contract during a holdout. These costs are captured in our extended model by assuming that holdout without work-to-rule is inefficient. Furthermore, if the union adopts a work-to-rule policy, then profitability is lower than in case the union would not work-to-rule. So, we explicitly distinguish two sources of inefficiency mentioned in Cramton and Tracy (1994a). As for the union the three strategic options are ranked with respect to the costs the firm has to bear. Holdout without work-to-rule inflicts the lowest costs, holdout with work-to-rule inflicts intermediate costs and strike inflicts the highest costs. In the empirical literature the associated efficiency loss due to holdouts is estimated to be around 2% for the Netherlands and 6% for the US, although no real distinction is made between holdouts and work-to-rule; see e.g., Van de Wijngaert (1994) and Cramton and Tracy (1992).

Summing up: In the simple wage bargaining model that will be characterized in section 8.3 holdouts are simply treated as production under the old contract that do not inflict any costs upon either party, i.e., holdout is efficient. This efficiency assumption is relaxed in 8.4 where we analyse the model with three types of industrial actions, i.e., strike, holdout with work-to-rule, and holdout without work-to-rule. All three types are assumed to impose costs on the players. So, this broader wage model captures several important aspects mentioned in the empirical literature. For convenience, throughout the remainder, we will dub holdouts without work-to-rule just as 'holdouts', and holdouts with work-to-rule as 'work-to-rule' .

8.2.2 The wage bargaining model

To structure our thoughts we briefly recall the notation chapter 2. Thus we focus on the triple $\langle S, \{A_i\}_{i=1,2}, \{d_i\}_{i=1,2}\rangle$ specifying the non-empty and convex set S of attainable utility pairs, player i's set A_i of 'threats' and player i's utility (or payoff) function d_i over the set of disagreement actions A. In particular, the firm's gross profits are normalized to $\Pi = 1$ in each period, and linear utility $u_i(x_i) = x_i$ is assumed for both players. Given this specification, it follows that

$$S = \left\{s \in \mathbb{R}^2_+ \mid s_1 + s_2 \leq 1\right\},$$

where s_1 denotes the union's payoff and s_2 denotes the firm's payoff.

The expired wage contract specifies the per-period expired wage w_0, $0 < w_0 < 1$. If the union decides to strike in case of disagreement, then the vector with per-period disagreement payoffs of strike is normalized to $(0, 0)$. Alternatively, the union may also choose to holdout or to work-to-rule. The vector with per-period payoffs under holdout is given by $(w_0, \alpha - w_0)$, with $w_0 < \alpha \leq 1$ an efficiency parameter. Similarly, the vector of per-period disagreement payoffs of work-to-rule are $((1-\gamma)\, w_0, \beta - w_0)$, with $0 \leq \gamma < 1$ the per-period costs of work-to-rule measured as a fraction of the expired wage and $\beta \leq \alpha$ the efficiency parameter of work-to-rule. We assume that production under either holdout or work-to-rule is profitable for the firm, i.e., $w_0 < \beta \leq \alpha$.

As already discussed in the previous subsection, holdout respectively work-to-rule induce some inefficiency which are captured by $1-\alpha$ and $1-\beta$. Note that the inefficiency of work-to-rule consists of two parts, namely the inefficiency $1-\alpha$ due to holdout and on top of that the inefficiency $\alpha-\beta$ due to deliberately work-to-rule. We will denote the union's disagreement action 'strike' by 's', 'holdout' by 'h', and 'work-to-rule' by 'wrt'. Note that the firm has only one disagreement action available, that is, to continue production, here denoted by 'c'. The disagreement game Γ specifying $d(a)$, $a \in A$, is now as follows.

	c
h	$w_0, \alpha - w_0$
wrt	$(1-\gamma)w_0, \beta - w_0$
s	$0, 0$

Note that the Nash equilibrium actions of Γ are given by $a^N = (h, c)$.[2]

Bargaining begins just after the expiration of the old contract at time $t = 0$ with the union making the initial proposal. As long as no agreement is reached the parties alternate in making wage offers with the union making offers in even rounds and the firm in odd rounds. In each round of disagreement, the union selects its threat, that is, decides to strike, or to adopt a work-to-rule policy or to holdout. If a proposed wage is accepted, then negotiations are over, and the new wage contract is assumed to hold thereafter. Thus, implicitly it is assumed that only a single new wage contract is negotiated. The bargaining procedure is thus as follows.

Bargaining Procedure 8.1.
Round t (t *even*, $t \geq 0$)*: The union proposes a wage* $w^t \in [0,1]$ *which the firm either accepts ('yes') or rejects ('no'). If the firm accepts, then the game ends and the (normalized) payoffs at round t of agreement are* w^t *for the union and* $1 - w^t$ *for the firm. However, if the firm rejects, then the disagreement game* Γ *is played, where the union selects its threat to hold out the negotiations, or to work-to-rule, or to execute a strike, inducing disagreement payoffs* $d((h,c)) = (w_0, \alpha - w_0)$, $d((wrt,c)) = ((1-\gamma)w_0, \beta - w_0)$, *respectively,* $d((s,c)) = (0,0)$. *Then bargaining continues to the next bargaining round.*

Round t (t *odd*, $t \geq 1$)*: The procedure is similar except that now the firm has the initiative to propose a wage* $w^t \in [0,1]$.

Recall that our wage bargaining model is a multi-stage game of (almost) complete information and, consequently, we will focus on subgame perfect equilibria (SPE).

The total payoffs of the firm and the union depend upon the disagreement payoffs before an agreement is reached (if reached at all) and the wage of the new agreement. Consider negotiations that are concluded at time T, $T \in \mathbb{N}$, with agreement upon w, $w \in [0,1]$, and the sequence of vectors $\{x^t\}_{t=0}^{T-1}$ that denote the payoff vector at round t, $x^t \in \{(0,0), (w_0, \alpha - w_0), ((1-\gamma)w_0, \beta - w_0)\}$ and $0 \leq t \leq T-1$. The corresponding vector of normalized discounted payoffs is given by

$$(1-\delta)\sum_{t=0}^{T-1} \delta^t \begin{bmatrix} x_1^t \\ x_2^t \end{bmatrix} + \delta^T \begin{bmatrix} w \\ 1-w \end{bmatrix}.$$

An innovative feature in our model is that the new wage contract is backdated. This means that the firm pays once an additional one-shot lump sum transfer to the workers on top of the newly agreed wage contract at the time the new agreement is reached. The size of this sum is equal to the foregone difference between the new and old wage contract times the number of rounds the contract is backdated. Formally, if w is the new wage contract agreed upon at time T and this contract is backdated for h_T, $0 \leq h_T \leq T$, periods, then the firm pays $w + h_T(w - w_0)$ at time T and w at time t, $t \geq T+1$. The union's utility of

State		Markov	
Union	proposes	$(w_0, 1 - w_0)$	always
	responds *yes*		if $w^t \geq w_0$
	action	h	always
Firm	proposes	$(w_0, 1 - w_0)$	always
	responds *yes*		if $w^t \leq w_0$
	action	c	always
Transitions		absorbing	

Table 8.1. The MPE strategies of proposition 8.2, where $((w_0, 1 - w_0), (w_0, 1 - w_0))$ is a fixed point of the function $p \times q$ for disagreement point $d((h, c)) = (w_0, 1 - w_0)$.

such an agreement at time T is given by

$$(1-\delta)\left[h_T(w-w_0)+\sum_{t=T}^{\infty}\delta^{t-T}w\right]=w_0+(1+(1-\delta)\,h_T)(w-w_0). \tag{8.1}$$

Similarly, the present value of the firm's profit at time T is given by

$$\begin{aligned}1-(1-\delta)\left[h_T(w-w_0)+\sum_{t=T}^{\infty}\delta^{t-T}w\right] &= 1-w_0- \\ &\quad (1+(1-\delta)\,h_T)(w-w_0).\end{aligned} \tag{8.2}$$

Backdating is not considered until section 8.4.3, where it is assumed that $h_T = T$. Different assumptions, for instance when backdating only applies to periods in which production takes place, would not qualitatively change our results.

8.3. Wage bargaining with efficient holdouts

Before characterizing the described wage bargainging game, we first turn to a simpler case. In this section a holdout is simply treated as normal production under the old wage contract that does not inflict any cost upon either party. That is, holdouts are efficient. Moreover, we assume that $\gamma = 0$ and $\alpha = \beta = 1$, indicating that effectively the union can only choose between a holdout ('h') and a strike ('s'). Again, the firm can only continue production ('c'). Hence, the simplified disagreement game Γ now looks as follows.

	c
h	$w_0, 1 - w_0$
s	$0, 0$

This specification brings us right back to the wage bargaining models as analysed in e.g., Fernandez and Glazer (1991) and Haller and Holden (1990). Note that the modified disagreement game has only one unique Nash equilibrium given by $a^N = (h, c)$. In the next subsection we briefly characterize this simplified wage bargaining game, proceeding step by step.

8.3.1 Markov perfect equilibrium

In this section the set of MPE strategies is characterized. By applying the concept of dynamic programming as introduced in chapter 3, we know that whatever actions the players choose in Γ, these will not affect their future utilities. Because of the history-independence property of MPE there is no threat of equilibrium switching in the next bargaining round and, therefore, deviations will simply not be punished. So, if $a \in A$ is a pair of MPE actions at the disagreement phase, then no player must have an incentive to unilateral deviate from his own action. But then these MPE disagreement actions must be NE actions of the disagreement game Γ, i.e. $a^N = (h, c)$, because otherwise some profitable (one-stage) deviation exists for one of the players when Γ is played.

Knowing that MPE strategies prescribe $a(t) = a^N = (h, c)$, for all $t \in \mathbb{N}$, this means that we can treat the endogenous disagreement actions as being fixed at (h, c). Then the results of section 3.3 immediately apply. Given the linear specification of S and $d(a^N) = (w_0, 1 - w_0)$ the fixed point of the function $p \times q$ yields $x^*(\delta, (h, c)) = (w_0, 1 - w_0)$ and $y^*(\delta, (h, c)) = (w_0, 1 - w_0)$. Note that the old contract already lies on the Pareto frontier, so that the effective surplus is zero. Therefore, the new contract wage is identical to the old one. In other words, in the MPE the old contract stays put. The following result is given without further proof.

Proposition 8.2. *Let* $((w_0, 1 - w_0), (w_0, 1 - w_0))$ *form be the unique fixed point of the function* $p \times q$ *for* $d(a^N) = (w_0, 1 - w_0)$, *where* $p \times q$ *is defined in (3.1). Then the strategies of table 8.1 are MPE strategies for all* $\delta \in [0, 1)$.

From this proposition it can be deduced that the derived MPE is unique, essentially because the Nash equilibrium of the disagreement game Γ is unique. This one-to-one relationship was also found when deriving the MPE of the variable-threat game with alternating offers in the previous chapter. Moreover, it is also obvious that the set of SPE payoffs $E(h(t), t)$ is non-empty for any history $h(t) \in H(t)$.

8.3.2 The minimum-wage and maximum-wage contract

In this subsection the minimum-wage and maximum-wage contracts as a function of the discount factor are derived. Again, the idea of equilibrium switching developed in chapter 5 is used in deriving the parties' worst equilibria.

State		*IA*	
Union	proposes	$(w^{max}(t), 1 - w^{max}(t))$	
	responds *yes*		if $w^t \geq w^{max}(t)$
	action	s	if rejection at t even
	action	h	if rejection at t odd
Firm	proposes	$(w^{max}(t), 1 - w^{max}(t))$	
	responds *yes*		if $w^t \leq w^{max}(t)$
	action	c	if rejection at t even
	action	c	if rejection at t odd
Transitions		Go to the strategies of proposition 8.2 if either $w^t \neq w^{max}(t)$, t even, or $a_1(t) \neq s$, t even, or union rejects $w^t = w^{max}(t)$, t odd.	

Table 8.2. A pair of SPE strategies that induce the maximum-wage contract. If union deviates equilibrium switching to the minimum-wage contract takes place. The latter state is absorbing.

This is to ensure that parties who might be tempted to deviate, stick to prescribed strategy profile.

Indeed, the minimum-wage contract is now easy to understand. In the minimum-wage equilibrium the union chooses the least costly option as long as no agreement is reached. In fact, the union can always secure the old wage w_0 by holding out the negotiations for indefinite time. Since playing a holdout is also the only Nash equilibrium of the disagreement game, the MPE described in the previous subsection actually supports the minimum wage contract, where agreement is immediately reached in every round on $(w^{min}(t), 1 - w^{min}(t)) = (w_0, 1 - w_0)$, for every $t \in \mathbb{N}$, in particular, for $t = 0$. Table 8.1 shows the strategies.

The maximum-wage contract is somewhat harder to derive. First, as we already learned from chapter 3, in exogenous alternating offers games only the disagreement payoffs of the responding player matter for the equilibrium outcome. Second, whenever strike is credible, the union alternates between holdout and strike in case of disagreement such that the costs it inflicts upon the firm are as large as possible. Hence, to derive the maximum-wage contract, the union should strike in even periods in which the firm responds to its offer, and the union should holdout whenever itself rejects an offer proposed by the firm. In this way the union manipulates the firm in its worst possible position at the bargaining table and by doing so the union is able to extract the highest wage. Recalling chapter 6 on alternating disagreement payoffs, these so-called 'alternating-strike', or 'stutter strike', strategies induce an effective disagreement point $d = (w_0, 0)$. Solving the fixed point problem of section 3.3.1. this yields

$$x^*(\delta, (w_0, 0)) = (w_0 + (1 - w_0)/(1 + \delta), \delta(1 - w_0)/(1 + \delta))$$

and

$$y^*(\delta, (w_0, 0)) = (w_0 + \delta(1 - w_0)/(1 + \delta), (1 - w_0)/(1 + \delta)).$$

Hence, the maximum wage is now specified by

$$w^{max}(t) = \begin{cases} w_0 + \frac{1-w_0}{1+\delta}, & \text{if } t \text{ even,} \\ w_0 + \frac{\delta(1-w_0)}{1+\delta}, & \text{if } t \text{ odd.} \end{cases}$$

Agreement will immediately be reached at $t = 0$.

But how to sustain this alternating-strike strategy profile if playing 'strike' is not an equilibrium action of the disagreement game? Strike does not only hurt the firm, it hurts the union just as well. Equilibrium switching takes care of the tempation to deviate. Potentially, the union has an incentive to deviate in even periods so as to increase its one-shot disagreement payoff from 0 to w_0. However, if the union is punished for its deviation by playing the minimum-wage contract equilibrium then it is obvious that for large enough discount factor the deviation will not be profitable. More precise, for t is even, it holds that

$$(1 - \delta)w_0 + \delta w^{min}(t + 1) \leq (1 - \delta)0 + \delta w^{max}(t + 1),$$

or, equivalently,

$$(1 - \delta^2)w_0 \leq \delta^2(1 - w_0),$$

if and only if $\delta^2 \geq w_0$.

Notwithstanding the fact that agreement is reached immediately in every round, the maximum-wage contract critically depends on the alternating strike profile although this pattern will never be observed in equilibrium. Still, the threat of executing a strike in even periods is credible for large enough discount factors. The strategies of table 8.2 support the maximum wage strategies for large enough discount factors; 'IA' means immediate agreement.

Without proof we state the following proposition. The case for t is odd is straightforward and therefore omitted.

Proposition 8.3. *Let t be even. For every $\delta \in [0, 1)$, the minimum wage $w^{min}(t)$ at round t as a function of δ is given by*

$$w^{min}(t) = w_0. \tag{8.3}$$

For $\delta^2 \geq w_0$, the maximium wage $w^{\max}(t)$ at round t as a function of δ is given by

$$w^{max}(t) = w_0 + \frac{1 - w_0}{1 + \delta}. \tag{8.4}$$

The SPE strategies which induce the minimum-wage and the maximum-wage contract are given in table 8.1, respectively, table 8.2. Moreover,

$$\lim_{\delta \to 1} (w^{min}(t), 1 - w^{min}(t)) = N(S, (w_0, 1 - w_0)),$$

and

$$\lim_{\delta \to 1} (w^{max}(t), 1 - w^{max}(t)) = N(S, (w_0, 0)).$$

8.3.3 Equilibria with lengthy strikes

Having constructed the minimum-wage and maximum-wage contracts, it is now easy to derive SPE which feature lengthy delay before agreement is reached during which the union is on strike. Consider the following payoff-relevant outcome path that consists of a round $T > 0$ in which agreement is reached upon wage contract $(w, 1-w)$ and a sequence of disagreement actions before agreement is struck, here $a(t) = (s, c)$, $0 \leq t < T$. For convenience, we assume T even. Let us denote

$$p_s(T) = \langle \{(s, c)\}_{t=0}^{t=T-1}, w \rangle.$$

Hence, equilibrium utilities take the form $(1 - \delta^T)d((s, c)) + \delta^T(w, 1 - w) = \delta^T(w, 1 - w)$. The idea is again to introduce suitable strategies that feature equilibrium switching and then derive equilibrium conditions for these strategies.

Loosely stated, the strategies prescribe unacceptable wage offers for both parties for the first T rounds during which the union strikes, and then, at round T, agreement on wage contract $(w, 1-w)$. If a player deviates from this strategy profile, it will be punished by playing its worst equilibrium. In principle, given these strategies, two opposing forces are at play. First, during a strike the union must be willing to forego additional income, available from immediate agreement, by expecting a sufficiently high settlement wage after the strike. This determines a lower bound on the settlement wage w. Second, the firm must not have an incentive to make an wage offer that the union cannot reject, i.e, by proposing the maximum-wage contract. This yields un upper bound on the settlement wage: profits afterwards must be high enough to make up for losses during the strike. In order to support an equilibrium, the settlement wage must at least balance these two opposing effects.

Given the specified outcome path $p_s(T)$, the associated continuation payoffs for the players at the start of round $t, t \leq T$, is denoted by $s(T - t, w, \delta)$ and given by

$$s\left(T - t, w, \delta\right) = \delta^{T-t} \begin{bmatrix} w \\ 1 - w \end{bmatrix}. \tag{8.5}$$

Note that both players' continuation payoffs strictly increases in t. Thus, if a party considers to deviate in offering a wage, she will do this as soon as possible, that is, the union at $t = 0$ and the firm at $t = 1$. Noticing that any player will always immediately accept an offer which gives him his highest equilibrium payoff, we must have in equilibrium

$$w^{min}(0) \leq s_1(T, w, \delta) \quad \text{and} \quad 1 - w^{max}(1) \leq s_2(T-1, w, \delta).$$

This yields

$$\delta^{-T} w_0 \leq w \leq 1 - \frac{\delta^{1-T}(1-w_0)}{1+\delta},$$

which implies that T cannot get too large, otherwise a feasible settlement wage w cannot be found. The maximum length T^* is found by solving

$$\delta^{-T} w_0 = 1 - \frac{\delta^{1-T}(1-w_0)}{1+\delta},$$

which gives

$$T^* = \frac{\log(w_0 + \delta) - \log(1+\delta)}{\log(\delta)}.$$

Clearly, in the limit, as the discount factor approaches unity, the maximum length of delay T^* approaches infinity. Obviously, as players become less and less impatient, the costs of delay go to zero. On the other hand, if for the firm the holdout payoff approaches the strike payoff, that is, $w_0 \to 1$, then the minimum wage approaches the maximum wage. For $w_0 \to 1$, the effective disagreement points $(w_0, 1-w_0)$ and $(w_0, 0)$ coincide and the wage bargaining game features a unique efficient equilibrium. Hence, delay is impossible, that is, $T^* = 0$.

We state the following proposition without proof.

Proposition 8.4. *Suppose $\delta^2 \geq w_0$. Then, for $T \geq 2$ and T even, $s(T-t, w, \delta)$ is a vector of equilibrium payoffs for $0 \leq t < T$ if and only if w and T satisfy*

$$\delta^{-T} w_0 \leq w \leq 1 - \frac{\delta^{1-T}(1-w_0)}{1+\delta},$$

and, at $t = T$, $w \in [w^{min}(T), w^{max}(T)]$. The strategies of table 8.3 support these equilbria.

8.3.4 Intermezzo: unequal discount factors

Assuming a common discount factor δ, we know from section 8.3.2 that the alternating-strike strategies induce the maximum equilibrium wage of the

State		Initial		A	
Union	proposes	$(1,0)$		w	
	responds	*yes*	if $w^t \geq w^{max}(t)$	*yes*	if $w^t \geq w$
	action	s		h	
Firm	proposes	$(0,1)$		w	
	responds	*yes*	if $w^t \leq w^{min}(t)$	*yes*	if $w^t \geq w$
	action	c		c	
Transition		Start in Initial state at $t = 0$, go to state A if $t \geq T$; Go to minimum-wage strategies of table 8.1 if union deviates, either, for $t < T$, if $w^t < 1$, t even, or $a_1(t) \neq s$, or for $t \geq T$, if $w^t \neq w$, t even, or rejection of $w^t = w$, t odd; Go to maximum-wage strategies of table 8.2 if firm deviates, either, for $t < T$, if $w^t < 1$, t odd, or for $t \geq T$, if $w^t \neq w$, t odd, or rejection of $w^t = w$, t even.			

Table 8.3. A pair of SPE strategies that induces a lengthy strike before agereement is reached on $(w, w - w)$ at $t = T$. If one of the players deviates then equilibrium switching is either to the minimum-wage contract or the maximum-wage contract equilibrium. These latter states are absorbing.

game. In this section the assumption of a common discount factor is relaxed. Let δ_u denote the individual discount factor of the union, and δ_f that of the firm. The main purpose is to show that the alternating-strike strategy profile does not support an equilibrium for $\delta_f < \delta_u$; see also Bolt (1995).

The alternating-strike strategy profile where the union strikes in odd rounds and holds out in even rounds, induces a sequence of alternating disagreement payoffs: $(0,0)$ in odd rounds, and $(w_0, 1 - w_0)$ in even rounds. For $\delta_u^2 \geq w_0$, it is now conjectured that the maximum equilibrium wage corresponds to the solution of the following (generalized) fixed point problem,

$$\begin{aligned} x_1 &= 1 - \delta_f(1 - y_1) \\ y_1 &= (1 - \delta_u)w_0 + \delta_u x_1. \\ x_2 &= 1 - x_1 \\ y_2 &= 1 - y_1. \end{aligned}$$

The fixed point $(x^*((\delta_u, \delta_f), (w_0, 0)), y^*((\delta_u, \delta_f), (w_0, 0)))$ of this system implies

$$x^*((\delta_u, \delta_f), (w_0, 0)) = \left(w_0 + \frac{(1 - \delta_f)(1 - w_0)}{1 - \delta_u \delta_f}, \frac{\delta_f(1 - \delta_u)(1 - w_0)}{1 - \delta_u \delta_f}\right),$$

and

$$y^*((\delta_u, \delta_f), (w_0, 0)) = \left(w_0 + \frac{\delta_u(1-\delta_f)(1-w_0)}{1-\delta_u\delta_f}, \frac{(1-\delta_u)(1-w_0)}{1-\delta_u\delta_f}\right).$$

The strategies induce immediate agreement on $x^*((\delta_u, \delta_f), (w_0, 0))$ at $t = 0$.

But does the alternating-strike strategy profile supports an SPE?. No! That is to say, it will depend on the relative values of δ_u and δ_f. By playing an alternative strategy, which is here called the *no-concession* strategy, the firm can increase its payoff. Informally, the firm's no-concession strategy is described by the following strategy: "reject all offers of the union in every even round and make 'non-acceptable' offers in every odd round."

By playing its no-concession strategy the firm makes sure that no agreement is ever reached and, therefore, the firm is able to secure the discounted stream of its disagreement payoffs. The normalized value of this stream is given by $(1-\delta_f)\sum_{t=0}^{\infty}\delta_f^t d_2(a(t))$, here dubbed the firm's *no-concession* payoff. It is clear that this no-concession payoff depends on the strategy of the union since the union influences the firm's disagreement payoff by its strike decision. In particular, given the alternating-strike strategy of the union, the firm's no-concession payoff amounts to $\delta_f(1-w_0)/(1+\delta_f)$. It is easy to show that

$$\frac{\delta_f(1-w_0)}{1+\delta_f} > x_2^*((\delta_u, \delta_f), (w_0, 0)) = \frac{\delta_f(1-\delta_u)(1-w_0)}{1-\delta_u\delta_f} \iff \delta_f < \delta_u.$$

Hence, given the alternating strike strategy of the union, if $\delta_f < \delta_u$ the firm is better off to play its no-concession strategy. So, clearly the alternating-strike strategies do not support a Nash equilibrium if $\delta_f < \delta_u$.

The intuition is clear. Because of its higher impatience and its low disagreement payoff in even rounds, the firm is manipulated into a bad bargaining position if it remains at the bargaining table. The firm would be willing to disagree forever, i.e., continue to make unacceptable offers and alternate between strikes and paying the old contract wage, rather than paying the maximum wage in the first round at $t = 0$. It can be shown that in the limit as the discount factors approach one, this result does not disappear. On the other hand, for $\delta_f \geq \delta_u$ the alternating-strike strategy profile does support the maximum equilibrium wage. Hence, the alternating-strike result only holds if the bargaining position of the firm is not too bad.

Let us define

$$(\bar{w}, 1-\bar{w}) = \left(\frac{1+\delta_f w_0}{1+\delta_f}, \frac{\delta_f(1-w_0)}{1+\delta_f}\right).$$

State		Even Agreement	
Union	proposes	$(\bar{w}, 1-\bar{w})$	
	responds *yes*		if $w^t \geq (1-\delta_u)w_0 + \delta_u \bar{w}$
	action	s	if rejection at t even
	action	h	if rejection at t odd
Firm	proposes	$(0, 1)$	
	responds *yes*		if $w^t \leq \bar{w}$
	action	c	if rejection at t even
	action	c	if rejection at t odd
Transitions		Go to the strategies of proposition 8.2 if union deviates either when $w^t \neq \bar{w}$, t even, or $a_1(t) \neq s$, t even,	

Table 8.4. A pair of modified alternating-strike strategies that restore equilibrium. If the union deviates, equilibrium switching to the minimum-wage contract takes place. The latter state is absorbing.

In order to restore equilibrium for the case $\delta_f < \delta_u$, consider the *modified* alternating-strike strategies in table 8.4. The next proposition states the main result.

Proposition 8.5. *For sufficiently large* $\delta_u < 1$,

i) if $\delta_f \geq \delta_u$, the alternating-strike strategies of table 8.2 support an SPE in which agreement on $x^*((\delta_u, \delta_f), (w_0, 0))$ is obtained at $t = 0$. This is the maximum-wage contract for the union.

ii) if $\delta_f < \delta_u$, the modified alternating-strike strategies of table 8.4 support an SPE in which agreement is reached only in even rounds. Holdouts occur in odd rounds as long as agreement has not been reached.

As a last remark, it can be shown that if the old contract wage w_0 is sufficiently low, the union can obtain a higher wage by threatening to strike in *every* round of dispute. Executing a strike is costly to both parties. However, if w_0 is small the costs of a strike are relatively low compared to the costs of a holdout. Moreover, by playing its 'always-strike' strategy the union reduces the firm's no-concession payoff from $\delta_f(1-w_0)/(1+\delta_f)$ to zero. Hence, if w_0 is small the threat of going on strike in every period of disagreement can improve the bargaining position of the union. This completely unravels the (modified) alternating-strike maximum-wage contract for $\delta_f < \delta_u$. To conclude, the analysis above that the assumption of a common discount factor is clearly not minor and that discount factors matter fot deriving SPE for bargaining games.

8.4. Dutch wage bargaining: an application

In this section we apply the wage bargaining model as specified in section 8.2 to the Dutch case of collective bargaining. First, as already mentioned, collective bargaining in the Netherlands is important since it covers the working conditions of around 75% of the labour force. Second, the Dutch wage negotiations process is time-consuming, it takes on average 7-8 months, although actual strike incidence is low. Moreover, and third, new wage contracts get often backdated to the expiration date of the old contract, so effectively workers get nominally compensated for the delay in the wage bargaining process. These three characteristics of Dutch collective bargaining have various strategic consequences which will be explored in this section.

8.4.1 Work-to-rule as a substitute for strike

In this section we characterize the minimum and maximum equilibrium wage as a function of the discount factor under the assumption that *no backdating* takes place. The method will probably be known by now, but here the aim is to derive conditions under which work-to-rule can be a substitute for strike.

As already discussed in the previous section the minimum-wage equilibriium corresponds to strategies in which the union chooses the least costly option, i.e. holdout, as long as no agreement is reached. Thus, the union refrains from work-to-rule or strike. Since holdout is also the action that inflicts the lowest costs upon the firm, holdout is the union's action with the lowest efficiency loss. Therefore, the Pareto improvement of any new contract is limited to $1 - \alpha$ and, consequently, the wage increase will be modest.

Whenever strike is credible, then the maximum-wage equilibrium strategies are identical to those described in section 8.3.2. Briefly recalling, the union alternates between holdout and strike in case of disagreement such that the costs it inflicts upon the firm are as large as possible. This is accomplished if the union strikes just after the firm has rejected a demand made by the union and it should holdout just after it rejected an offer made by the firm. Equilibrium switching ensures that the threat of a strike is credible, by prescribing an immediate switch to the equilibrium that induces the lowest equilibrium wage whenever the union fails to carry out such a strike threat. So, at the first occasion in which the union does not carry out its threat of strike, the minimum wage equilibrium strategies prescribe the continuation in the game from that point in time onwards. If strike is not considered credible, i.e., $\delta^2 < w_0/\alpha$ (see below), then the union can use the threat of work-to-rule similarly as just described with respect to strike (read work-to-rule instead of strike every time strike is mentioned).[3]

The next theorem precisely characterizes the minimum and maximum wage at round t, denoted by $w^{\min}(t)$, respectively, $w^{\max}(t)$, for t is even. The economic interpretation is that the maximum equilibrium wage is achieved if the union adopts the option that inflicts the *highest* costs upon the firm among the options that are *credible*. We do not explicitly state the equilibrium wages at t is odd, because it consists of w_0 plus δ times the equilibrium wage increases at t is even.

Theorem 8.6. *Let t be even. The wage $w^{min}(t)$ at round t as function of δ is given by*

$$w^{min}(t) = w_0 + \frac{1-\alpha}{1+\delta}. \tag{8.6}$$

If $\gamma < \frac{\alpha-\beta}{\alpha-w_0}$, then the wage $w^{\max}(t)$ at round t as function of δ is given by

$$w^{\max}(t) = \begin{cases} w^{\min}(t), & \text{if } \delta^2 < \gamma w_0 / (\alpha - \beta + \gamma w_0), \\ w_0 + \frac{1-\beta}{1+\delta}, & \text{if } \delta^2 \in [\gamma w_0 / (\alpha - \beta + \gamma w_0), w_0/\alpha), \\ w_0 + \frac{1-w_0}{1+\delta}, & \text{if } \delta^2 \geq w_0/\alpha. \end{cases} \tag{8.7}$$

Similarly, if $\gamma \geq \frac{\alpha-\beta}{\alpha-w_0}$, then the wage $w^{\max}(t)$ at round t is given by $w^{\min}(t)$ if $\delta^2 < w_0/\alpha$ and $w_0 + \frac{1-w_0}{1+\delta}$ otherwise.

The results of section 8.3, i.e., $\alpha = \beta = 1$ and $\gamma = 0$, belong to the case $\gamma \geq (\alpha - \beta)/(\alpha - w_0)$, which shows that these results are robust to including more types of industrial action. Furthermore, strike (work-to-rule) is credible if the union's costs w_0 (γw_0) of this action do not exceed the net gain of this action that comes in the form of a future wage increase, i.e., investment in such an action should be profitable. Note that γ does not enter $w^{\max}(t)$, because work-to-rule is only used in every even round in which only the firm's disagreement payoff $\beta - w_0$ matters.

Theorem 8.6 makes it possible to answer the question to what extent work-to-rule can be used as a substitute for strike. It is easy to see that the maximum wage increase corresponding to work-to-rule is a factor $\lambda := (1-\beta)/(1-w_0)$ times the wage increase associated with strike. Obviously, $\beta = 1$ corresponds to $\lambda = 0$. Furthermore, work-to-rule is an imperfect substitute for strike, i.e., $\lambda < 1$, iff $\beta - w_0 > 0$. The latter inequality should be read as: Production under the work-to-rule yields a higher profit than strike does, or equivalently, the firm's costs of strike exceed those of work-to-rule. However, there is a situation in which work-to-rule serves as a substitute for strike, namely in case the union's costs of work-to-rule are small and work-to-rule is credible while the more effective strike is not available as a credible option, that is, $\gamma \in [0, (\alpha - \beta)/(\alpha - w_0))$ and $\delta^2 \in [\gamma w_0 / (\alpha - \beta + \gamma w_0), w_0/\alpha)$.

The results in this section enable us to briefly comment on a closely related issue of independent interest, namely the special case in which the union fails strike as a strategic weapon and it has to resort to holdout or work-to-rule. This is the relevant case for professions such as the police, the army, customs and firemen for which strike is simply forbidden by law. Also, in the Netherlands strike is forbidden by law if the coverage of workers that are willing to strike is too low. Finally, this is the relevant case if there are other compelling non-economic reasons, as for instance ideological reasons, for why it is simply taboo for individual employees to go on strike. From Theorem 8.6 it immediately follows that for this special case $w^{\min}(t)$ is not affected and that $w^{\max}(t)$ at t even is simply given by

$$w^{\max}(t) = \begin{cases} w^{\min}(t), & \text{if } \delta^2 < \gamma w_0 / (\alpha - \beta + \gamma w_0), \\ w_0 + \frac{1-\beta}{1+\delta}, & \text{if } \delta^2 \geq \gamma w_0 / (\alpha - \beta + \gamma w_0). \end{cases}$$

As a last remark, one essential variable that is absent in the wage bargaining model is employment. If the wage bargaining model with backdating would be modified such that the firm's employment adjusts to wage increases and the union cares about wages and employment, then the maximum wage increase in such an extended model would be lower than the maximum wage increase in Theorem 8.6. The intuition is simple. The union faces a trade off between a higher wage and a lower level of employment and it therefore sacrifices some of the wage increase in order to make the deterioration of employment less. Thus, the absence of employment considerations in our model leads to a systematic bias toward higher wage increases and, consequently, toward a systematic higher prediction of the dampening effect of holdouts on wage increases.

8.4.2 Equilibria with lengthy work-to-rule

Dutch wage negotiations often feature lengthy delay without strike activity before agreement is reached. The question arises whether this pattern of wage determination can be supported within the bargaining model under investigation. In this section an affirmative answer to this question is given. Since holdout can be regarded as a special case of work-to-rule, i.e. $\beta = \alpha$ and $\gamma = 0$ only equilibria with lengthy work-to-rule are considered. As in section 8.3.3 where we considered lengthy strikes, we will first derive necessary and sufficient equilibrium conditions for lengthy work-to-rule before the negotiations are concluded. Second, we derive limit results for such equilibria if the time between proposals vanishes.

The strategies show much resemblance to the strategies of section 8.3.3. Loosely stated, the strategies with work-to-rule for the first T rounds (without loss of generality we assume T is even) are as follows: at an even round t, $t < T$, the union demands a wage equal to 1; the firm (obviously) rejects such offer, after which the union works to rule. At round T the union demands w and the firm accepts every wage not exceeding w. At an odd round t, $t < T$, the firm offers the wage w_0, which the union rejects followed by work-to-rule. As soon as the union does not make the prescribed demand at even rounds t, $t \leq T$, this party is punished by an immediate switch to the minimum wage equilibrium of Theorem 8.6. Similar, if the firm does not make the prescribed offer at odd rounds before T this party is punished by an immediate switch to the maximum-wage equilibrium of Theorem 8.6. Obviously, these strategies induce T rounds of work-to-rule followed by agreement upon w. Let us denote the corresponding payoff-relevant outcome path by

$$p_h(T) = \langle\{(wrt, c)\}_{t=0}^{t=T-1}, w\rangle.$$

Given this outcome path $p_h(T)$, the associated continuation payoff vector at the start of round t, $t \leq T$, is denoted by $s^*(T-t, w, \delta)$ and given by

$$s^*(T-t, w, \delta) = \left(1-\delta^{T-t}\right)\begin{bmatrix}(1-\gamma)\,w_0 \\ \beta - w_0\end{bmatrix} + \delta^{T-t}\begin{bmatrix} w \\ 1-w\end{bmatrix}. \tag{8.8}$$

Note that the firm's continuation payoff strictly decreases in t if and only if $1 - w < \beta - w_0$, i.e. work-to-rule generates higher profits than the new wage.

The presence of decreasing continuation payoffs is the more interesting case from both a theoretical as from an empirical point of view. From a theoretical point of view this case includes $\alpha = \beta = 1$ and $\gamma = 0$ which was analysed in the previous section. From an empirical point of view this case reflects the estimate of the efficiency parameter of 0.98 for the Netherlands and 0.94 for the US, see e.g., Van de Wijngaert (1994) and Cramton and Tracy (1992).

As in deriving equilibria with lengthy strikes, here also the equilibrium settlement wage must balance two opposing forces. First, because the union anticipates on a sufficient high settlement wage after the delay, it is willing to give up additional current income which would be available from immediate agreement. This yields a lower bound on the settlement wage. Second, the firm must not have an incentive to make an offer that the union cannot reject, i.e., by offering the union the maximum equilibrium wage. This yields an upper bound on the settlement wage; profits afterwards must be sufficient to compensate for losses taken on during the delay. The next theorem states this result more formally. Strategies which support these equilibria with delay are of similar structure to those derived in the previous section.

Theorem 8.7. *Suppose $\beta > \frac{1+\delta w_0}{1+\delta}$ and $\delta^2 \geq w_0/\alpha$. Then, for $T \geq 2$ and T even, $s^*(T-t, w, \delta)$ is a vector of equilibrium payoffs for $0 \leq t < T$ if and only if w and T satisfy*

$$w_0 + \frac{\left(\frac{1-\alpha}{1+\delta} + \left(1-\delta^T\right)\gamma w_0\right)}{\delta^T} \leq w \leq w_0 + \frac{1-w_0}{1+\delta} - \frac{(1-\beta)(1-\delta)}{\delta},$$

and, at $t = T$, $w \in \left[w^{\min}(T), w^{\max}(T)\right]$.

The upper bound upon the settlement wage is independent of the length of the holdout period, while the lower bound upon the settlement wage is increasing in the length of the work-to-rule period.[4] So, these bounds cannot unambiguously explain the negative relation between length of the holdout period and wage increases observed for the Netherlands. Of course, the multiplicity of equilibria implies that it is not hard to find two pairs (w, T) and (w', T') such that $T < T'$ and $w > w'$. However, doing so is not convincing because the opposite, i.e. $T < T'$ and $w < w'$, can also easily be achieved.

Finally, we mention that the interval of wages is not empty if and only if

$$\delta^T \geq \frac{\delta(1 - w_0) + \delta(1 + \delta)\gamma w_0 - (1 - \delta^2)(1 - \beta)}{\delta(1 - \alpha + (1 + \delta)\gamma w_0)}, \tag{8.9}$$

i.e., the length of the equilibrium work-to-rule cannot become too large.

We continue by characterizing the limit set of equilibrium payoffs corresponding to equilibria with lengthy work-to-rule as time between proposals vanishes. This limit set is denoted as S^* and it is given by

$$S^* = \{s \in Co\{(\underline{s}_1, 1 - \underline{s}_1), (1 - \underline{s}_2, \underline{s}_2), ((1 - \gamma) w_0, \beta - w_0)\}\}, \tag{8.10}$$

where $\underline{s}_1 = w_0 + \frac{1}{2}(1 - \alpha)$, $\underline{s}_2 = \frac{1}{2} - \frac{1}{2}w_0$ and $Co\{\dots\}$ refers to the convex hull. Denote Δ, $\Delta > 0$, as the time between every two consecutive bargaining rounds, r as the rate of time preference and ℓ, $\ell \geq 0$, as the length of the work-to-rule phase measured in continuous time. It is standard to take $\delta = e^{-r\Delta}$. Every $s \in S^*$ uniquely determines a wage $w(s)$ and a delay $\ell(s)$ measured in real time (to made precise later). Hence, given $s \in S^*$ and $\Delta > 0$, the number of rounds featuring work-to-rule is $T(s, \Delta) = \ell(s)/\Delta$, which goes to infinity as Δ goes to 0. Note that $\underline{s}_1 = \lim_{\Delta \to 0} w^{\min}(t)$ and $\underline{s}_2 = \lim_{\Delta \to 0} 1 - w^{\max}(t)$ in the definition of S^*.

The following theorem states that S^* is the limit set of equilibrium payoffs and specifies the wage $w(s)$ and length of work-to-rule $\ell(s)$ for every $s \in S^*$. The proof is deferred to the appendix.

Theorem 8.8. *Every payoff vector $s \in S^*$ is an equilibrium payoff vector corresponding to an equilibrium with work-to-rule for*

$$\ell(s) = r^{-1}\left|\ln \frac{s_2 + s_1 - \beta + \gamma w_0}{1 - \beta + \gamma w_0}\right| \tag{8.11}$$

length of time and agreement upon the wage

$$w(s) = w_0 + \frac{\gamma w_0 (1 - w_0 - s_2) + (s_1 - w_0)(1 - \beta)}{s_1 + s_2 - \beta + \gamma w_0}. \tag{8.12}$$

Note that condition $\delta^2 \geq w_0/\alpha$, which is imposed in Theorem 8.7, is automatically satisfied for sufficiently small $\Delta > 0$. As is the case in Theorem 8.7

the condition $\beta > \frac{1}{2}(1 + w_0)$ is the necessary and sufficient condition that ensures equilibria with decreasing continuation payoffs for the firm are present. For completeness we mention that this theorem also holds for $\beta \leq \frac{1}{2}(1 + w_0)$. For the special case $\alpha = \beta = 1$ and $\gamma = 0$, as considered in the previous section, the set S^* is a line piece on the Pareto frontier with endpoints $(\underline{s}_1, 1 - \underline{s}_1), (1 - \underline{s}_2, \underline{s}_2)$. The length of $\ell(s)$ is a measure of the degree of inefficiency; if s is relatively close to the Pareto-frontier, then $\ell(s)$ is relatively close to 0.

8.4.3 Backdating

In this section we first show that the union's minimum and maximum utility of Theorem 8.6 are not affected if backdating is incorporated into the model. Therefore, the aspect of backdating does not effect the parties strategic opportunities in terms of utilities, which confirms the commonly held point of view that backdating is only a minor detail of wage negotiations. However, this theorem also states that lengthy work-to-rule in the presence of backdating have a dampening effect on the equilibrium wage. Let $s_1^{\max}(h_t, t)$, respectively, $w^{\max}(h_t, t)$ denote the union's maximum equilibrium utility, respectively, the maximum equilibrium wage at period t after h_t periods of production under the old contract. Similarly, $s_1^{\min}(h_t, t)$ and $w^{\min}(h_t, t)$ refer to the minimum equilibrium values.

Theorem 8.9. *Let $w^{\max}(t)$ and $w^{\min}(t)$ be given as in Theorem 8.6. Then $s_1^{\min}(h_t, t) = w^{\min}(t)$ and $s_1^{\max}(h_t, t) = w^{\max}(t)$ and the corresponding wages are given by $w^{\max}(h_t, t) = w_0 + \frac{w^{\max}(t) - w_0}{1 + (1-\delta) h_t}$ and $w^{\min}(h_t, t) = w_0 + \frac{w^{\min}(t) - w_0}{1 + (1-\delta) h_t}$.*

The dampening effect of holdouts on the wage increase is relatively small. This can be seen as follows. Rewriting the expression for $w^{\max}(h_t, t)$ yields

$$w^{\max}(t) - w^{\max}(h_t, t) = (1 - \delta) h_t (w^{\max}(h_t, t) - w_0) \geq 0 \qquad (8.13)$$

and the term $(1 - \delta) h_t$ is relatively small for 'realistic' values of δ and h_t. For example, $(1 - \delta) h_t \approx 0.029$ if $\Delta = 1$ (one bargaining round lasts a day), $h_t = 210$ (roughly 7 months) and $\delta = e^{-r\Delta} \approx r$ with $r = 14 \cdot 10^{-5}$ (an annual rate of 5.11%). Thus, neglecting backdating yields a prediction of the maximum wage increase $w^{\max}(h_t, t) - w_0 = \frac{w^{\max}(t) - w_0}{1 + (1-\delta) h_t}$ that overshoots the prediction of the model with backdating (by about 2.9% in the example). Empirical evidence for this theoretical small effect is reported in Van Ours and Van de Wijngaert (1996), which finds a 0.1% negative effect on new wages per two months of production under the expired wage contract for the Netherlands.

The equilibria of the previous section can be easily extended to incorporate backdating. Backdating simply means that we have to distinguish between utilities and wages. The relation between wage w and utility s_1 after T periods of holdout is straightforward: $s_1 = w_0 + (1-\delta) T (w - w_0)$. Hence, backdating has a dampening effect. This result also holds in the limit as Δ goes to 0, provided the length of the holdout in real time is kept constant. Let $s \in S^*$, then $w(s)$ given by (8.12) has to be interpreted as the union's utility of the agreement that includes backdating after $\ell(s)$ time of work-to-rule, where $\ell(s)$ is given in (8.11). Denote the settlement wage including backdating as $w^B(s, \ell(s))$. The following theorem states that the negative relation between the wage $w^B(s, , \ell(s))$ and the length of work-to-rule $\ell(s)$. Hence, backdating unambiguously explains the empirical findings as observed for The Netherlands.

Theorem 8.10. *Every $s \in S^*$ is a vector of equilibrium utilities and the limit wage $w^B(s, \ell(s)) = w_0 + \frac{w(s)-w_0}{1+r\ell(s)}$, where $w(s)$ respectively $\ell(s)$ are given in (8.12) and (8.11).*

For every $s \in S^*$ it holds that the limit discrepancy between the union's utility and the level of the settlement wage level is given by

$$w(s) - \lim_{\Delta \to 0} w^B(s, \Delta, T(s, \Delta)) = \frac{r\ell(s)}{1 + r\ell(s)} \cdot (w(s) - w_0), \qquad (8.14)$$

which increases the larger $\ell(s)$ becomes. The implication for empirical work is evident: If production under the old contract and backdating are observed in the data, then the union's utility and the level of the wage should be clearly distinguished and a modification is necessary.

The bargaining model can easily be extended in order to let the parties propose whether or not to backdate wage contracts, i.e., endogenous backdating. From above we have that both the firm and the union are indifferent between the wage $w(s)$ without backdating and the wage $w^B(s, \Delta, T(s, \Delta))$ at every period t. But then all the equilibrium strategies derived thus far constitute one of the SPE's in the extended model with endogenous backdating. Furthermore, the (limit) set of equilibrium payoffs will not change. Thus, a richer model can explain the equilibrium behaviour derived in this section, i.e., lengthy work-to-rule and backdating.

The interesting case is the extension to different discount factors, i.e., $\delta_u \neq \delta_f$. First, suppose the firm is more patient than the union, that is, $\delta_f > \delta_u$. Then the reduction in future wage level that the union will require in order to obtain backdating is less than what the firm would be willing to offer. This means that there is room for Pareto improvement by backdating. Formally, consider the wage contract $w^B > w_0$ after T periods of production, then the sum of the parties utilities is equal to $1 + T\left(w^B - w_0\right)\left(\delta_f - \delta_u\right) > 1$ and the parties will backdate new wage contracts. Recursive relations for the union's maximum equilibrium $s_1^{\max}(h_t, t)$ and $w^{\max}(h_t, t)$ can easily be given simply by replacing δ by either δ_u or δ_f in the proof of Theorem 8.9, but its solution is very cumbersome. Therefore, it remains an open question whether the immediate agreement result in the union's best and worst SPE found for $\delta_u = \delta_f$ also holds for $\delta_f > \delta_u$, because backdating and lengthy production under the old contract (which causes delay) enlarge the surplus. For the opposite case, apart from the problems pointed to in section 8.3.4, we do not expect backdating, because it reduces the size of the surplus.

One last remark should be made with respect to equilibria in which the union strikes in all periods before a new settlement wage is agreed upon. Since backdating only applies to periods in which the union held out and these equilibria do not involve holdouts it is obvious that an analysis of such equilibria in our model simply boils down to the analysis presented in section 8.3. In fact, a minor modification is needed in order to take into account the efficiency parameter of holdout.

8.5. Related literature

Economic theory has trouble explaining strikes in wage bargaining. In fact, this comes back to what Kennan (1986) calls the 'Hicks paradox', which states that rational parties should always be able to agree to an efficient outcome in advance, thereby avoiding the costs of delay and share the increased gains from trade. Since delay in bargaining is often costly to both players in the process, the common rationalization of this apparent inefficiency is that one of the parties (or both parties) is poorly informed. A poorly informed union may resort to use its strike weapon in order to discriminate among firms with different profitability. That is, strikes reflect temporal wage discrimination in which the union compensates for its uncertainty about the firm's profitability by exploiting the greater impatience of a high-profit firm to reach an wage agreement quickly. However, the empirical evidence on this theoretical relation is ambiguous, see e.g., Kennan and Wilson (1989). Moreover, most of these asymmetric information wage bargaining models suffer from the *Coase conjecture*. That is, as the length of a bargaining round becomes arbitrarily small, so does the real time of delay; a result that is rigorously proven in Gul and Sonnenschein (1988).

The analysis in Fernandez and Glazer (1991), Haller (1991), and Haller and Holden (1990) makes an important contribution to the theory of strikes by showing that strikes may occur in a simple bargaining framework with complete and symmetric information. This is a remarkable theoretical result, because virtually all previous work on explaining strikes asserted that asymmetric information was a necessary condition for strikes to occur in equilibrium. Further, they show that their wage bargaining model has a great multiplicity of subgame perfect equilibria, including some which are Pareto efficient and others which involve strikes. The analysis of the perfect information wage bargaining game shows that delay may occur in equilibrium. However, the analysis provides no real explanation for its occurrence and duration. In fact, the strategies that support delay in equilibrium rule out any compromises during the course of the negotiations, while the aspect of compromise lies at the heart of any bargaining situation. A player who wants to make a compromise during a bargaining impasse in the negotiations by offering an acceptable wage will be regarded as 'weak' and punished by imposing his worst wage contract. This is a considerable drawback of these perfect information models, but they may serve as a benchmark model and show that asymmetric information is not a necessary ingredient to describe the occurrence of delay in equilibrium. Further, the practical relevance of the derived 'stutter strikes' may be questioned, since, for example, the union's strategy of striking in every other period is an unfair labour practice in the United States. Labour law in the U.S. explicitly forbids stutter strikes, in which the union strikes for a short period, then returns to work, strikes again and so on. However, this type of industrial action is allowed in, for instance, the Netherlands; see Van de Wijngaert (1994) for a thorough investigation of the Dutch system of collective bargaining. Moreover, the characterization results do not really depend on stutter strikes being allowed or not. If unions are forbidden to engage in stutter strikes, and are bounded by law only to strike until agreement is reached, then this other institutional environment will have no real consequences for the derivation of inefficient equilibria. Only the upper bound on equilibrium wages will be affected.

In this chapter the asymmetric case is considered where only the union is able to engage in industrial action. In real life wage bargaining the firm may threaten to lockout employees. It can be shown that if the union's payoffs during a lockout are sufficiently small then the firm is able to press wages down and avoid wage increases. As similar result is found in Holden (1994), where, using a more complex macroeconomic setup, a zero wage inflation can be obtained if employment is low, so that credible lockout threats by the firm are used to prevent demands for wage increases.

Finally, in Houba and Lomwel (2001) the new wage contract is of finite length and the parties negotiate an infinite sequence of wage contracts, with the understanding that the negotiations for a new contract take place after the expiration of the current contract. In this setting where profit growth is autonomous, the authors show that the extreme SPE utility levels of both parties are not affected. The maximum-wage contract shows some interesting wage dynamics, including a generic area of parameter values for which the maximum equilibrium wage involves a *wage decrease.*

Notes

1 Several explanations of these costs' sources are mentioned in Cramton and Tracy (1992), (1994a) and (1994b), Holden (1989), (1994) and (1997), Moene (1988) and Van Ours and Van de Wijngaert (1996).

2 As an aside, for the limiting case $\gamma = 0$, (wrt, c) is also Nash, however, this not a generic case.

3 The results in Haller (1991) can be applied directly in order to determine the highest equilibrium wage that can be obtained by the threat of work-to-rule.

4 The two conditions of theorem 8.7 are only imposed for explanatory reasons. Condition $\beta > \frac{1+\delta w_0}{1+\delta} = w^{\max}(0)$ is the necessary and sufficient condition that ensures equilibria with decreasing continuation payoffs for the firm are present. Without this condition only Case 1 in the proof has to be considered and nothing changes if $\beta \in \left[\frac{w_0+\delta}{1+\delta}, \frac{1+\delta w_0}{1+\delta}\right]$ and for $\beta < \frac{w_0+\delta}{1+\delta}$ condition (A.10) in the proof becomes the upper bound upon w. Condition $\delta^2 \geq w_0/\alpha$ is imposed in order to restrict the number of cases to be considered, because the analysis in case of $\delta^2 \in [\gamma w_0/(\alpha - \beta + \gamma w_0), w_0/\alpha)$ would be similar to the one in Case 1 in the proof and only a minor modification is needed with respect to the relevant maximum equilibrium wage.

Chapter 9

THE POLICY BARGAINING MODEL

9.1. Introduction

The wage bargaining model of the previous chapter treats the parties involved in the negotiations asymmetrically, because only one of the two parties could use threats. In this chapter we consider the symmetric situation in which both parties have threats at their disposal according to the full model introduced in chapter 2 and the analysis relies on many of the results derived in the previous chapters. The aim of this chapter is to pursue the most general mathematical results, rather than an application to one particular application. Despite the emphasis on mathematics this analysis will serve as a solid base for applications.

In this chapter first the equilibria in Markov strategies are characterized and it is shown that at least one such equilibrium exists. Next, each player's worst equilibrium is characterized. Recall from the wage bargaining models in chapter 8 that the firm's worst equilibrium is not an equilibrium in Markov strategies and this is also the case in this chapter. However, in this chapter the firm also chooses among its disagreement actions and, similar as the union in the previous chapter, the firm may be tempted to deviate from the described disagreement action. This causes the following problem: How to construct the firm's worst equilibrium strategies such that the firm does not deviate from the described disagreement action (when called upon to do so) given that in the firm's worst equilibrium we fail the threat of equilibrium switching to some pair of equilibrium strategies that are strictly worse for the firm, i.e., logic dictates that in equilibrium the firm cannot do strictly worse than its worst equilibrium strategies. In this chapter this problem is solved and this makes the full characterization of each player's worst equilibrium possible. This includes deriving the necessary and sufficient conditions for a unique pair of equilibrium utilities.

The analysis continues by considering a special class of bargaining problems, called *policy bargaining problems.* In these problems we assume that the players' economic environment is modelled as an infinitely repeated game and that the contract space consists of all joint policies in this infinitely repeated game, where the standard assumption of binding agreements is maintained. Furthermore, playing the infinitely repeated game up to the round a joint agreement is agreed upon are then the disagreement actions in the negotiations.

9.2. Subgame perfect equilibria

The characterization of the limit set of SPE payoffs starts with deriving the relatively simple MPE strategies and it is shown that at least one of these equilibria always exists. By means of an example we discuss the structure of each player's worst SPE before the general case is considered. The general case involves several technical problems that are tackled one by one in three sections, including a section devoted to the necessary and sufficient conditions for a unique pair of SPE utilities. The full characterization of the set of SPE utilities is then relatively easy, because it applies techniques that are already explained in previous chapters. Part of the analysis includes the limit results as the risk of breakdown vanishes.

9.2.1 Markov perfect equilibrium

In this section the set of MPE strategies is characterized, which is almost a trivial exercise if we apply the dynamic programming as introduced in section 3.3.2. To see this, consider the history independent value functions $V^i = (V_1^i, V_2^i) \in S$, $i = 1, 2$, that represent the players' MPE utilities from the start of the next bargaining round onward whenever player i is the proposing player. Then during disagreement one round earlier, whatever actions the players choose in Γ, these will not affect their future utilities given by the history-independent value functions. This means that when the players choose their disagreement actions there is no threat of equilibrium switching in the next round and, therefore, deviations will simply not be punished. So, if $a \in A$ is a pair of MPE actions at the disagreement phase, then no player must have an incentive to unilateral deviate from his own action. But then these MPE disagreement actions must be NE actions of the disagreement game Γ, i.e., $a \in A^N$, because otherwise some profitable (one-stage) deviation exists for one of the players when Γ is played.

Having established that MPE strategies prescribe $a(t) = a^N$, $a^N \in A^N$, for all $t \in \mathbb{N}$ means that we can treat the endogenous disagreement actions as if being fixed at some a^N, $a^N \in A^N$, in the remainder of the analysis. Then the results of section 3.3 immediately apply for $d = d(a^N)$, $a^N \in A^N$. In order to express that the fixed point of the function $p \times q$ in section 3.3.1 now depends

State		Markov	
Player 1	proposes	$x^*(a^N)$	always
	responds *yes*		if $s_1^t \geq y_1^*(a^N)$
	action	a_1^N	always
Player 2	proposes	$y^*(a^N)$	always
	responds *yes*		if $s_2^t \geq x_2^*(a^N)$
	action	a_2^N	always
Transitions		absorbing	

Table 9.1. The MPE strategies of proposition 9.1, where $(x^*\left(a^N\right), y^*\left(a^N\right))$, $a^N \in A^N$, is a fixed point of the function $p \times q$ with disagreement point $d\left(a^N\right)$.

upon $d\left(a^N\right)$ we write $x^*\left(a^N\right)$ and $y^*\left(a^N\right)$. The following two results are given without further proof.

Proposition 9.1. *Let $a^N \in A^N$ and $(x^*(a^N), y^*(a^N))$ be the unique fixed point of the function $p \times q$ for the disagreement point $d(a^N)$ in lemma 3.2. Then the strategies of table 9.1 are MPE strategies.*

Corollary 9.2. *For every $h(t) \in H(t)$ the set of SPE utilities $E(h(t), t) \subseteq S^{IR}$ is nonempty.*

This proposition implies that there are as many MPE strategies as there are Nash equilibria in Γ. Furthermore, this result is similar to the one in proposition 7.13 for bargaining procedure 7.11, which can be explained as follows. In procedure 7.11 the proposals and responses during the negotiations cannot be conditioned upon past play of disagreement actions in Γ, because the bargaining game simply ends when the negotiations break down and the disagreement actions in Γ are chosen once. In the model of this chapter past disagreement actions can be incorporated into the players' strategies, but the Markov strategies discard this possibility. The latter means that the Markov strategies in this chapter are thereby similar to the Markov strategies in procedure 7.11. This similarity in Markov strategies is reflected in the similarity of results obtained.

9.2.2 Worst SPE strategies: Example

In this section the issues involved in deriving player 1's worst SPE strategy are discussed by means of the following example. Consider the bargaining problem (S, A, d) in utility representation with $S = \left\{s \in \mathbb{R}_+^2 \,\middle|\, s_1 + s_2 \leq 1\right\}$ and the utility pairs (or payoffs) in the disagreement game Γ given by

	L_2	R_2
L_1	$\frac{1}{2}, \frac{1}{2}$	$-\frac{1}{6}, \frac{2}{3}$
R_1	$\frac{2}{3}, -\frac{1}{6}$	$0, 0$

Note that the disagreement game is a prisoners' dilemma and each player's minmax value is 0.

Recall that the main issue is how to deter deviation from the prescribed disagreement actions in any subgame the players are called upon to play the disagreement game, especially for the player who is held down at his worst SPE. The strategies that support player 1's worst SPE prescribe the actions (R_1, R_2) with payoffs $(0,0)$ in the disagreement phase of every odd period (i.e., player 2 made the last proposal) and the actions (L_1, R_2) with payoffs $\left(-\frac{1}{6}, \frac{2}{3}\right)$ at the disagreement phase of every even period. Obviously, at even periods player 1 foregoes the payoff $\frac{1}{6}$ by not optimally responding to the action R_2 played by player 2. Therefore, in equilibrium player 1 has to be compensated in the next period for this forgone payoff, which can be achieved as follows. If player 1 deviates from L_1 at an even period, then the strategies continue in such a way that player 1 receives his worst SPE payoff from the next (even) period onward whereas this player would get a compensation for the foregone payoff on top of his worst SPE payoff if he plays L_1. The amount $\frac{1-\delta}{6\delta}$ suffices to compensate player 1 one period later. At an odd period player 2's equilibrium proposal leaves either player 1's worst payoff or the latter payoff plus compensation to player 1 dependent upon the disagreement action played by player 1 in the previous even period.

The complete strategies that support player 1's worst SPE payoff are as follows. At every even period player 1 proposes $x^m = \left(\frac{1}{2(1+\delta)}, \frac{1+2\delta}{2(1+\delta)}\right)$, player 2 accepts every proposal s^t made by player 1 if and only if $s_2^t \geq x_2^m$ and the players play (L_1, R_2) in case of disagreement. At every odd period player 2 proposes $y^* = \left(\frac{\delta}{2(1+\delta)} + \frac{1-\delta}{6\delta}, \frac{2+\delta}{2(1+\delta)} - \frac{1-\delta}{6\delta}\right)$ if player 1 did play L_1 and the partitioning $y^m = \left(\frac{\delta}{2(1+\delta)}, \frac{2+\delta}{2(1+\delta)}\right)$ if not. Player 1 accepts every proposal s^t made by player 2 if and only if $s_1^t \geq \widehat{y}_1^m$, provided player 2 made the 'right' proposal at this period. At the disagreement phase the strategies prescribe the actions (R_1, R_2). If player 2 deviates, then this player is punished by equilibrium switching to this player's worst SPE strategies, which are strategies similar as above with the players' roles reversed. As we will see in section 9.2.3 the SPE proposals x^m and y^* can be computed by taking $\left(\frac{1}{6\delta}, \frac{2}{3}\right)$ as the disagreement point in the functions $p \times q$. The first component of the constructed disagreement point is related to player 1's compensation and the second component is player 2's payoff at odd periods. The 'disagreement' point depends upon δ and converges to $\left(\frac{1}{6}, \frac{2}{3}\right)$ as δ tends to 1. Furthermore, all three SPE proposals converge to $\left(\frac{1}{4}, \frac{3}{4}\right)$, which maximizes the Nash product $\left(s_1 - \frac{1}{6}\right)\left(s_2 - \frac{2}{3}\right)$.

Is it really necessary to go through all the trouble of introducing the compensation? The answer is affirmative, without compensation we would have $a^* = (R_1, R_2) = a^N$ with payoffs $(0,0)$. The latter means that we would end up in the unique pair of MPE strategies for this example, which is clearly a result different than the one just discussed.

Next, we mention that in the above example the disagreement point lies below the Pareto frontier of S. However, this is no general property. Consider for example the set of feasible payoffs S that is equal to the set of feasible payoffs in the infinitely repeated prisoners' dilemma, which means that the Pareto frontier consists of the two line pieces given by $s_2 = \frac{5}{8} - \frac{1}{4}s_1$, $s_1 \in \left[0, \frac{1}{2}\right]$, and $s_2 = \frac{5}{2} - 4s_1$, $s_1 \in \left[\frac{1}{2}, \frac{5}{8}\right]$. Then similar as before, the SPE proposals x^m and y^* can be found by solving for the fixed point of $p \times q$, provided $\left(\frac{1}{6\delta}, \frac{2}{3}\right)$ is again taken as the disagreement point. Now, the latter point $\left(\frac{1}{6\delta}, \frac{2}{3}\right)$ lies above the Pareto frontier. This is caused by the fact that player 1's net gain of $\frac{1}{6}$ corresponding to the optimal deviation from the actions (L_1, R_2) is unrelated to player 2's payoff from these actions. This observation means that the results in section 6.3.2, which in turn require a good knowledge of the results in chapter 5, will play an important role in the analysis. For completeness, we mention that the SPE proposals supporting player 1's worst SPE, which are adapted to the 'new' Pareto frontier, become $x^m = y^m = \left(0, \frac{5}{8}\right)$ and $y^* = \left(\frac{1-\delta}{6\delta}, \frac{5}{8} - \frac{1-\delta}{24\delta}\right)$.

We conclude this section by comparing the limit set of SPE payoffs in both examples. In the first example this set is equal to $\left\{s \in \mathbb{R} | s_1 + s_2 \leq 1, s_1, s_2 \geq \frac{1}{4}\right\}$. For the second example it follows that the limit set of SPE payoffs is equal to the whole set of individually rational payoffs. These limit sets are supported with strategies that apply the idea of equilibrium switching.

9.2.3 Worst SPE strategies: General case

The structure of the punishment strategies is already mentioned in the previous section and table 9.2 (on the next page) summarizes the key ideas by stating the general structure of these strategies.

From table 9.2 it follows that there are two states and state i, $i = 1, 2$, refers to the state in which player i is punished. The initial state could be either of the two states with the understanding that whenever the initial state is state i, $i = 1, 2$, then it means that the strategies prescribe player i's worst SPE strategies. For explanatory reasons we mainly discuss the case in which the initial state is state 1. For some histories the strategies prescribe a switch from one state to the other. Note that the strategies do not have an absorbing state, which is a novel feature that does not alter the techniques employed. It simply means that we can no longer split the equilibrium analysis into a sequence of separate steps by first deriving the equilibrium conditions for each of the absorbing states and, then, turning to the initial state. Here, both states have to be considered simultaneously and one has to be aware that some deviations, given some particular history that induces the second state, might cause a switch back to the first state. Obviously, this is similar as the by now familiar shifts from the first to the second column encountered before.

For the first state where player 1 is punished we specify three proposals in S, one representing the proposal made by player 1, denoted as x^m, and two representing different proposals for player 2 to make, denoted as y^m and y^*. The latter two proposals are conditional upon how player 1 played Γ the last time. The superscript m stands for minimum, because player 1's worst SPE payoff is equal to x_1^m and y_1^m at time t is even respectively t is odd. The key idea behind the prescribed behaviour in the first column is that both players play the normal form game Γ according to the minmax actions a^1 with payoffs $\left(m_1, d_2\left(a^1\right)\right)$ in the disagreement phase of every odd period and the actions a^* with payoffs $d\left(a^*\right)$ at the disagreement phase of every even period.

At this stage we simply mention that the proposals x^m and y^* will be linked to the fixed points of the function $p \times q$ for $\left(m_1 + (1-\delta)\,\delta^{-1} g_1\left(a^*\right), d_2\left(a^*\right)\right)$, where the function $g_1 : A \to \mathbb{R}$ refers to player 1's net gain of switching from a_1^* to his optimal response against a_2^*. Recall from lemma 3.3 that x_1^m is increasing in the first component of the disagreement point and decreasing in the second one.

The rationale behind the minmax actions a^1 in odd rounds is twofold. Player 1 as a responding player rejects every proposal that is short of his threshold level $y_1^m = p_1\left(x^m\right)$ and for arbitrary $a \in A$ the first component of the disagreement point is equal to $d_1\left(a\right) + (1-\delta)\,\delta^{-1} g_1\left(a\right)$. Then the lower $d_1\left(a\right)$ is, the more player 1 has to give in to player 2 and the more severe player 1's punishment is. Since deviations from a_1 are not punished this imposes the equilibrium condition that player 1 does not have an incentive to deviate from the prescribed action, i.e., $g_1\left(a\right) = 0$. Thus, in player 1's worst equilibrium $d_1\left(a\right)$ should simultaneously be minimal and robust against deviation. This is accomplished with the minmax actions a^1. With respect to the disagreement actions to be played at even rounds the story is quite complicated, because a^* enters both components of the constructed disagreement point with opposed effects and player 2 must trade off the two opposed effects of $a^*\left(\delta\right)$. We defer this issue to section 9.2.4.

As mentioned before, at even periods player 1 foregoes a payoff $g_1\left(a^*\right)$ by not optimally responding to the action a_2^*. Therefore, in equilibrium player 1 has to be compensated in the next period for this forgone payoff, which can be achieved as follows. If player 1 deviates from a_1^* at an odd period, then the strategies continue in such a way that player 1 receives his minimum SPE payoff from the next (even) period onward whereas this player would get a compensation for the foregone payoff on top of this minimum payoff if he plays a_1^* that is incorporated in the proposal y^*. In equilibrium player 1 should comply to a^* at odd rounds and in order for player 1 not to deviate we must

State		1		2	
	proposes	x^m	always	x^{**}	if $a_2(t-1) = a_2^{**}$
				x^M	if $a_2(t-1) \neq a_2^{**}$
Player 1	responds	*yes*	if $s_1^t \geq y_1^m$	*yes*	if $s_1^t \geq y_1^M$
	action	a_1^*	if t is odd	a_1^2	if t is odd
		a_1^1	if t is even	a_1^{**}	if t is even
	proposes	y^*	if $a_1(t-1) = a_1^*$	y^M	always
		y^m	if $a_1(t-1) \neq a_1^*$		
Player 2	responds	*yes*	if $s_2^t \geq x_2^m$	*yes*	if $s_2^t \geq x_2^M$
	action	a_2^*	if t is odd	a_2^2	if t is odd
		a_2^1	if t is even	a_2^{**}	if t is even
Transition		Go to state 2, if either $s_1^t < y_1^*$ and for t even $a_1(t-1) = a_1^*$, or $a_2(t-1) \neq a_2^*$, t odd, or $a_2(t-1) \neq a_2^1$, t even.		Go to state 1, if either $s_2^t < x_2^{**}$ and for t odd $a_2(t-1) = a_2^{**}$, or $a_1(t-1) \neq a_1^2$, t odd, or $a_1(t-1) \neq a_1^{**}$, t even.	

Table 9.2. The strategies that represent player 1's worst SPE, where the initial state is state 1, the actions a^i, $i = 1, 2$, minmax player i, the actions a^* and a^{**} are characterized in section 9.2.4, the proposals x^m, y^m and y^* correspond to (9.2) and x^M, y^M and x^{**} correspond to (9.3). These strategies also represent player 2's worst SPE, but then the initial state becomes state 2.

have

$$(1-\delta)\, d_1\left(a^*\right) + \delta y_1^* \geq (1-\delta)\left[d_1\left(a^*\right) + g_1\left(a^*\right)\right] + \delta y_1^m. \tag{9.1}$$

Solving for y_1^* yields

$$y_1^* \geq y_1^m + (1-\delta)\delta^{-1} g_1(a^*)$$

and the equal sign suffices in equilibrium. Thus, the amount $(1-\delta)\,\delta^{-1} g_1\left(a^*\right)$ suffices to compensate player 1 one period later at an odd period and player 2's equilibrium proposal at an odd period leaves either the minimum payoff or the minimum payoff plus compensation to player 1 dependent upon the action played by player 1 in the previous even period.

Let us now turn to the question why player 1's worst SPE can be linked to the fixed points of $p \times q$ for $d = \left(m_1 + (1-\delta)\,\delta^{-1} g_1\left(a^*\right), d_2\left(a^*\right)\right)$. First, all proposals should be Pareto efficient. To see this, if x^m is not Pareto efficient player 1 would deviate by proposing some proposal s^t that is better for both players knowing that i) he cannot be punished for this deviation and ii) that it will be accepted by player 2, because player 1's worst SPE coincides with player 2's best SPE for reasons similar as in section 5.4.2. According to the strategies in the first state player 2's threshold of offers he can credibly reject, which takes into account player 1's compensation, is equal to $(1-\delta)\, d_2\left(a^*\right) +$

δy_2^*. Thus, in equilibrium the proposal x^m just offers the latter threshold to player 2, i.e. $x_2^m = (1-\delta)\, d_2(a^*) + \delta y_2^*$. Now, since y^* is part of player 2's threshold it has to be Pareto efficient, because otherwise there would be some slack for increasing this threshold without deteriorating player 1's payoff y_1^* but forcing $x_1^m = f_1(x_2^m)$ further down. Next, the strategies imply that in any even round immediately following a deviation from a_1^* in the previous round, player 1 should be kept at his threshold induced by the strategies and this threshold $y_1^m = (1-\delta)\, m_1 + \delta x_1^m$. Finally, also the proposal y^m should be Pareto efficient in order to give player 2 an incentive to punish player 1 that is as large as possible.

To summarize, for the behaviour under the first state to be player 1's worst SPE the following three necessary conditions have to hold:

1 The proposals y^* and y^m are such that $y_1^* = y_1^m + (1-\delta)\delta^{-1} g_1(a^*)$.

2 The proposals x^m, y^m and y^* are Pareto efficient.

3 Player 1 and player 2 in the role of responding players are indifferent between accepting and rejecting the SPE proposals y^m respectively x^m.

The last condition does not have to hold for the SPE proposal y^*, because in SPE the threat of a change in the state forces player 2 to make the proposal y^*. Finally note that substitution of the expression for y_1^m into the expression for y_1^* yields

$$y_1^* = (1-\delta)\left(m_1 + \delta^{-1} g_1(a^*)\right) + \delta x_1^m.$$

So, it is as if player 1's disagreement payoff at odd periods is equal to $m_1 + \delta^{-1} g_1(a^*)$.

Combining all the above insights and making use of the function $p \times q$ as defined in section 3.3.1 implies the following equilibrium conditions

$$\begin{array}{ll} x_2^m = p_2(y^*) = (1-\delta) d_2(a^*) + \delta y_2^* & x_1^m = p_1(y^*) = f_1(x_2^m) \\ y_1^* = q_1(x^m) = (1-\delta)(m_1 + \delta^{-1} g_1(a^*)) + \delta x_1^m & y_2^* = q_2(x^m) = f_2(y_1^*) \\ y_1^m = \delta x_1^m & y_2^m = f_2(y_1^m) \end{array} \tag{9.2}$$

Thus, (x^m, y^*) is a fixed point of the function $p \times q$ for the constructed disagreement point $\left(m_1 + \delta^{-1} g_1(a^*), d_2(a^*)\right)$ and y^m follows once x^m is known. As already mentioned in the example, the disagreement point may lie above the Pareto frontier, which may cause the function $p \times q$ to admit multiple fixed points (x^m, y^*), which is similar as in section 6.3.2. Recall that if the latter is the case, then proposition 6.13 applies to both x^m and y^* meaning that $x_1^m < y_1^*$.

Let us make one remark concerning the interpretation of the disagreement point. In Nash's variable-threat game of section 7.2.1 the notion of a threat point has two interpretations, namely as the disagreement point and as each player's threat to use disagreement actions that either strengthen his own position or hurt the other player's position in case of a disagreement. Obviously, the disagreement point $\left(m_1 + \delta^{-1} g_1\left(a^*\left(\delta\right)\right), d_2\left(a^*\left(\delta\right)\right)\right)$ cannot represent the 'true' payoffs in case of perpetual disagreement (unless $a^*\left(\delta\right) = a^1$), but they can be regarded in terms of a threat. Player 2 can successfully increase his bargaining position by the use of the actions $a^*\left(\delta\right)$ at the even periods and the threat to hurt player 1 by the use of the minmax actions a^1 in odd periods. However, the use of the actions $a^*\left(\delta\right)$ is not without costs for player 2, because player 1 has to be compensated.

To conclude this section, if the initial state in table 9.2 is state 2, then we would start in player 2's worst SPE. Then similar as above, the SPE proposals x^M, x^{**} and y^M satisfy the system

$$\begin{aligned} x_2^{**} &= (1-\delta)(m_2 + \delta^{-1} g_2(a^{**})) + \delta y_2^M & x_1^{**} &= f_1(x_2^{**}) \\ y_1^M &= (1-\delta) d_1(a^{**}) + \delta x_1^{**} & y_2^M &= f_2(y_1^M) \\ x_2^M &= \delta y_2^M & x_1^M &= f_1(x_2^M) \end{aligned} \tag{9.3}$$

Moreover, the proposals x^M and y^M correspond to the fixed points of the function $p \times q$ for disagreement point $d = \left(d_1\left(a^{**}\right), m_2 + \delta^{-1} g_2\left(a^{**}\right)\right)$. The proposals x^M and y^M represent player 2's minimum SPE payoff at even respectively odd periods.

9.2.4 Optimal disagreement actions

Player 1's worst SPE strategies depend upon the disagreement actions $a^* \in A$ and in this section these actions are characterized. For given $\delta < 1$ such actions exist. However, in general these actions also depend upon δ and, therefore, the broader issue is whether or not a^* as a correspondence of δ is well defined. For bargaining problems with a piecewise linear Pareto frontier the latter indeed holds, but the general case is still unsolved.

Before we turn to these issues we derive some properties of the net-gain function g_1 that are needed in the analysis. Formally, we define the net-gain function $g_1 : A \to \mathbb{R}$ as $g_i(a) = \max_{\hat{a}_i \in A_i} d_i(a \backslash \hat{a}_i) - d_i(a)$, $i = 1, 2$, where $a \backslash \hat{a}_i$ means that player i unilaterally deviates from the pair of actions $a \in A$ by playing $\hat{a}_i$. Obviously, $g_i(a) \geq 0$, $i = 1, 2$, and $d_i(a) + g_i(a) \geq m_i$ for all $a \in A$. Furthermore, $g_i(a) = 0$, $i = 1, 2$, if and only if $a_i \in A_i$ is a best response to $a_j \in A_j$, $j = 1, 2$, $j \neq i$. Hence, $a^N \in A^N$ if and only if $g_1(a^N) = g_2(a^N) = 0$. We make the following additional assumption with respect to Γ.

Assumption 9.3. *The normal form game* Γ *is such that* g_1 *and* g_2 *are continuous functions in* a.

This assumption is satisfied in case Γ is a bimatrix game and A represent its set of *mixed* strategies, e.g., Farrel and Maskin (1989).

This assumption is important, because it implies two properties that are sufficient for the analysis below. The continuity of g_1 in a means that the function $p \times q$ for $\left(m_1 + (1-\delta)\,\delta^{-1} g_1(a), d_2(a)\right)$ is continuous in both a and δ. This is important, because it in turn guarantees that the fixed points of $p \times q$ form a upper semi-continuous correspondence in a and δ.

We are now ready to consider the optimization problem that characterizes a^* for each $\delta < 1$. For convenience, we write $x(\delta, a)$ and $y(\delta, a)$ for the fixed points of $p \times q$ and $a^*(\delta)$ instead of a^* from here on.

The actions $a^*(\delta)$ that constitute player 1's worst SPE for each $\delta < 1$ have to minimize this player's SPE utility in odd and even rounds. Since $y_1^m(\delta, a^*) = (1-\delta)\, m_1 + \delta x_1^m(\delta, a^*)$ it is immediately clear that we can restrict attention to the minimization by which we obtain $x_1^m(\delta, a^*)$. To make this statement precise, define the correspondence $F_{x_1} : [0,1] \times A \to [0, s_1^{max}]$ as the correspondence of all components $x_1(\delta, a)$ corresponding to these fixed points, i.e.

$$F_{x_1}(\delta, a) = \left\{ x_1(\delta, a) \,\middle|\, \begin{array}{c} (x(\delta, a), y(\delta, a)) \text{ is a fixed point of } p \times q \\ \text{for } \left(m_1 + (1-\delta)\,\delta^{-1} g_1(a), d_2(a)\right) \end{array} \right\}.$$

Then player 1's worst SPE payoff corresponds to

$$x_1^m(\delta, a^*(\delta)) = \min_{a \in A} \left\{ \min_{x_1(\delta, a)} F_{x_1}(\delta, a) \right\}.$$

Denote the set of minimizers in A for each $\delta < 1$ as $A^*(\delta)$. The following proposition states that for each δ at least one minimizing $a \in A$ exists.

Proposition 9.4. *If assumption 9.3 holds, then*

$$A^*(\delta) = \arg\min_{a \in A} \left\{ \min_{x_1^m(\delta, a)} F_{x_1}(\delta, a) \right\} \neq \emptyset. \tag{9.4}$$

In the above proposition we kept δ fixed, but formally, $A^*(\delta)$ is a correspondence $A^* : [0,1) \to A$ and we can only derive the limit δ goes to 1 if this correspondence is well behaved. So, we need to investigate the correspondence $A^*(\delta)$ a little further. First, note that in general $A^*(\delta)$ will change with δ. This can be easily seen as follows if we assume for explanatory simplicity that all functions are continuously differentiable and that $F_{x_1}(\delta, a)$ consists of a unique element. Application of the implicit-function theorem to all four equations in the fixed point problem $(x, y) = (p(y), q(x))$ and, then, solving for the derivative of $x_1(\delta, a)$ with respect to a yields

$$\tfrac{\partial}{\partial a} x_1(\delta, a) = 0, \quad \Leftrightarrow \quad \tfrac{\partial}{\partial a} d_2(a) + f_2'(y_1(\delta, a)) \tfrac{\partial}{\partial a} g_1(a) = 0. \tag{9.5}$$

Hence, if f_2 is nonlinear, then every element in $A^*(\delta)$ depends upon δ through $y_1(\delta, a)$. Only for the special case of a linear Pareto frontier there exists a set $A^* \subseteq A$ such that $A^*(\delta) = A^*$ for all $\delta \in [0, 1)$.

With respect to the minimization problem (9.4) two technical problems arise. These two problems concern whether the limit set $\lim_{\delta \to 1} A^*(\delta)$ is well defined and the characterization of the limit values $\lim_{\delta \to 1} x^m(\delta, a^*(\delta))$ and $\lim_{\delta \to 1} y^*(\delta, a^*(\delta))$ in case these exist. Consider $A^*(\delta)$ first. There is no a problem in case the function $p \times q$ admits a unique fixed point for all $(\delta, a) \in [0, 1) \times A$, because then the correspondence $F_{x_1}(\delta, a)$ is a continuous function in both δ and a that guarantees that $A^*(\delta)$ is an non-empty and upper semi-continuous correspondence in δ and, hence, $\lim_{\delta \to 1} A^*(\delta)$ exists. From section 6.3.2 we know $p \times q$ admits a unique fixed point if the Pareto frontier is linear. However, from the same section we also know that under a non-linear Pareto frontier multiple fixed points may exist meaning that $\min_{x_1(\delta,a)} F_{x_1}(\delta, a)$ may be a discontinuous function and then it is not known what the properties of the correspondence $A^*(\delta)$ are.

Second, we consider the limit values of the SPE proposals $x^m(\delta, a^*(\delta))$ and $y^*(\delta, a^*(\delta))$ as δ goes to 1. Unlike theorem 6.10 the constructed disagreement point $\left(m_1 + \delta^{-1} g_1(a^*(\delta)), d_2(a^*(\delta))\right)$ depends upon δ directly and also indirectly through $a^*(\delta)$ and, therefore, this theorem needs to be generalized in order to apply to the current case. One technical problem to solve is the following. In order to establish notation, let $F(\delta, a)$ and $T(a)$ denote the set $F(\delta)$ and $T(S, d)$ of theorem 6.10 with disagreement points, respectively, $\left(m_1 + \delta^{-1} g_1(a^*), d_2(a^*)\right)$ and $d = (m_1 + g_1(a^*), d_2(a^*))$. Similar as before, the correspondence $F(\delta, a^*(\delta)) \subseteq S$ consists of all $x^m(\delta, a^*(\delta))$ and $y^*(\delta, a^*(\delta))$ and for each the common limit (provided it exists) corresponds to some point $\widehat{s}$ on the Pareto frontier such that the Nash product curve

$$(s_1 - m_1 - g_1(\lim_{\delta \to 1} a^*(\delta)))(s_2 - d_2(\lim_{\delta \to 1} a^*(\delta)))$$

through $\widehat{s}$ is tangent to the Pareto frontier. The question is whether or not the one pair of $x^m(\delta, a^*(\delta))$ and $y^*(\delta, a^*(\delta))$ corresponding to (9.4) converges to the point s^* with the minimal first component in the set of tangent points $T(\lim_{\delta \to 1} a^*(\delta))$. Such convergence would make life easier, because then $\lim_{\delta \to 1} a^*(\delta)$ can be calculated by minimizing the smallest $s_1^*(a)$ of points $s^*(a) \in T((m_1 + g_1(a), d_2(a)))$ over the set A.

These technical difficulties can be overcome for bargaining problems with a piecewise linear Pareto frontier. The latter implies that the interval $[m_1, s_1^{\max}]$

can be partitioned into a finite number of non-degenerate subintervals and the derivative of the function f_2 is constant on the interior of each interval. Consider some $\delta < 1$ and suppose $\widehat{a} \in A^*(\delta)$ is one of the minimizers in (9.4). Then for δ' sufficiently close to δ we have that $y_1^*(\delta', \widehat{a})$ lies in the same subinterval as $y_1^*(\delta, \widehat{a})$ and, therefore, (9.5) implies that $\widehat{a} \in A^*(\delta')$. This implies that a non-degenerate range of δ's exists for which $a^*(\delta)$ is constant on this range. The next proposition states that, generically, the set $A^*(\delta)$ is constant on the non-degenerate range $(\overline{\delta}, 1)$ for some $\overline{\delta} < 1$ and this result suffices to prove that $\lim_{\delta \to 1} A^*(\delta)$ exists. Moreover, generically, we have convergence of the set F to the set T. For a proof and for the nongeneric conditions for no convergence we refer to Houba (1997).

Proposition 9.5. *Let the Pareto frontier of S be piecewise linear. Then there exists a non-empty set $A^* \subseteq A$ and a $\overline{\delta} < 1$ such that for all $\delta > \overline{\delta}$ the set $A^*(\delta) = A^*$. Moreover, for the generic case* , $\lim_{\delta \to 1} F(\delta, a^*) = T(a^*)$, $a^* \in A^*$.

The last result can be used to construct an algorithm that enables us to compute the set A^*. This algorithm goes roughly as follows. Suppose that there are k line pieces of f_2 that describe the Pareto frontier of S^{IR}. Then $[m_1, s_1^{\max}]$ can be partitioned into k subintervals and the derivative of f_2 on each subinterval is constant. Since f_2 is piecewise linear and concave, there exists k numbers α_ℓ, $\ell = 1, \ldots, k$, such that $\alpha_1 < \ldots < \alpha_k$ and $f_2'(s_1) = -\alpha_\ell$ for s_1 in the ℓ-th subinterval. The number ℓ also counts the number of iterations in the algorithm and it starts with $\ell = 1$.

Iteration ℓ Solve (9.5) with $f_2'(s_1) = -\alpha_\ell$, which yields the set $A^*(\alpha_\ell)$. Next, for each $a^* \in A^*(\alpha_\ell)$ calculate the point $s^*(\alpha_\ell) \in T(a^*(\alpha_\ell))$ with the minimal first component. If the obtained $s_1^*(\alpha_\ell)$ belongs to the ℓ-th subinterval, then the algorithm stops and $A^* = A^*(\alpha_\ell)$. Otherwise, proceed to $\ell + 1$.

In Houba (1997) it is shown that this algorithm always yields an element of A^* within a finite number of iterations.

Some final remarks are in place. First, note that in general every Pareto frontier of S can be approximated by some piecewise linear frontier and, by doing so, an approximation of the set $A^*(\delta)$ is obtained. Therefore, there is not much loss in generality if we assume from here on that $\lim_{\delta \to 1} A^*(\delta)$ exists and that $\lim_{\delta \to 1} F(\delta, a^*(\delta)) = T(\lim_{\delta \to 1} a^*(\delta))$ in the remainder of the analysis.

Second, the prisoners' dilemma already showed that compensation matters. In general this can be seen as follows. No compensation would boil down to imposing the extra restriction $g_1(a^*) = 0$ in finding the optimal a^* that constitutes player 1's worst SPE. Adding an additional constraint usually deteriorates the optimum and, therefore, taking compensation into account is necessary.

9.2.5 Conditions for uniqueness

In this subsection the necessary and sufficient conditions for a unique limit SPE payoff are derived, i.e., $\lim_{\delta \to 1} x^m(\delta, a^*(\delta)) = \lim_{\delta \to 1} x^M(\delta, a^{**}(\delta))$. First, it is shown that the NBS corresponding to the standard bargaining problem (S^{IR}, m), $m = (m_1, m_2)$, always belongs to the limit set of SPE payoffs. Second, this result is used to derive the conditions such that $\lim_{\delta \to 1} x^m(\delta, a^*(\delta))$ and $\lim_{\delta \to 1} x^M(\delta, a^{**}(\delta))$ coincide with this NBS.

The first lemma states that in case of uniqueness there is only one candidate for the limit point and this candidate is always contained in the limit set of SPE payoffs.

Lemma 9.6. *Let $N(S, m)$ be the NBS of (S, d) with $d = m$. Then*

$$N(S, m) \in \lim_{\delta \to 1} E(\delta).$$

Lemma 9.6 implies that the derivation of the necessary and sufficient conditions for uniqueness boils down to deriving necessary and sufficient conditions such that both $x^m(\delta, a^*(\delta))$ and $x^M(\delta, a^{**}(\delta))$ converge to $N(S, m)$. We distinguish two exclusive cases, namely whether or not the Pareto frontier is differentiable in this NBS. First, the conditions are derived in case of differentiability.

Lemma 9.7. *Let $A^* = \lim_{\delta \to 1} A^*(\delta)$, $A^{**} = \lim_{\delta \to 1} A^{**}(\delta)$ and the Pareto frontier be smooth in $N(S, m)$. Then both $x^m(\delta, a^*(\delta))$ and $x^M(\delta, a^{**}(\delta))$ converge to $N(S, m)$ as δ goes to 1 if and only if $d(a^N) = m$, $a^N \in A^N$, and $A^N \subseteq A^* \cap A^{**}$.*

This result can be directly linked to the second geometrical interpretation of section 4.2.2 for $N(S, m)$. To see this, recall that every disagreement point $d' \in S$ that lies on the line through m and $N(S, m)$ yields a NBS $N(S, d')$ equal to $N(S, m)$. Requiring uniqueness means that with respect to the disagreement points $(m_1, d_2(a^N))$ and $(m_1 + g_1(a^*), d_2(a^*)))$, $a^* \in A^*$, we impose that each of these two points lies on the line through m and $N(S, m)$. This can only be accomplished if $d(a^N) = m$ and $(m_1 + g_1(a^*), d_2(a^*)))$ lies on this line. The latter translates into that a^N is also a minimizer of $\lim_{\delta \to 1} x_1^m(\delta, a^*(\delta))$. Note that these conditions are similar to those derived in section 7.3.6, because $N(S, m)$ is the unique SPE agreement of bargaining procedure 7.4 and uniqueness in the current model has to be achieved through Markov strategies.

In case the Pareto frontier is not differentiable in $N(S, m)$, then each of the points $(m_1, d_2(a^N))$ and $(m_2 + g_1(a^*), d_2(a^*)))$, $a^* \in A^*$, has to lie on one of the infinitely many lines through $N(S, m)$ with slope equal to the absolute value of some of the subdifferentiables. Thus, if the Pareto frontier is nondifferentiable in $N(S, m)$, then the sufficient and necessary conditions for uniqueness of lemma 9.7 become less restrictive and are formulated as follows.

Lemma 9.8. *Let $a^* \in \lim_{\delta\to 1} A^*(\delta)$, $a^{**} \in \lim_{\delta\to 1} A^{**}(\delta)$ and the Pareto frontier be nondifferentiable in $N(S,m)$. Both $x^m(\delta, a^*(\delta))$ and $x^M(\delta, a^{**}(\delta))$ converge to $N(S,m)$ as δ goes to 1 if and only if*

1 $N(S,(d_1(a^N),m_2)) = N(S,(m_1,d_2(a^N))) = N(S,m)$ *for all* $a^N \in A^N$,

2 $N(S,m) \in T(a^*)$ *and* $N_1(S,m) = \min_{s_1}\{s_1 \mid s \in T(a^*)\}$,

3 $N(S,m) \in T(a^{**})$ *and* $N_2(S,m) = \min_{s_2}\{s_2 \mid s \in T(a^{**})\}$.

Note that the first condition is weaker than the condition $d(a^N) = m$ stated in lemma 9.6, because the pairs of Nash equilibrium utilities do not have to coincide with the point m. The first condition of lemma 9.8 allows the Nash equilibrium utilities to differ from m up to the line where a further shift of the Nash equilibrium utilities would shift the Nash bargaining solution $N(S,(d_1(a^N),m_2))$ away from $N(S,m)$. Formally, in terms of the set $D(s^*)$ as defined in section 4.2.2, the first condition requires that $d(a^N)$ belongs to the set $D(s^*)$ with $s^* = N(S,m)$. Thus, as long as alternating between minmaxing one of the players and the Nash equilibrium does not yield an agreement that is different from $N(S,m)$ the pairs of Nash equilibrium utilities remain within the bounds that guarantee that $N(S,(d_1(a^N),m_2))$ and $N(S,(m_1,d_2(a^N)))$ coincide with $N(S,m)$. The last two conditions of lemma 9.8 require that $N(S,m)$ simultaneously corresponds to player 1's worst and best SPE agreement.

9.2.6 Characterization of SPE utilities

In this subsection the set of SPE payoffs is characterized. First, the method in Shaked and Sutton (1984) is applied in order to show that the punishment strategies of table 9.2 are indeed the bounds upon the players' SPE payoffs. Next, it is shown that the punishment strategies are SPE. Finally, the limit set of SPE payoffs is characterized as the risk of breakdown vanishes.

Recall that the set of equilibrium payoffs at the start of each bargaining round is denoted as $E(h_1(t),t)$, $t \in \mathbb{N}$ and $h_1(t) \in H_1(t)$. Similar as in lemma 3.17 it can be shown that there exists two sets E_1 and E_2 such that $E(h_1(t),t) = E_1$ for all even t and $h_1(t) \in H_1(t)$ and $E(h_1(t),t) = E_2$ for all odd t and $h_1(t) \in H_1(t)$, because the bargaining problem with endogenous disagreement actions is stationary. As in section 3.4.1, we define $\underline{s}_i^j(\delta)$ and $\overline{s}_i^j(\delta)$, $i,j = 1,2$, as the lower and upper bound upon player i's equilibrium payoffs at the start of every bargaining round player j makes the proposal. The first theorem relates these bounds to the strategies of table 9.2. This then implies that these strategies can indeed be regarded as each player's worst SPE.

Proposition 9.9. *For each $\delta < 1$ the bounds upon the sets $E_1(\delta)$ and $E_2(\delta)$ are given by*

$$\begin{aligned} \underline{s}_1^1(\delta) &= x_1^m(\delta, a^*(\delta)), & \overline{s}_2^1(\delta) &= f_2(\underline{s}_1^1(\delta)), \\ \overline{s}_1^1(\delta) &= x_1^M(\delta, a^{**}(\delta)), & \underline{s}_2^1(\delta) &= f_2(\overline{s}_1^1(\delta)), \\ \underline{s}_2^2(\delta) &= y_2^M(\delta, a^{**}(\delta)), & \overline{s}_1^2(\delta) &= f_1(\underline{s}_2^2(\delta)), \\ \overline{s}_2^2(\delta) &= y_2^m(\delta, a^*(\delta)), & \underline{s}_1^2(\delta) &= f_1(\overline{s}_2^2(\delta)). \end{aligned}$$

This last proposition characterizes the equilibrium bounds, but this does not yet mean that we have shown that the strategies of table 9.2 are indeed SPE strategies. Now, the arguments presented thus far in this chapter can be used to prove that this is indeed the case. The only deviations not yet discussed are deviations that trigger a switch from one state to the other. In the first state, i.e., player 1 is punished, there are three types of subgames for which this is the case. In the first type of subgame the strategies prescribe that player 2 proposes $y^*(\delta, a^*(\delta))$. Deviations are punished and yield player 2 the payoff $y_2^M(\delta, a^*(\delta))$. Such deviation is not profitable if and only if $y_2^M(\delta, a^{**}(\delta)) \leq y_2^*(\delta, a^*(\delta))$. The other two type of subgames refer to the playing of the disagreement game Γ. Player 2 does not deviate from the minmax actions a^1 a condition similar to (9.1) holds and rewriting yields

$$(1-\delta)\,\delta^{-1} g_2(a^1) \leq x_2^M(\delta, a^{**}(\delta)) - x_2^m(\delta, a^*(\delta)).$$

As δ goes to 1 the left-hand side goes to 0, whereas the limit of the right-hand side is positive in case of multiplicity. Similar, player 2 does not deviate from the actions $a^*(\delta)$ in Γ if and only if

$$(1-\delta)\,\delta^{-1} g_2(a^*(\delta)) \leq y_2^M(\delta, a^{**}(\delta)) - y_2^*(\delta, a^*(\delta)).$$

In case of multiplicity of SPE payoffs all three conditions are satisfied for sufficiently large $\delta < 1$. This leads to the following proposition.

Proposition 9.10. *If the conditions of lemma 9.7 and lemma 9.8 do not hold, then there exists a $\overline{\delta} < 1$ such that for all $\delta > \overline{\delta}$ the strategies of table 9.2 are SPE strategies.*

In all the previous chapters multiplicity of SPE payoffs also implies that the closure of the set of SPE payoffs is the set $\{s \in S^{IR} |\, s_1 \geq \underline{s}_1, s_2 \geq \underline{s}_2\}$, where $\underline{s}_i = \min\{s_i | s \in T(S,d)\}$. A similar results holds also here and we state it without a proof, because the arguments are familiar by now.

Theorem 9.11. *The limit set of SPE utility pairs is given by*

$$\lim_{\delta \to 1} E_j(\delta) = \{s \in S^{IR} |\, s_1 \geq \underline{s}_1,\ s_2 \geq \underline{s}_2\}, \quad j = 1, 2,$$

where $\underline{s}_1 = \lim_{\delta \to 1} x_1^m(\delta, a^*(\delta))$ *and* $\underline{s}_2 = \lim_{\delta \to 1} x_2^M(\delta, a^{**}(\delta))$.

We conclude this section with a simple example in which the theory derived is applied to one of the bargaining models analyzed in the previous chapter.

Example 9.12. *Consider the wage bargaining problem of section 8.3 once more. The disagreement game* Γ *is given by*

	Continue
Holdout	$w_0, 1 - w_0$
Strike	$0, 0$

Then $d(a^N) = d(a^1) = (w_0, 1 - w_0)$, $d(a^2) = (0,0)$ *and* $g_2(a) = 0$ *for all* $a \in A$. *It is easy to see that* $A^* = A^{**} = \{a^N\}$. *Thus,*

$$\left(m_1 + (1-\delta)\,\delta^{-1} g_1(a^*), d_2(a^*)\right) = (w_0, 1 - w_0)$$

and

$$\left(d_1(a^{**}), m_2 + (1-\delta)\,\delta^{-1} g_2(a^{**})\right) = (w_0, 0).$$

These are exactly the constructed disagreement points obtained in section 8.3.

9.3. Policy Bargaining

Policy bargaining problems refer to economic situations in which two economic agents interact strategically with each other over time and simultaneously have the opportunity to negotiate over some joint policy which will be implemented if agreed upon. For example, recall the discussion of the WTO negotiations in section 1.2 (of chapter 1), where the subject of the negotiations are the nations' trade barriers and, simultaneously, each nation can use its import barriers as a threat, i.e., a trade war, in the ongoing negotiations. So, roughly speaking, the joint policy concerns the future use of threats. To give another example, most EC nations fish in the territorial EC waters and, in the past, these countries negotiated quotas, i.e., a joint policy, for each member. Negotiations for fish quota are deferred to chapter 10.

The class of policy bargaining problems are a special case of the bargaining problems analyzed in the previous section, because the Pareto frontier of the set S coincides with the Pareto frontier, expressed in utility pairs, of the disagreement game. So, with respect to the equilibrium analysis we are there yet. However, this class of bargaining problems deserves some attention on its own, especially because this class of problems provides the framework to study for example the negotiations within the WTO. This framework also enables us to address the questions of renegotiation and nonbinding agreements in section 9.3.3 and 9.3.4.

9.3.1 The policy bargaining model

Policy bargaining problems have the following stylized features: First, each party's decisions concerning economic variables impose external effects on the other player. Second, the subject of the negotiations is a binding *joint policy* for the parties in order to internalize the externalities of 'competition'. Third, the economy and the negotiations proceed simultaneously. Fourth, the parties are not able to commit in advance on their decisions concerning the economy in case of disagreement. The bargaining models analyzed in chapter 7 do not fit all stylized features, but the bargaining model with endogenous threats introduced in chapter 2 does. In terms of the latter model the idea is that the disagreement game Γ represents the economic environment in which both parties operate and that the contract space consists of joint policies in the infinitely repetition of Γ. So, policy bargaining problems impose a mild additional restriction upon the contract space and, thereby, it is slightly less general in structure than the model analyzed in the previous section. Given the economic relevance of policy bargaining problems these problems deserve some attention on their own.

The economic environment in which the two parties negotiate over a joint policy is represented by the standard infinitely repeated game corresponding to Γ. A *joint policy* c^t proposed at t, $t \in \mathbb{N}$, is defined as an *infinite* sequence of actions $a(t+\tau) \in A$, $\tau \in \mathbb{N}$. Thus, $c^t \in C = \prod_{\tau=0}^{\infty} A$. An *agreement* c^t agreed upon at time t, $t \in \mathbb{N}$, is a joint policy agreed upon by both parties at time t and, similar as imposed in section 2.2.1, it is assumed to be *binding* and immediately becomes effective. The latter means that once the parties agree upon $c^t \in C$ that they implement the specified actions in the current and all future periods.

Assumption 9.13. *An agreement $c \in C = \prod_{\tau=0}^{\infty} A$ reached at t, $t \in \mathbb{N}$, is binding and specifies a joint policy of infinite length starting immediately at t.*

This assumption begs the question what the set of utility pairs is that can be attained by all nonstationary paths as agreements, which we now address. The normalized utility or payoff to player i, $i = 1, 2$, of the joint policy $c^t \in C$ accepted at time t is assumed to be $u_i\left(c^t\right) = (1-\delta)\sum_{\tau \in \mathbb{N}} \delta^{\tau} d_i(a(t+\tau))$. The associated set of attainable utility pairs $d(A) \subset \mathbb{R}^2$ in Γ is defined by $d(A) = \{s \in \mathbb{R}^2 |\ \exists a \in A \text{ s.t. } s = d(a)\}$ and this set may not be a convex set. Fortunately, the set of attainable utility pairs in the infinitely repeated game associated with Γ is always the convex hull of the set $d(A)$ for $\delta \geq \frac{1}{2}$. The latter is first shown in Sorin (1985), where it is shown that for $\delta \geq \frac{1}{2}$ in a two-player game it holds that every s in the convex hull of $d(A)$ is the normalized payoff of some infinite stream of at most two distinct payoffs $\widehat{s}, \tilde{s} \in d(A)$, e.g., see Farrel and Maskin (1989) and Fudenberg and Maskin (1986) for similar convexity results. This result implies that the set of attainable utility pairs S in the policy bargaining problem is the convex hull of the set $d\,(A)$ for $\delta \geq \frac{1}{2}$ and, therefore, the policy bargaining problem is a *convex* bargaining problem for every normal form game Γ. Since the limit δ goes to 1 is of main interest the condition $\delta \geq \frac{1}{2}$ is automatically met sufficiently near the limit and there is no loss in generality to work with a convex S as the set of feasible payoffs.[1]

In what follows we will work in the utility representation. For games Γ with utility functions d_i that are strict quasiconcave in a_i the Pareto frontier of S will consists of stationary contracts $c^t = \left(a^{PE}, a^{PE}, \ldots\right)$, where a^{PE} is Pareto efficient pair of actions in Γ. However, for the class of bimatrix games Γ every $s \in S$ is likely to represent multiple utility-equivalent joint policies in C. Recall from section 9.2.4 that we additionally assumed a piecewise linear Pareto frontier in order to overcome several technical problems. In the context of the policy bargaining model we mention that every game Γ can be arbitrarily close approximated by some bimatrix game Γ' and every bimatrix game admits a piecewise linear Pareto frontier. So, it is without much loss of generality to assume that Γ is a bimatrix game.

The bargaining model is rich enough to analyze renegotiation of agreements. Renegotiation has to be considered, because it is unlikely that players will refrain from bargaining after they have agreed upon a joint policy that is not Pareto efficient. However, in subsection 9.3.3 it is shown that the presence of renegotiation does not affect the set of SPE utility pairs. So, the following assumption is innocuous.

Assumption 9.14. *No renegotiation of existing agreements takes place.*

9.3.2 Characterization of SPE utilities

Since the policy bargaining problem is a special case of the bargaining model analyzed in section 9.2 it is immediately clear that the characterization result of that section applies to the policy bargaining model and we will not repeat it here. Instead we want to point at a novel feature in the policy bargaining model: In case of multiplicity of SPE utility pairs it is possible to construct SPE strategies with perpetual disagreement.

The strategies that support perpetual disagreement are based upon equilibrium switching and are similar as the strategies introduced in section 5.4.4 and 7.3.5 with two exceptions. First, we have to add suitable disagreement actions and, second, the column with the state A that deals with agreement at $t(\sigma^*)$ can be discarded because perpetual disagreement means $t(\sigma^*) = \infty$. The behaviour in the initial state is given in table 9.3 and the state transitions to state 1 and 2 refer to these states in table 9.2. The strategies prescribe the behaviour in which player i, $i = 1, 2$, always demands his utopia payoff $s_i^{\max}$ over and over again and his opponent rejects. In case of disagreement the players play Γ according to the actions $\widehat{a} \in A$ over and over again. If one of the players deviates from this behaviour, this player is punished with his worst SPE strategies. This

State		Initial	
Player 1	proposes	$(s_1^{\max}, s_2^*)$	always
	responds *yes*		if $s_1^t \geq s_1^{\max}$
	action	$\hat{a}_1$	always
Player 2	proposes	$(s_1^*, s_2^{\max})$	always
	responds *yes*		if $s_2^t \geq s_2^{\max}$
	action	$\hat{a}_2$	always
Transition		Go to state 1, if $s_1^t < s_1^{\max}$, t even, or $a_1(t-1) \neq \hat{a}_1$	
		Go to state 2, if $s_2^t < s_2^{\max}$, t odd, or $a_2(t-1) \neq \hat{a}_2$	

Table 9.3. Perpetual disagreement SPE strategies, where 'state 1' and 'state 2' refer to table 9.2.

behaviour results in an infinite stream of per-period utilities associated with $\widehat{a}$ and its normalized discounted utility is equal to $(1-\delta)\sum_{t=0}^{\infty}\delta^t d(\widehat{a}) = d(\widehat{a})$.

The equilibrium analysis is rather simple. For sufficiently large $\delta < 1$ the subgame perfectness of the behaviour in state 1 and state 2 is already established. Furthermore, player 1 does not deviate from playing $\widehat{a}_1$ if a condition similar to (9.1) holds in odd and even periods and when combined and rewritten imply

$$(1-\delta)\, g_1(\widehat{a}) \leq \delta\left[\min\left\{x_1^m(\delta, a^*(\delta)), y_1^m(\delta, a^*(\delta))\right\} - d_1(\widehat{a})\right].$$

It is easy to see that the latter inequality holds for sufficiently large $\delta < 1$ in case $d_1(\widehat{a}) > \underline{s}_1$. Similar, player 2 does not deviate from $\widehat{a}_2$ for sufficiently large $\delta < 1$ if $d_2(\widehat{a}) > \underline{s}_2$. This yields the following result.

Proposition 9.15. *Let $\widehat{a} \in A$ be such that $d_i(\widehat{a}) > \underline{s}_i$, $i = 1, 2$. Then the strategies of table 9.3 are SPE strategies for sufficiently large $\delta < 1$. Moreover, perpetual disagreement is a SPE outcome.*

This result is similar as the one stated in corollary 7.22. Furthermore, it can be easily extended to nonstationary outcome paths $p(\sigma)$ associated with perpetual disagreement as long as the continuation utilities $U_i^k(p(\sigma);\delta)$ from each round $k \in \mathbb{N}$ onward satisfy $\lim_{\delta \to 1} U_i^k(p(\sigma);\delta) > \underline{s}_i$, $i = 1, 2$.

9.3.3 Renegotiation of Agreements

In this subsection it is shown that dropping the assumption that no renegotiation of (inefficient) agreements takes place does not affect the set of SPE payoffs as derived in theorem 9.11. Renegotiation of agreements can be incorporated in a very natural way. In case the two players have agreed upon an agreement, called the *existing agreement*, the bargaining continues as before and the assumption of binding agreements means that as long as the players do not reach a new agreement both players implement the actions specified by the existing agreement.

Any subgame in which only renegotiation takes place boils down to the modified alternating offer model with exogenously given disagreement payoffs. The results of section 6.3.2 (for alternating disagreement actions) already imply that the latter model may admit multiple SPE's due to the non-linearity of the Pareto frontier. However, even without being able to derive unique SPE payoffs for the subgame with renegotiation, the set of SPE payoffs can be characterized. First, since every agreement in the strategies of table 9.1 and 9.2 is Pareto efficient it is obvious that these strategies are also SPE strategies in the model with renegotiation. Second, the strategies in the model with renegotiation not only prescribe the player's behaviour as long as they do not agree but also prescribe the players' behaviour during the renegotiation. Let superscript R denote the utilities of a proposed joint policy taking into account future renegotiation. As in the proof of proposition 9.9 denote $\underline{s}_1^{1,R}$ as the infimum of player 1's SPE payoffs at $t+2$ even. Applying the arguments in the proof of proposition 9.9 yields that player 1 can secure $(1-\delta)\, m_1 + \delta \underline{s}_1^{1,R}$ at $t+1$ and, therefore, player 2 can obtain at most $f_2((1-\delta)\, m_1 + \delta \underline{s}_1^{1,R})$. In order to sustain $a^R \in A$ as the disagreement actions in any odd period (the reasoning is similar as in the proof of proposition 9.9) player 2's supremum payoff is at most $f_2((1-\delta)\left[m_1 + \delta^{-1} g_1(a^R)\right] + \delta \underline{s}_1^{1,R})$ and player 1 cannot do better than

$$\underline{s}_1^{1,R} \geq f_1\left((1-\delta) d_2(a^R) + \delta f_2\left((1-\delta)\left[m_1 + \delta^{-1} g_1(a^R)\right] + \delta \underline{s}_1^{1,R}\right)\right).$$

As in the proof of theorem 9.9, this implies $\underline{s}_1^{1,R} = x_1^m(\delta, a^*(\delta))$. Hence, player 1's worst SPE payoff does not change by allowing for renegotiation. The following theorem summarizes the main conclusion of this subsection.

Theorem 9.16. *The limit set of SPE utilities under renegotiation is equal to limit set of SPE utilities given in theorem 9.11.*

9.3.4 Nonbinding agreements

In several real-life policy bargaining situations agreements are not binding, i.e., there is no third authority that enforces that both parties implement the agreement. This is the case in the past WTO agreements, but also in case of (secret) illegal cartel agreements. In this section nonbinding agreements in policy bargaining are briefly investigated by means of a simple example with common interest and a finite horizon of one single bargaining round. For this simple example it is shown that nonbinding agreements are devastating to the previous results.

For explanatory reasons we assume that the game Γ is played just once and that there is one single round of bargaining in which player 1 proposes a joint policy c how to play Γ and player 2 responds with *yes* or *no*. After the response

Player 1	proposes	(L_1, L_2)	
	action	R_1	for every h_3
Player 2	responds *yes*		if $c \in \{(L_1, L_2)\}$
	action	R_2	for every h_3

Table 9.4. Strategies that support an SPE in which both players agree upon the Pareto efficient joint policy (L_1, L_2) but both players deviate by playing the inefficient (R_1, R_2).

of player 2 the game Γ is played once. We consider the common interest game Γ given by

	L_2	R_2
L_1	$2,2$	$0,0$
R_1	$0,0$	$1,1$

The set of pure joint policies $C = \{(L_1, L_2), (L_1, R_2), (R_1, L_2), (R_1, R_2)\}$ and this game has two pure NE, namely the Pareto efficient equilibrium (L_1, L_2) and the inefficient equilibrium (R_1, R_2).

Subgame perfectness requires that always one of the two NE of Γ is played. Formally, let H_3 denote the set of all histories up to the playing the game Γ. Then subgame perfectness of the strategies σ^* only imposes that $\sigma^*(h_3) \in \{(L_1, L_2), (R_1, R_2)\}$, $h_3 \in H_3$. However, if the history h_3 implies that the players agreed upon (L_1, L_2), then not only implementing this agreement is a possible equilibrium continuation but also using the actions of the Pareto dominated NE is an equilibrium continuation. So, there exists a SPE outcome path in which player 1 proposes $c = (L_1, L_2)$, player 2 agrees, the players implement the agreement and (R_1, R_2) are the threats if no agreement is reached. This SPE can be interpreted as if the players negotiate their way out of the inefficient NE and this conforms to our intuition about common interests bargaining problems.

However, there also exist SPE strategies in which both players agree upon $c = (L_1, L_2)$ and then both players deviate from their agreement by choosing R_i, $i = 1, 2$, instead of L_i. The strategies in table 9.4 support the latter SPE. At first glance this result seems counter intuitive, because one would expect that negotiations in a common interest bargaining problem not only result in an agreement on (L_1, L_2) but also that this agreement is carried out. However, in every SPE each player will choose an optimal action in Γ given the beliefs (or expectations) about the opponent's behaviour and if both players believe that the opponent ignores any outcome of the bargaining phase and that (R_1, R_2) is going to be played, then their behaviour is sensible. For instance, in many war situations the parties agree upon a cease fire (which is a non-binding agreement), but often each of them does not trust (read: believe) the opponent to act according to this agreement and continuing the fight results in this case, as for example happened many times in the early nineties in Bosnia. Then continuing the fight is a self-fulfilling prophecy.

The discussion above indicates that negotiations with nonbinding contracts yields the same set of equilibrium utility pairs as the standard normal-form game Γ without any negotiations. The reason is that every NE of Γ can be translated into a SPE of the policy bargaining model in which the strategies prescribe that this particular NE is played when Γ is played and to 'ignore' what happened at the negotiation table. This insight easily extends to infinitely repeated games and the policy bargaining model with nonbinding agreements and it then reads as follows.

Theorem 9.17. *The limit set of SPE utility pairs in the policy bargaining game with nonbinding agreements is equal to the limit set of SPE utility pairs in the standard infinitely repeated game associated with the disagreement game and, therefor,* $\lim E(\delta) = S^{IR}$.

Thus, every individual rational payoff belongs to the set of SPE utilities in the policy bargaining model with nonbinding agreements and this result is equal to the folk theorem in infinitely repeated games, e.g., Fudenberg and Maskin (1986). The negative implication is clear: Nonbinding agreements are devastating to the results obtained earlier and imply that no reduction in the set of SPE utility pairs can be expected if negotiations over joint policies are added to a standard infinitely repeated game. This has some far reaching consequences for the literature on renegotiation-proof equilibria for (in)finitely repeated games not only with two players but with an arbitrary number of players, e.g., Asheim (1991), Benoît and Krishna (1993), Bernheim, Peleg, and Whinston (1987), Bernheim and Ray (1989), Farrel and Maskin (1989), Ray (1994) Rubinstein (1980) and Schultz (1994).[2]

The particular example discussed is an adapted version of the example in Farrel (1988),[3] where the additional assumption is made that each player believes that if one of the NE is agreed upon as the joint policy that then his opponent will carry out the agreement. In this reference this assumption is motivated by assuming that the players use a common language. However, the assumption of rational players means that both players fully understand the game tree, i.e., speak each others language. Therefore, the nature of the problem is the self-fulfilling beliefs of the players as they choose their action in Γ rather then whether or not the players have a common language. Nevertheless, under this additional assumption the policy bargaining model admits the unique pair of SPE utilities $(2, 2)$.

The renegotiation concepts all presuppose that some modelled negotiations with nonbinding agreements prior to the repeated game refrain the players from playing dominated equilibria. So, these prior negotiations between the players are thought off as a means for the players to negotiate their way out of every dominated equilibria. Obviously, the implicit assumption is that all parties believe that agreed upon equilibria will be followed. If renegotiation is modelled as the policy bargaining model with nonbinding agreements, then theorem 9.17 implies that the equilibrium set of utility pairs still admits an awful lot of dominated equilibrium pairs. Hence, this result questions the crucial foundation of renegotiation-proof equilibria.

The discussion above indicates that the renegotiation-proof concepts can only be salvaged in a policy bargaining model with nonbinding agreements if it is explicitly assumed that there is a set of SPE strategies that are 'renegotiation' proof and all players believe that agreements upon such renegotiation-proof equilibria will be followed. For finite horizon versions of such policy bargaining problem we conjecture that the results in Benoît and Krishna (1993) will be obtained and for the infinite horizon case the results will be similar to those in Ray (1994). Moreover, without going into details, we mention that it can be shown that the set of strong perfect equilibria analyzed in Rubinstein (1980) always belong to the set of SPE utility pairs in such policy bargaining problems. In particular, for the prisoners' dilemma the latter result and the results in Van Damme (1989) immediately imply that the entire Pareto efficient frontier of S^{IR} can be sustained as SPE in the policy bargaining model.

9.4. Numerical Examples

This section contains several interesting examples. The first two examples briefly state the formal results for the examples discussed in section 9.2.2. In the third example uniqueness is established. These examples show the variety of results that can possibly be obtained.

Example 9.18. *Consider the bargaining problem (S, A, d) in utility representation with $S = \{ s \in \mathbb{R}^2_+ \mid s_1 + s_2 \leq 1 \}$ and the utility pairs (or payoffs) in the disagreement game Γ given by the prisoners' dilemma of section 9.2.2. Then $A^N = a^1 = a^2 = \{(R_1, R_2)\}$. The set $A^* = \{(L_1, R_2)\}$ of proposition 9.4 is found in the first step of the algorithm given in section 9.2.4 and each element $a^* \in A^*$ maximizes $d_2(a) - g_1(a)$ on A. Thus, $d(a^*) = \left(-\frac{1}{6}, \frac{2}{3}\right)$ and $g_1(a^*) = \frac{1}{6}$. The disagreement point becomes $\left(m_1 + \delta^{-1} g_1(a^*), d_2(a^*)\right) = \left(\frac{1}{6\delta}, \frac{2}{3}\right)$. Then solving system (9.2) yields*

$$x^m = \left(\frac{1}{2(1+\delta)}, \frac{1+2\delta}{2(1+\delta)}\right), \qquad y^m = \left(\frac{\delta}{2(1+\delta)}, \frac{2+\delta}{2(1+\delta)}\right),$$

$$y^* = \left(\frac{\delta}{2(1+\delta)} + \frac{1-\delta}{6\delta}, \frac{2+\delta}{2(1+\delta)} - \frac{1-\delta}{6\delta}\right).$$

Furthermore, the proposals $x^m(a^*, \delta)$, $y^m(a^*, \delta)$ and $y^*(a^*, \delta)$ converge to $\left(\frac{1}{4}, \frac{3}{4}\right)$ as δ goes to 1, which maximizes the Nash product $(s_1 - \frac{1}{6})(s_2 - \frac{2}{3})$ on *the Pareto frontier. By symmetry,* $A^{**} = \{(R_1, L_2)\}$ *and*

$$x^M = \left(\tfrac{2+\delta}{2(1+\delta)}, \tfrac{\delta}{2(1+\delta)}\right), \qquad y^M = \left(\tfrac{1+2\delta}{2(1+\delta)}, \tfrac{1}{2(1+\delta)}\right),$$
$$x^{**} = \left(\tfrac{2+\delta}{2(1+\delta)} - \tfrac{1-\delta}{6\delta}, \tfrac{\delta}{2(1+\delta)} + \tfrac{1-\delta}{6\delta}\right).$$

These proposals converge to $\left(\frac{3}{4}, \frac{1}{4}\right)$ *as* δ *goes to 1. Hence, theorem 9.11 implies that the limit set is equal to* $\left\{ s \in S |\, s_i \geq \frac{1}{4}, i = 1, 2\right\}$.

Example 9.19. *Consider the policy bargaining problem* (S, A, d) *in utility representation associated with the game* Γ *given by the prisoners' dilemma of section 9.2.2. Then* $A^N = a^1 = a^2 = \{(R_1, R_2)\}$, *the set* S *is equal to the convex hull of the four utility pairs in* Γ *and* S^{IR} *consists of all* $s \in \mathbb{R}^2_+$ *such that*

$$s_2 \leq \begin{cases} \frac{5}{8} - \frac{1}{4}s_1, & s_1 \in \left[0, \frac{1}{2}\right], \\ \frac{5}{2} - 4s_1, & s_1 \in \left(\frac{1}{2}, \frac{5}{8}\right]. \end{cases}$$

The set $A^*(\delta) = \{(L_1, R_2)\}$ *of proposition 9.4 is found in the first step of the algorithm given in section 9.2.4 and each element* $a^* \in A^*$ *maximizes* $d_2(a) - \frac{1}{4}g_1(a)$ *on* A. *Thus,* $d(a^*) = \left(-\frac{1}{6}, \frac{2}{3}\right)$ *and* $g_1(a^*) = \frac{1}{6}$. *Then the disagreement point* $\left(m_1 + \delta^{-1}g_1(a^*), d_2(a^*)\right)$ *is equal to* $\left(\frac{1}{6\delta}, \frac{2}{3}\right)$ *and solving system (9.2) yields*

$$x^m = y^m = \left(0, \tfrac{5}{8}\right) \quad \text{and} \quad y^* = \left((1-\delta)\tfrac{1}{6\delta}, \tfrac{5}{8} - (1-\delta)\tfrac{1}{24\delta}\right).$$

The disagreement point $\left(m_1 + \delta^{-1}g_1(a^*), d_2(a^*)\right) = \left(\frac{1}{6}\delta^{-1}, \frac{2}{3}\right)$ *lies above the Pareto frontier of* S^{IR} *and, hence, proposition 6.13 applies, i.e.,* $x_1^m < y_1^m$. *Furthermore, the proposals* x^m, y^m *and* y^* *converge to* $\left(0, \frac{5}{8}\right)$ *as* δ *goes to 1, which maximizes the Nash product* $(s_1 - \frac{1}{6})(s_2 - \frac{2}{3})$ *on the Pareto frontier. By symmetry,* $A^{**} = \{(R_1, L_2)\}$ *and*

$$x^{**} = \left(\tfrac{5}{8} - (1-\delta)\tfrac{1}{24\delta}, (1-\delta)\tfrac{1}{6\delta}\right) \quad \text{and} \quad x^M = y^M = \left(\tfrac{5}{8}, 0\right).$$

These proposals converge to $\left(\frac{5}{8}, 0\right)$ *as* δ *goes to 1. Hence, theorem 9.11 implies that the limit set is equal to* S^{IR}. *Since every individually rational payoff can be sustained as an SPE payoff this result is very similar to the folk theorem in standard repeated games, e.g., Fudenberg and Maskin (1986).*

Example 9.20. *Let* $\alpha \in \left(1\frac{1}{2}, 3\frac{1}{2}\right)$ *and consider the policy bargaining model* (S, A, d) *associated with an adapted version of the advertisement game* Γ *treated in Farrel and Maskin (1989) given by*

	L_2	M_2	R_2
L_1	$\left(\frac{1}{3}, \frac{1}{3}\right)$	$(1, 0)$	$\left(-\frac{7}{3}, \frac{2}{3}\alpha\right)$
M_1	$(0, 1)$	$(0, 0)$	$(0, 0)$
R_1	$\left(\frac{2}{3}\alpha, -\frac{7}{3}\right)$	$(0, 0)$	$(0, 0)$

Then $A^N = a^1 = a^2 = (R_1, R_2)$ *and* $d(a^N) = 0$ *imply that the set* $S^{IR} = \{s \in \mathbb{R}^2_+ \mid s_1 + s_2 \leq 1\}$. *The set* $A^* = A^{**} = A^N$. *Thus, the conditions of lemma 9.7 for uniqueness apply. Therefore, the MPE of proposition 9.1 is the unique SPE and the limit set of SPE payoffs is* $\{(\frac{1}{2}, \frac{1}{2})\}$ *being the NBS of* $(S^{IR}, 0)$. *Note that solving system (9.2) yields*

$$x^m = \left(\tfrac{1}{1+\delta}, \tfrac{\delta}{1+\delta}\right) \quad \textit{and} \quad y^m = y^* = \left(\tfrac{\delta}{1+\delta}, \tfrac{1}{1+\delta}\right).$$

Two conclusions can be drawn from this example. First, the policy bargaining model may predict a different set of SPE payoffs than the folk theorem in the standard infinitely repeated game, e.g., Fudenberg and Maskin (1986). Second, in Okada (1991b) the limiting average criterion is assumed (which can be more or less regarded as $\delta = 1$*) and it is shown that the set of SPE payoffs is equal to the set* S^{IR}. *Hence, the limit set of theorem 9.11 does not necessarily coincide with* S^{IR} *as in Okada (1991b). The reason is that in the policy bargaining model time is valuable because of discounting, whereas in the model with the limiting average criterion time is valueless, meaning that* $s \in S$ *at time* $t \in \mathbb{N}$ *has the same value as* s *at* $t = 0$. *A similar result holds for the standard alternating offers procedure where a unique outcome results in case* $\delta < 1$ *whereas for* $\delta = 1$ *every* $s \in S^{PE}$ *is a fixed point of the function* $p \times q$.

Finally, if $\alpha \in (\frac{1}{2}, 1\frac{1}{2})$, *then* $A^* = \{(M_1, L_2)\}$, $A^{**} = \{(L_1, M_2)\}$ *and the limit set of SPE payoffs of theorem 9.11 is equal to* $\{s \in S^{IR} \mid s_1, s_2 \geq \frac{1}{3}\alpha\}$. *Hence, a result 'in between' uniqueness and a folk theorem is also possible.*

9.5. Related literature

The characterization and existence of MPE strategies with endogenous threats was first shown in Okada (1991a). The characterization of the set of SPE utility pairs was conducted in Okada (1991b) under the restriction that the players do not face any risk of breakdown. Then the set of SPE utility pairs coincides with the set S^+, as defined in section 6.2.1. Example 9.20 shows that a vanishing positive risk of breakdown yields a different (limit) set of SPE payoffs. The reason is that the model in Okada (1991b) lacks the time-is-valuable property, i.e., $\delta < 1$, and without this property all other bargaining models in the literature would predict the set S^+ as the set of SPE payoffs, including the standard alternating offers procedure.

The characterization of the optimal punishment strategies with compensation are first proposed in Busch and Wen (1995) for bargaining problems with a linear Pareto frontier. The presentation in this chapter is based upon Houba (1997), in which the generalization to nonlinear Pareto frontiers is analyzed. In many bargaining applications assuming a linear Pareto frontier is without loss of generality, but the analysis in this chapter illustrates one of the very rare occasions that linearity is a very special case. Nevertheless, the basic idea underlying each player's worst SPE strategy as proposed in Busch and Wen (1995) extends to the general case. The analysis in this chapter implies one modification of the results in the latter reference. Note that the results in Bolt (1995) imply that caution should be taken if different discount factors are assumed. In Busch and Wen (1995) the disagreement point is identified with $(m_1, d_2(a^*) - g_1(a^*))$ and also this disagreement point does not represent the true payoffs from perpetual disagreement. However, the results in this chapter reveal that the latter result needs some revision. Finally, in Busch and Wen (2001) the assumption of observably of mixed disagreement actions is dropped. In this reference it is shown that the continuation utilities in each player's worst SPE might be inefficient.

The use of disagreement actions is very flexible, because players can alter these at will in every round. Holden (1994) is the first where less flexible disagreement actions are considered in the context of wage bargaining and it is shown that choosing actions that remain constant for two rounds periods is sufficient to obtain a unique SPE outcome. In Furusawa and Wen (2001) it is assumed that the disagreement actions are chosen once in every $M \geq 1$ bargaining rounds and this limits the set of SPE outcomes. In particular, for every M the limit set of SPE utilities, as the discount factor goes to 1, consists of a single Pareto efficient contract that corresponds to the unique Pareto efficient agreement in Nash's variable-threat game of section 7.2. Furusawa and Wen (2002) apply the latter model to negotiations in an international trade context in which only one country faces restrictions in choosing disagreement actions.

Notes

1 Note that if we would allow the players to correlate on the actions $a \in A$ then the set $d(A)$ is always a convex set. In that case it sufficies to consider stationary paths only. For economic application this route might disguish the problem of a nonconvex set of attainable utilities.

2 The literature on renegotiation-proof equilibria mostly considers games Γ with a finite set of pure actions, such as bimatrix games. In Schultz (1994) it is shown that the set of equilibrium utilities corresponding to several of the renegotiation-proof equilibrium concepts is empty in case the normal-form game has continuous pure actions, as for example in the Cournot duopoly game. Thus, the latter results is an impossibility result, whereas our criticism is methodological.

3 There is one minor difference with Farrel (1988) and that is that the response of player 2 is not part of the model in this reference.

Chapter 10

DESTRUCTIVE THREATS

10.1. Introduction

In the bargaining models analyzed thus far we made the implicit assumption that the use of threats does not affect the contract space. In several real life situations this assumption is violated. For example, in negotiations over fish quota a country's threat to encourage more catches by its national fleet in the current fishing season has a negative effect upon the future stock of fish and the latter determines the future gains of cooperation. Adding intertemporal linkages between past threats and future gains increases the complexity of the bargaining problem.

To this end, we slightly modify the policy bargaining model introduced in the last chapter in such a way that the infinitely repeated game is replaced by a difference game. A *difference game* is a dynamic game in discrete time in which i) each player's intertemporal objective functions depend on the chosen actions and on several variables related to *physical stocks*, and ii) these stocks change over time under the influence of the actions of the players. For instance, investments change the capital stock, depletion changes the resource stock and emissions change the stock of pollutants. In difference games the past matters in two different ways: First, as before, the players can condition their current and future actions upon the observable history of the game. Second, the used actions of the past, i.e., the history, affect the current and future stocks and, by doing so, the current and future gains depend upon the history of the game. The analysis in this chapter restricts attention to two classes of difference games that can be regarded as 'exponential-logarithmic' and 'linear-quadratic' in nature.

The first class is applied to policy bargaining in the context of the 'Tragedy of the Commons'. Loosely stated, the joint exploitation of a common resource leads to inefficient overexploiting of the resource and it is often the cause of disputes between national governments. Disputes may be ended if the governments involved manage to negotiate a joint policy. In section 10.3 we explicitly consider negotiations between two governments over quota concerning catches of fish and the regeneration of fish is modeled by simple biological dynamics. Nevertheless, we regard the underlying economic problem to be more general.

The equilibrium analysis mainly focuses on interior and linear MPE strategies and we show that just one such MPE exists. A novel feature from the perspective of bargaining theory is that both the efficient frontier and the countries' disagreement utilities depend upon the countries' time-preferences. Numerical solutions show that patience is weakness, which is in contrast with the results derived in section 6.3.1. This means that the bargaining position of a country deteriorates if it attaches more importance to future catches. The analysis is concluded with characterizing the set of SPE utility pairs and this result is then related to the strategies discussed in section 9.2.3.

The application to negotiations over fish quota assumes a single species of fish, i.e., one stock variable. The presence of several stock variables is investigated for a different class of difference games, namely those called linear-quadratic. These games have quadratic utility functions and the function governing the evolution of the stock variables is linear. Such functions can be seen as Taylor approximations of more general functional forms and this explains part of the popularity of this class of games in economics. Another reason is that this class of games admits a MPE that is analytically tractable and that has a quadratic structure, meaning the associated value functions are quadratic in the stock variables. With respect to analytically tractable solutions a negative result is obtained for policy bargaining problems, namely a breakdown of the quadratic structure in case of multiple stock variables. Tractability only survives for the case of a single stock variable. For explanatory reasons we conduct the analysis only for a *finite* horizon.

This chapter is organized as follows. Since not all readers are familiar with the framework of difference games we briefly introduce these first, including the two classes of difference games analyzed. Then the analysis proceeds with deriving equilibria for each of these two classes. We conclude with a brief discussion about the general case.

10.2. Difference games

Policy bargaining in the context of difference games requires some additional notation that slightly extends the notation of chapter 2. In this section the general framework of difference games is briefly introduced and how it is incorporated into the policy bargaining model. Next, we discuss the two classes of difference games analyzed in this chapter and derive the MPE strategies for the ordinary difference game without negotiations. We conclude this section with a discussion how to represent the set of Pareto efficient joint policies by a set of 'bargaining' or 'social welfare' weights.

10.2.1 General framework

Difference games introduce a novel feature to the policy bargaining problem and this is the presence of stock or *state* variables. For example, the capital stock or the accumulated pollutants such as carbon-dioxide. The vector of state variables at time $t \in \mathbb{N}$ is denoted by $x(t)$ and it is an element of the set of feasible state vectors $X \subseteq \mathbb{R}^k$ for some $k \geq 1$. (Note that infinitely repeated games can be considered as the special case $k = 0$.) At the start of the negotiations there is an initial state x_0, that is exogenously given. The set of feasible actions for player i, $i = 1, 2$, at time t, $t \in \mathbb{N}$, becomes history dependent through the state and is now defined as $A_i(x(t)) \subseteq \mathbb{R}^{n_i}$ for some $n_i \geq 1$. Similar as before, $A(x(t)) = A_1(x(t)) \times A_2(x(t))$. Furthermore, the union of all sets of disagreement actions $A_i(x(t)), i = 1, 2$, over $x(t) \in X$ is denoted as $A_i(X)$ and, similar, $A(X) = A_1(X) \times A_2(X)$.

The current state and actions determine the state for the next round. The function $f : X \times A(X) \to X$ that governs this relation is called the state transition and it is given by

$$x(t+1) = f(x(t), a(t)), \quad t \in \mathbb{N}, \quad x(0) = x_0,$$

where $a(t) \in A(x(t))$. The players's utility functions are assumed to be additive separable with respect to time and autonomous, i.e., independent of time. Slightly modifying our notation we denote player i's per-period utility at time t as $d_i(x(t), a(t))$, $i = 1, 2$. Overall discounted utility (where we drop the normalization) then becomes

$$\sum_{t \in \mathbb{N}} \delta_i^t d_i(x(t), a(t)), \quad i = 1, 2,$$

where $\delta_i \in (0, 1)$ denotes player i's discount factor. We note that it is customary within the literature on difference games that discounted utilities are not normalized and the policy bargaining models in Houba and De Zeeuw (1995) and Houba, Sneek, and Várdy (2000) follow this tradition. Incorporating normalization is not difficult from a methodological point of view, but since derivations of analytical solutions are tedious we rather avoid redoing all the calculations in these references.

Note that the utility functions have the property that the remaining per-period utilities at time t do not depend directly upon past actions $a(0), \ldots, a(t-1)$, but the remaining utilities may depend indirectly on these actions through the state variable $x(t)$. In other words, the past is sunk with respect to direct utilities of used actions, but not indirectly because the past actions determine the current state. Furthermore, the utility functions from period t onwards will be the same for different histories which result in the same state vector $x(t)$. Markov strategies in difference games are strategies that are a function of the state vector $x(t)$ and the associated value function $V_i : \mathbb{N} \times X \to \mathbb{R}$ is denoted as $V_i(t, x)$, $i = 1, 2$, given the state x at t, $t \in \mathbb{N}$, where we sometimes drop the time index if it does not cause confusion. Obviously, the history of the game is much broader than the physical stock variables, because the stock variables are only a derivative of the full history. This means that even if different histories result in the same level of the state variables then still differences within these histories may matter, provided the players choose it to matter through their strategies. In Markov strategies the players never let it matter, but in previous chapters we have seen SPE strategies where the history did matter.

The question is how to imbed the difference game in the policy bargaining model. The answer is very simple. Similar as for infinitely repeated games, one round of play in the difference game corresponds to one round in the negotiation process with the understanding that the negotiation round takes place before the disagreement actions of that round are chosen. So, similar as in chapter 2, at every round t, $t \in \mathbb{N}$, first one of the players proposes a joint policy followed by his opponent's response and, next, in case of disagreement, the disagreement actions at bargaining round time t correspond to the actions $a(t)$ in the difference game. We maintain the convention that player 1 proposes at even rounds and player 2 at odd rounds.

Consider round t, $t \in \mathbb{N}$, and the state $x(t) = x$. Then a *feasible* joint policy $c^t(x) = (a(t), a(t+1), \dots)$ proposed at round t consists of an infinite stream of actions $a(t+\tau)$, $\tau \in \mathbb{N}$, such that $a(t+\tau) \in A(x(t+\tau))$ for $x(t+\tau+1) = f(x(t+\tau), a(t+\tau))$ and $x(t) = x$, where x can be regarded as the initial state having arrived at round t. The latter conditions impose that the joint policy is feasible and we take it for granted that only feasible joint policies are proposed. Note that we do not impose that these feasible policies are Pareto efficient, because (failure of) this property should follow from the equilibrium analysis. As in section 9.3.1 an agreed upon joint policy $c^t(x)$, $t \in \mathbb{N}$, at round t is binding and becomes immediately effective. The utility of the joint policy $c^t(x)$, evaluated at round t and given the state $x(t) = x$, is denoted as

$$u_i\left(x, c^t(x)\right) = \sum_{\tau \in \mathbb{N}} \delta_i^{\tau} d_i\left(x(t+\tau), a(t+\tau)\right), \quad i = 1, 2.$$

We maintain the assumption of perfect recall and perfect observability of all past actions as made in assumption 2.16. Since each player is assumed to know the structure of the game, which includes the state transition, this means that each player also perfectly knows the current state $x(t)$ at the start of round t.

Then the policy bargaining maintains the property of almost perfect information and the SPE concept remains the appropriate equilibrium concept.

Finally, in order to distinguish the value functions at different stages of the policy bargaining model we add the superscript b to the value functions at the start of each bargaining round. Similar, we add the superscript d for the value functions just after the rejection of a proposal and before the disagreement actions are chosen. Thus, for each round $t \in \mathbb{N}$ and the state $x(t) = x$ at round t we distinguish the value functions $V_i^b(t,x)$ and $V_i^d(t,x)$, $i = 1, 2$.

10.2.2 The great fish war

In the application to negotiations over fish quota we assume that the difference game underlying the policy bargaining problem is the great fish war model in Levhari and Mirman (1980). In this game two countries share a common pool filled with a single species of fish. The one-dimensional state $x(t) \in X \subset \mathbb{R}_+$, $t \in \mathbb{N}$, denotes the amount of fish in the common pool at the beginning of period t. The state is normalized to stay within the interval $[0,1]$, i.e., $X = [0,1]$, and the initial state x_0, $x_0 \in [0,1]$, is exogenously given. Furthermore, the set of admissible actions for country i at t given $x(t)$ is the set $A_i(x(t)) = [0, x(t)]$, $i = 1, 2$. The action $a_i(t) \in [0, x(t)]$ denotes the *planned* catch of country i at t. The population of country i actually catches or consumes the amount $c_i(t)$ given by

$$c_i(t) = \begin{cases} a_i(t), & \text{if } a_1(t) + a_2(t) \leq x(t), \\ \\ \frac{a_i(t)}{a_1(t)+a_2(t)} x(t), & \text{otherwise.} \end{cases} \tag{10.1}$$

The additional variable $c_i(t)$ is needed to guarantee that the total amount of fish caught does not exceed the amount available, i.e. $c_1(t) + c_2(t) \leq x(t)$. After fishing at t, the amount of fish $x(t) - c_1(t) - c_2(t) \geq 0$ regenerates. The simple biological dynamics governing the regeneration follow the state transition given by

$$x(t+1) = (x(t) - c_1(t) - c_2(t))^{\alpha}, \quad t \in \mathbb{N}, \tag{10.2}$$

where $\alpha \in (0,1]$ is a measure of the rate of regeneration. The speed of regeneration slows down if α increases and for the extreme case $\alpha = 1$ it becomes a nonrenewable resource such as an oil well.

The effect of current catches upon the future stock of fish can be seen from two different angles. The first angles regards fishing as a public investment problem. Since catches have a negative effect upon the future stock it can be said that each country unilaterally 'invests' in the future stock of fish by restraining its current period catches. Obviously, such investment is public by nature, because part of the gains accrue to the other country. Thus, 'investments' cause positive externalities that give countries an incentive to free ride by lowering their own investment. This argument provides a simple explanation for why Tragedies of the Commons occur.

The second angle regards catches as threats. The negative effect of threats upon future stocks means that threats are destructive. The issue then becomes how can each country use its destructive threat to obtain a more favourable outcome in the negotiations. Similar as in the modified variable threat game of section 7.3.1 massive destruction of the future surplus is not something that each country prefers, because it will face its consequences if carried out. So, each country should trade off the benefits of using its threat in the negotiations and facing the consequences when called upon to carry out its threat.

The players' per-period utility functions do not depend upon the state, only on catches. These functions are given by

$$d_i\left(x\left(t\right),a\left(t\right)\right)=\ln c_i\left(t\right),i=1,2,$$

where $c_i\left(t\right)$ depends upon $a\left(t\right)$ according to (10.1). Because utilities depend upon the per-period catches $c_i\left(t\right)$ we will write $c^t\left(x\right)=\left(c\left(t\right),c\left(t+1\right),\ldots\right)$ for convenience instead of $c^t\left(x\right)=\left(a\left(t\right),a\left(t+1\right),\ldots\right)$. The intertemporal utility for country i, $i=1,2$, corresponding to the joint policy $c^t\left(x\right)$ at round t, $t\in\mathbb{N}$, given state $x\left(t\right)=x$ is given by

$$u_i(c^t)=\sum_{\tau=0}^{\infty}\delta_i^{\tau}\ln c_i\left(t+\tau\right),\quad i=1,2.$$

This utility function is the logarithmic form of a Cobb-Douglas utility function with infinitely many commodities.

The fish war model admits an interior and linear MPE that is analytically tractable, e.g., Levhari and Mirman (1980). By *linear* we mean that the catches are linear functions of $x\left(t\right)$ and *interior* refers to $c\left(x\left(t\right)\right)>0$ and $c_1\left(x\left(t\right)\right)+c_2\left(x\left(t\right)\right)<x\left(t\right)$ for all $t\in\mathbb{N}$. Thus, $a\left(t\right)=c\left(t\right)$ in terms of (10.1). The associated value functions have the logarithmic form. The linear and interior MPE is given in the following proposition, which is given without proof.

Proposition 10.1. *The MPE disagreement catch at round t for state* $x\left(t\right)=x$ *at t is given by*

$$c_i\left(x\right)=\frac{\delta_j\left(1-\alpha\delta_i\right)}{\delta_1+\delta_2-\alpha\delta_1\delta_2}x,\quad i,j=1,2,\quad i\neq j,$$

the state evolves as

$$x\left(t+1\right)=\left(\frac{\alpha\delta_1\delta_2}{\delta_1+\delta_2-\alpha\delta_1\delta_2}\right)^{\alpha}x^{\alpha}$$

and the value function $V_i(t,x) = \ln \tilde{A}_i + \frac{1}{1-\alpha\delta_i} \ln x$, *where*

$$\tilde{A}_i = \left(\frac{\delta_j (1-\alpha\delta_i)(\alpha\delta_1\delta_2)^{\frac{\alpha\delta_i}{1-\alpha\delta_i}}}{(\delta_1+\delta_2-\alpha\delta_1\delta_2)^{\frac{1}{1-\alpha\delta_i}}} \right)^{\frac{1}{1-\delta_i}} \in (0,1).$$

This proposition implies that $c_1(x) \le c_2(x)$ if and only if $\delta_1 \ge \delta_2$. Thus, the impatient country 'invests' less in the future surplus than the more patient country because doing so is more costly for the impatient country. Furthermore, it can be shown that this MPE is Pareto inefficient meaning overexploiting takes place and there is room for negotiations.

10.2.3 Linear-quadratic difference games

The analysis of multi-dimensional stock variables in policy bargaining problems is for explanatory reasons restricted to finite-horizon *linear-quadratic* (LQ) difference games. Such games are defined as difference games with quadratic utility functions, a linear state transition and a finite time horizon. Similar as for the great fish war, this class of LQ games is also analytically tractable, which is demonstrated at the end of this section.

The finite time horizon means that for some $t_f < \infty$ the players choose their actions in the difference game in rounds $t \in T = \{0, \dots t_f\}$, where t_f stands for the final round where the players are *active* in the game. The state variables are defined as the stocks at the beginning of each round and, similar to there being an initial state, there is also a final state that results after all actions are taken. The final state is associated with the state at time $t_f + 1$ meaning that $x(t) \in X$ for $t \in T \cup \{t_f + 1\}$. We do not impose restrictions upon each player's feasible set of actions, which means that $A_i(x(t)) = \mathbb{R}^{n_i}$, $i = 1, 2$, for all $t \in T$ and $x(t)$. Similar, there are no restrictions upon the set of feasible states, i.e. $X = \mathbb{R}^k$, $t \in T$. The state transition is defined as

$$x(t+1) = Ax(t) + B_1 a_1(t) + B_2 a_2(t), \qquad t \in T, \qquad x(0) = x_0,$$

where A is a $k \times k$ matrix A and B_i, $i = 1, 2$, is a $k \times n_i$ matrix.

The per-period utility function $d_i : \mathbb{R}^{k+n_1+n_2} \to \mathbb{R}$, $i = 1, 2$, is given by the quadratic function given by

$$d_i(x(t), a(t)) = -\tfrac{1}{2}\left[a_i^\top(t) R_i a_i(t) + x^\top(t) Q_i x(t)\right], \qquad t \in T,$$

where the $n_i \times n_i$ matrix R_i is positive definite and the $k \times k$ matrix Q_i is semi-positive definite. The product of $a_i(t)$, $i = 1, 2$, with either $x(t)$ or $a_j(t)$ $j \neq i$ (or both) can be easily included in the utility functions, but at the cost of more complex expressions for the MPE strategies. Each player also derives some utility from the final state given by

$$-\tfrac{1}{2}x^\top(t_f+1)Q_i x(t_f+1), \quad i=1,2.$$

The feasible joint policy $c^t(x)$ proposed at time t given the state $x(t)=x$ simply becomes $c^t(x)=(a(t),\ldots,a(t_f))$, where $a(t+\tau)\in\mathbb{R}^{n_i}$ for $\tau=0,\ldots,t_f-t$. So, as time goes by, the dimension of the set of available joint policies shrinks. This means that the sooner the players start with the joint use of available instruments the more they can control the economic system to their joint benefit. Clearly, this gives incentives to both parties to reach an early agreement. The overall utility functions for $i=1,2$ are given by

$$u_i(x,c^t)=\sum_{\tau=0}^{t_f-t} d_i(x(t+\tau),a(t+\tau))-\tfrac{1}{2}x^\top(t_f+1)Q_i x(t_f+1).$$

Discounting can be included, but in order to minimize the expressions we assume no discounting, i.e., $\delta_1=\delta_2=1$.

In Houba and De Zeeuw (1995) the policy bargaining game with the above LQ game is analyzed with mild assumptions upon the matrices at the costs of having to solve several technical problems that we do not want to discuss in this chapter. Therefore, we impose the following simplifying assumption.[1]

Assumption 10.2. *The matrices A, B_i, R_i and Q_i, $i=1,2$, are all square, diagonal and positive definite.*

Note that all matrices have positive elements on the diagonal and zeros of the diagonal.

We conclude this section by deriving the unique pair of MPE strategies and the associated value functions are quadratic in the state variables. This shows that LQ games are analytically tractable. In fact, due to the finite horizon, it is also the unique SPE. A quadratic value function means that $V_i(t,x)=-\tfrac{1}{2}x^\top K_i(t)x$, $t\in T$, where $K_i(t)$ is a positive-definite $n\times n$ matrix, given the state $x(t)=x$ at round t. The MPE with quadratic value functions is derived in the following proposition.

Proposition 10.3. *The unique MPE strategies at round t, $t\in T$, given the state $x(t)=x$ prescribe the MPE actions*

$$a_i(t,x)=-R_i^{-1}B_i^\top K_i(t+1)E(t)^{-1}Ax, \quad i=1,2,$$

the evolution of the state

$$x(t+1)=E(t)^{-1}Ax$$

and quadratic value functions

$$V_i(t,x)=-\frac{1}{2}x^\top K_i(t)x,$$

where

$$E(t) = I + B_1 R_1^{-1} B_1^\top K_1(t+1) + B_2 R_2^{-1} B_2^\top K_2(t+1),$$

$$\begin{aligned} K_i(t) =& Q_i + A^\top E(t)^{-1} K_i(t+1) E(t)^{-1} A \\ &+ A^\top E(t)^{-1} K_i(t+1) B_i R_i^{-1} B_i^\top K_i(t+1) E(t)^{-1} A \end{aligned}$$

and $K_i(t_f+1) = Q_i$. *Moreover,* $K_i(t)$ *is positive definite.*

The following example illustrates this proposition.

Example 10.4. *Consider the LQ game with* $T = \{0,1\}$, *dimension* $n = k_1 = k_2 = 1$ *for all vectors, the players' utility function given for player 1 by*

$$-\tfrac{1}{2} \sum_{t=0,1} [a_1^2(t) + x^2(t)] - \tfrac{1}{2} x^2(2),$$

for player 2 by

$$-\tfrac{1}{2} \sum_{t=0,1} [a_2^2(t) + 2x^2(t)] - x^2(2)$$

and the state transition is given by

$$x(t+1) = x(t) + a_1(t) + a_2(t), \qquad t \in T, \qquad x(0) = 1$$

and $a_i(t) \in \mathbb{R}$, $i = 1,2$, $t \in T$. *Substitution of* $A = B_1 = B_2 = 1$, $R_1 = Q_1 = R_2 = 1$ *and* $Q_2 = 2$ *into the expressions of proposition 10.3 for* $t = 1$ *and the state* $x\,(1) = x$ *yields*

$$a(1,x) = (-0.25x, -0.5x), \quad x(2) = 0.25x$$

and the value functions

$$V_1(1,x) = -0.5625x^2, \quad V_2(1,x) = -1.1875x^2.$$

Then a second substitution into proposition 10.3 for $t = 1$ *and the initial state* $x\,(0) = 1$ *yields*

$$a(0,1) = (-0.2632, -0.5263), \quad x(1) = 0.2105$$

and the value functions

$$V_1(0,1) = -0.5596, \quad V_2(0,1) = -1.1911.$$

The complete MPE path follows if we substitute $x(1) = 0.2105$ *into the expressions for* $a(1,x)$ *and* $x\,(2)$ *in order to obtain*

$$a(1, 0.2105) = (-0.0526, -0.1052), \quad x(2) = 0.0526.$$

10.2.4 Pareto efficient joint policies

The MPE strategies of the policy bargaining model derived in later sections of this chapter have some properties in common with the MPE strategies in section 3.3. To be specific, the proposed MPE joint policy maximizes the proposing player's utility under the restriction that it just meets the responding player's threshold utility level. Thus, the MPE proposal is Pareto efficient. In this section we reformulate the associated proposing player's maximization problem and obtain an equivalent optimization problem in which a weighted sum of the players' utility functions is maximized. The advantage is that we reinterpret the subject of the negotiations as a bargaining problem over the proposing player's weight in the weighted sum of utilities. By doing so we reduce the dimension of the contract space from infinity to a one-dimensional compact and nonempty interval.

From section 2.2.2 on Pareto efficient contracts we know that such contracts are found by solving (in this chapter's notation)

$$\max_{c^t(x)} u_1\left(x, c^t\left(x\right)\right), \text{ s.t. } u_2\left(x, c^t\left(x\right)\right) \geq \widehat{u}_2,$$

where $\widehat{u}_2 \in \left[\min_{c^t(x)} u_2\left(x, c^t\left(x\right)\right), \max_{c^t(x)} u_2\left(x, c^t\left(x\right)\right)\right]$ and $c^t\left(x\right)$ is a feasible joint policy (which imposes additional constraints including the state transitions). Under the hypothesis that the constraint is binding we can maximize the associated Lagrangian. Formally, Pareto efficient contracts $c^t\left(x\right)$ at round t given state $x\left(t\right) = x$ are found by solving

$$\max_{c^t(x)} u_1\left(x, c^t\left(x\right)\right) - \lambda\left(-u_2\left(x, c^t\left(x\right)\right) + \widehat{u}_2\right), \tag{10.3}$$

where $\lambda \geq 0$ denotes the Lagrange multiplier.

Maximization problem (10.3) can be solved in several ways. One two-step procedure that is often implicitly applied is the following. First, calculate $c^t\left(x\right)$ as a function of λ, denoted as $c^t\left(x, \lambda\right)$, by solving all equations associated with the partial derivatives of the Lagrangian with respect to the actions of $c^t\left(x\right)$ and, second, compute the only positive value of λ that solves $\left(\widehat{u}_2 - u_2(x, c^t(x, \lambda))\right) = 0$. It is easy to verify that the equations that are solved in the first step of this two-step procedure also correspond to the first-order conditions of the maximization problem given by

$$\max_{c^t(x)} \tfrac{1}{1+\lambda} u_1\left(c^t\left(x\right)\right) + \tfrac{\lambda}{1+\lambda} u_2\left(c^t\left(x\right)\right). \tag{10.4}$$

where the Lagrange multiplier $\lambda \geq 0$ is simply treated as a parameter. The latter maximization problem is very popular within economics, because it is interpreted as a *social welfare* function and $1/(1+\lambda) \in [0,1]$ is player 1's weight the social planner gives to player 1. Similar, $\lambda/(1+\lambda) \in [0,1]$ is player 2's weight.[2] Obviously, $\lambda = 0$ and (taking the limit) $\lambda = \infty$ correspond to the extreme cases in which player 2 respectively player 1 is completely ignored by the social planner. Note that player 1's utility decreases in λ, whereas player 2's utility increases in λ.

In the remainder of this chapter the set of Pareto efficient contracts is either solved by maximizing the Lagrangian and the Lagrange multiplier then plays an important role in the analysis, or, alternatively, the weighted average of utilities is maximized as a function of the vector of weights. In the latter case, we think in terms of a reformulated bargaining problem in which the contract space of joint policies is replaced by the one-dimensional contract space $[0, 1]$. This interval represents the proposing player i's weight $\lambda_i \in [0, 1]$, $i = 1, 2$, with the understanding that the opponent's weight $\lambda_j = 1 - \lambda_i$, $j \neq i$. Furthermore, the content of the negotiations are the players' weights with the understanding that an agreement upon the weight λ_i specifies the unique Pareto efficient joint policy associated with this weight. So, for example player 1's proposal $\lambda_1 \in [0, 1]$ means the joint policy that solves (10.4) for $\lambda = (1 - \lambda_1)/\lambda_1$.

The above is illustrated by deriving the set of Pareto efficient quota for the great fish war as a function of $\lambda_i \in [0, 1]$, $i = 1, 2$, and $x(t) = x$ under the additional restriction that the set of feasible quota specify time-independent fractions of the amount of fish in the pool for each round.[3] Formally, a joint policy $c^t = (c(t), c(t+1), \ldots)$ specifies the infinite stream of consumptions $c(t+\tau) = (\gamma_1 x(t+\tau), \gamma_2 x(t+\tau))$, $\gamma_1 + \gamma_2 \leq 1$ and $\tau \in \mathbb{N}$, with the understanding that $x(t+\tau+1) = (1 - \gamma_1 - \gamma_2)^\alpha x^\alpha(t+\tau)$. In the remainder, the term 'Pareto efficiency' in the fish quota negotiations therefore refers to optimal allocations within the set of time-independent linear 'Markovian' paths. Furthermore, for convenience later on we solve for the Pareto efficient joint policy associated with the vector of weights $\lambda = (\lambda_1, \lambda_2)$, $\lambda_1, \lambda_2 \geq 0$ and $\lambda_1 + \lambda_2 = 1$, without making the substitution $\lambda_j = 1 - \lambda_i$, $i, j = 1, 2$ and $j \neq i$. The following proposition characterizes Pareto efficient paths in the great fish war.

Proposition 10.5. *The Pareto efficient joint policy $c^t(x)$, $t \in \mathbb{N}$ and $x(t) = x$, corresponding to the vector of weights $\lambda = (\lambda_1, \lambda_2)$, $\lambda_1, \lambda_2 \geq 0$ and $\lambda_1 + \lambda_2 = 1$, is given by*

$$c_i^b(x(t+\tau)) = \frac{\lambda_i}{1 + \lambda_1 \frac{\alpha\delta_1}{1-\alpha\delta_1} + \lambda_2 \frac{\alpha\delta_2}{1-\alpha\delta_2}} x(t+\tau), \quad i = 1, 2, \quad \tau \in \mathbb{N},$$

the state evolves as

$$x(t+\tau+1) = \left(\frac{\lambda_1 \frac{\alpha\delta_1}{1-\alpha\delta_1} + \lambda_2 \frac{\alpha\delta_2}{1-\alpha\delta_2}}{1 + \lambda_1 \frac{\alpha\delta_1}{1-\alpha\delta_1} + \lambda_2 \frac{\alpha\delta_2}{1-\alpha\delta_2}} \right)^\alpha x^\alpha(t+\tau), \quad \tau \in \mathbb{N}$$

and country i's utility at round t is $\ln A_i^b + \frac{1}{1-\alpha\delta_i} \ln x$, *where*

$$A_i^b = \left(\frac{\lambda_i \left(\lambda_1 \frac{\alpha\delta_1}{1-\alpha\delta_1} + \lambda_2 \frac{\alpha\delta_2}{1-\alpha\delta_2} \right)^{\frac{\alpha\delta_i}{1-\alpha\delta_i}}}{\left(1 + \lambda_1 \frac{\alpha\delta_1}{1-\alpha\delta_1} + \lambda_2 \frac{\alpha\delta_2}{1-\alpha\delta_2} \right)^{\frac{1}{1-\alpha\delta_i}}} \right)^{\frac{1}{1-\delta_i}}, \quad i = 1, 2. \tag{10.5}$$

Note that if λ_i, $i = 1, 2$, increases (decreases), then country i is allowed to catch more (less) fish in every round, because $\frac{\partial c_i^b(x(t+\tau))}{\partial \lambda_i} > 0$. If one of the countries becomes more patient, then both countries' quota are reduced and, therefore, the 'investment' in future amounts of fish increases, i.e., $\frac{\partial c_i^b(x(t+\tau))}{\partial \delta_k} < 0$ and $\frac{\partial x^b(t+\tau+1)}{\partial \delta_k} > 0$ for $k = 1, 2$.

10.3. Negotiations for quota

The analysis in this section characterizes *interior* and *linear* MPE strategies in policy negotiations over fish quota and, then, it is shown that the model admits just one such equilibrium. The MPE weights are not analytically tractable and we resort to numerical methods to investigate the MPE. This reveals a 'patience is weakness' result. Finally, we characterize the set of SPE utility pairs.

Before we start it is important to note that interior and linear MPE strategies correspond to value functions that are logarithmic. The latter can only hold if the constants A_i^b, $i = 1, 2$, in (10.5) do not depend upon the state. This already imposes that the MPE weights in an interior and linear MPE are independent of the state. Formally, the weight $\lambda_j : X \to [0, 1]$, $j = 1, 2$, as a function of the state x is constant in x, i.e., $\frac{\partial \lambda_j(x)}{\partial x} = 0$. This particular form is postulated from the start of the MPE analysis and we will discuss its consequences at the end of the first section.

10.3.1 Optimal disagreement catches

In the derivation of the interior and linear MPE with logarithmic value functions we start with deriving the disagreement actions.

Consider round t and the state $x(t) = x$. Then dynamic programming requires that the disagreement catches $c^d(x) \in A(x)$ and the corresponding value functions $V_i^d(x)$, $i = 1, 2$, at this round are derived from the Pareto efficient MPE utilities one round later. This leads to the following result.

Proposition 10.6. *Let the vector of weights* (λ_1, λ_2), $\lambda_1 + \lambda_2 = 1$ *and* $\lambda_1, \lambda_2 \geq 0$, *represent next round's Pareto efficient MPE agreement of proposition 10.5. The MPE disagreement catch at round t,* $t \in \mathbb{N}$, *given the state* $x(t) = x$ *is* *given by*

$$c_i^d(x) = \frac{\delta_j(1-\alpha\delta_i)}{\delta_1+\delta_2-\alpha\delta_1\delta_2}x, \quad i,j=1,2, \quad i\neq j,$$

the state evolves as

$$x(t+1) = \left(\frac{\alpha\delta_1\delta_2}{\delta_1+\delta_2-\alpha\delta_1\delta_2}\right)^{\alpha} x^{\alpha}$$

and the value function $V_i^d(x) = \ln A_i^d + \delta_i \ln A_i^b + \frac{1}{1-\alpha\delta_i}\ln x$, *where* A_i^b *is given in equation (10.5) and*

$$A_i^d = \frac{\delta_j(1-\alpha\delta_i)(\alpha\delta_1\delta_2)^{\frac{\alpha\delta_i}{1-\alpha\delta_i}}}{(\delta_1+\delta_2-\alpha\delta_1\delta_2)^{\frac{1}{1-\alpha\delta_i}}} \in (0,1).$$

Note that the linear MPE disagreement *catches* are independent of next round's MPE agreement, i.e., independent of (λ_1, λ_2). Hence, the MPE disagreement catches are the same in odd and even rounds. Moreover, these catches are equal to the interior and linear MPE catches of proposition 10.3. As we will see in section 10.4.2, the latter result is rather special. Nevertheless, as before, $c_1^d(x) \leq c_2^d(x)$ if and only if $\delta_1 \geq \delta_2$.

Why are the disagreement actions the same as in proposition 10.3? Partly, this result is a direct consequence of the combined effect of linearity, because linearity implies that next round's MPE pair of weights (λ_1, λ_2) is independent of $x(t+1)$ and, therefore, $\frac{\partial\lambda_i}{\partial x(t+1)} = 0, i = 1, 2$. Without the latter property the first-order conditions for MPE disagreement actions are no longer analytically tractable, which can be seen from the proof of proposition 10.6 that is presented in such a way that linearity is only imposed at the end of it and up to that point the arguments allow for $\frac{\partial\lambda_i}{\partial x(t+1)} \neq 0, i = 1, 2$.

Another partial explanation of this result can be explained by the specific functional forms. The great fish war belongs to a class of difference games that are *state separable* in terms of Dockner, Feightinger, and Jørgensen (1985). The value functions for this class of difference games always consist of a term with the state part and a constant term. Linear MPE disagreement catches are therefore analytically tractable and only the term with the state variable matters in deriving the equilibrium catches. As just mentioned, without the assumption of linearity in MPE strategies, state separability is lost and the nonlinear MPE disagreement catches are no longer tractable. To summarize the discussion: The independence of next round's MPE agreement in proposition 10.6 is the combined effect of the specific linear MPE solved for and a peculiarity of the specific functional forms.

10.3.2 Optimal proposals

The second step in deriving interior and linear MPE strategies consists of characterizing the responding player's optimal response and the optimal proposal for the proposing player at each bargaining round. The derivation of both is similar to the derivation given in section 3.3 and we will briefly go through the derivation in this section.

Consider round t, $t \in T$, and the state $x(t) = x$ at this round. Subgame perfectness requires that the responding player in round t accepts every joint policy that yields a strictly higher utility than the MPE utility in case of disagreement and rejects if the proposed joint policy yields strictly less utility. When the responding player is indifferent between accepting and rejecting, then both are optimal but for similar reasons as in section 3.3.4 it holds that: In any MPE the responding player accepts in case he is indifferent between accepting and rejecting. Thus, the responding player j, $j = 1, 2$, accepts the proposed joint policy $c^t(x)$ by player i, $i = 1, 2$ and $i \neq j$, if his utility of doing so is at least his disagreement utility $V_j^d(x)$ that he can obtain after rejecting the current proposal. The proposing player i can always secure his disagreement utility $V_i^d(x)$ by proposing an unacceptable joint policy, but in any MPE this player has no incentive to do so. Thus, the proposing player's best joint policy maximizes this player's utility over the set of joint proposals that gives the responding player j a utility of at least $V_j^d(x)$. Formally, player i's optimal joint policy is found by solving the optimization problem given by

$$\arg\max_{c^t} u_i\left(x, c^t\right), \quad \text{s.t.} \quad u_j\left(x, c^t\right) \geq V_j^d\left(x, \lambda_j\right), \tag{10.6}$$

where we write $V_j^d(x, \lambda_j)$ in order to express that the disagreement utility of the responding player j depends upon this player's MPE proposal λ_j in the next round. The constraint is binding in the optimum for every state $x > 0$. This implies that the resulting solution is Pareto efficient and that the proposing player i's weight λ_i solves $V_j^b(x, \lambda_i) = V_j^d(x)$, $j \neq i$, for all $x \in (0, 1]$. In the next theorem this result is combined with the expressions of proposition 10.5 and 10.6 and the uniqueness of a pair of weights λ_1 and λ_2 is shown. The MPE strategies are given in table 10.1.

Theorem 10.7. *The strategies of table 10.1 constitute the* unique *interior and linear MPE, where* $\lambda_1, \lambda_2 \in [0, 1]$ *are the* unique *solution for* $i, j = 1, 2$, $i \neq j$, *to*

$$\frac{(1-\lambda_i)^{1-\alpha\delta_j}(\Lambda_i)^{\alpha\delta_j}}{\Lambda_i + 1} - \left[A_j^d\right]^{(1-\alpha\delta_j)(1-\delta_j)} \left(\frac{\lambda_j^{1-\alpha\delta_j}(\Lambda_j)^{\alpha\delta_j}}{\Lambda_j + 1}\right)^{\delta_j} = 0, \tag{10.7}$$

State			Markov
Country 1	proposes	λ_1	always
	responds *yes*		if $u_1\left(x, c_1^t(x)\right) \geq V_1^d(x, \lambda_1)$
	catches	$\frac{\delta_2(1-\alpha\delta_1)}{\delta_1+\delta_2-\alpha\delta_1\delta_2}x_t$	always
Country 2	proposes	λ_2	always
	responds *yes*		if $u_2\left(x, c_2^t(x)\right) \geq V_2^d(x, \lambda_2)$
	catches	$\frac{\delta_1(1-\alpha\delta_2)}{\delta_1+\delta_2-\alpha\delta_1\delta_2}x_t$	always
Transition		absorbing	

Table 10.1. The MPE strategies of theorem 10.7.

where $\Lambda_i = \frac{\alpha\delta_j}{1-\alpha\delta_j} + \left(\frac{\alpha\delta_i}{1-\alpha\delta_i} - \frac{\alpha\delta_j}{1-\alpha\delta_j}\right)\lambda_i$. *Moreover,* $\lambda_i > 1 - \lambda_j$, $j = 1, 2$ *and* $j \neq i$.

Equation 10.7 states the implicit solution for the equilibrium weights that determine all the value functions. This solution is not analytically tractable. The property $\lambda_i > 1 - \lambda_j$ is the first-mover advantage of section 3.4.3.

To conclude, we discuss interior but nonlinear MPE's in C^1-functions, which implicitly impose that $\lambda_1(x)$ and $\lambda_2(x)$ are C^1-functions as well. The generalization of theorem 10.7 involves six equations with six unknown *functions* as its 'variables', all six with domain the state space $X = [0, 1]$. These functions are the four nonlinear disagreement catches $c_i^{even}(x)$ and $c_i^{odd}(x)$, $i = 1, 2$, and the two weights $\lambda_1(x)$ and $\lambda_2(x)$. The derivatives of the unknown functions $\lambda_1(x)$ and $\lambda_2(x)$ are also present in four of these six equations, namely the four equations that determine the disagreement catches (see the proof of proposition 10.6 for details). These six relations constitute a highly nonlinear system of first-order *differential* equations. Theorem 10.7 implies that this system admits one set of functions as its solution with the property $\lambda_1'(x) = \lambda_2'(x) = 0$. Whether there is a solution with the property $\lambda_1'(x) \neq 0$ and $\lambda_2'(x) \neq 0$ remains an open question.

10.3.3 Numerical Solutions

The pair of MPE weights of theorem 10.7 is not analytically tractable. In order to obtain more insight numerical solutions for the weights λ_1 and λ_2 are calculated on a grid of the parameter space with $\alpha = 0.5$ and $\delta_1, \delta_2 = 0.15, 0.35, \ldots, 0.95$.

The numerical solutions for the MPE weight λ_1, proposed by country 1, are reported in table 10.2. County 2's MPE proposal λ_2 can be found at the position (δ_2, δ_1). This table illustrates the first-mover advantage, i.e., $\lambda_1 +$

$\alpha = 0.50$	$\delta_2 \rightarrow$				
$\downarrow \delta_1$	0.15	0.35	0.55	0.75	0.95
0.15	0.51	0.74	0.84	0.90	0.93
0.35	0.27	0.51	0.66	0.76	0.83
0.55	0.17	0.36	0.51	0.63	0.71
0.75	0.11	0.25	0.39	0.51	0.59
0.95	0.07	0.18	0.29	0.41	0.50

Table 10.2. MPE weights λ_1 of theorem 10.7 for $\alpha = 0.50$ and $\delta_1, \delta_2 = 0.15, 0.35, \ldots, 0.95$.

$\lambda_2 \geq 1$, where the equal sign for some combinations of δ_1 and δ_2 is due to rounding off. The weight is almost symmetric on the diagonal, but it rapidly decreases if δ_1 increases. In the lower-left corner country 2 almost becomes a monopolist. Therefore, small asymmetries in the discount factors already imply large asymmetries in equilibrium weights.

The results in table 10.2 contrast with the result obtained in section 6.3.1, where a player's bargaining position is strengthened when he becomes more patient. If we take the weight a country can assure itself as a measure of its strength, then country 2 gets stronger at the expense of country 1 when country 1 becomes more patient. In Houba, Sneek, and Várdy (2000) it is reported that the same result holds in utility terms and that this result depends upon the state.[4]

The explanation of these numerical results is as follows. Proposition 10.6 implies that country 1 voluntarily catches less fish today the higher its δ_1 is, in order to invest in future consumption. However, country 2 behaves as a free rider, immediately extracting part of that investment in the current round by increasing its catch, i.e., $\partial c_2^d(x) / \partial \delta_1 > 0$. Obviously,[5]

$$\frac{\partial (1-\alpha\delta_1)^{-1} \ln x}{\partial \delta_1} < 0, \quad \frac{\partial \ln A_1^d}{\partial \delta_1} < 0 \quad \text{and} \quad \frac{\partial \ln A_2^d}{\partial \delta_1} > 0$$

imply that country 1's disagreement utility decreases, while at the same time country 2's disagreement utility increases. Clearly, such a shift improves country 2's bargaining position at the expense of 1. The economic intuition goes somewhat deeper. Contrary to the standard alternating offer model, the players can no longer delay the division of the pie by vetoing an agreement. This is because in the disagreement phase, endogenously determined fishing will take place even in the absence of any agreement. The patient player has therefore lost the strategic advantage he had in the alternating offer model. Moreover, as he cares more about the future than the impatient player, he has more to lose from an inefficient exploitation of the renewable resource than the impatient player. The option of delaying an agreement, leading to a fish war and an inefficient exploitation of the stock of fish for at least another round, has therefore become a strategic weapon in the hands of the impatient player.

State			1
Country 1	proposes	0	always
	responds *yes*		if $u_1(x, c^t(x)) \geq 0$
	catches	0	if t is odd
		x	if t is even
Country 2	proposes	1	if $a_1(t-1) = 0 \cdot x(t-1)$
	responds *yes*		if $u_2(x, c^t(x)) \geq V_2^d(x, 1)$
	catches	$(1-\alpha\delta_2)x$	if t is odd
		x	if t is even
Transition		absorbing	

Table 10.3. Worst SPE strategies of proposition 10.8.

We report without presenting tables that the rapid decline in both country 1's weight and utility is accelerated when the regeneration speed is lower, i.e., α is higher. The intuition is that the biological system needs more time to 'repair' the damage from a round of overexploiting the higher α is. The damage is permanent when $\alpha = 1$, i.e., the case of a nonrenewable resource.

10.3.4 The set of SPE utility pairs

The point made in this section is that the set of SPE utilities can be quite large. This result is based upon the unrealistic assumption that each country has the ability to empty or destroy the entire pool of fish in a single season by choosing $a_i(x) = x$, $i = 1, 2$. Unrealistic as this may be, there is a direct parallel, both in model and results, with the bargaining procedure in Ponsati and Sákovics (1998), where in a modified alternating offers procedure both players can simultaneously leave the negotiations for ever in between bargaining rounds. In the latter context, a player who leaves automatically destroys the entire surplus and we could say that $a_i(t)$ is either 0 or x. So, the fish war model is somewhat richer, because it allows for the continuous variable $a_i(t) \in [0, x]$. There is a second reason for analyzing SPE strategies in this model. The equilibrium analysis in this chapter mainly concerns MPE strategies and in this section we briefly address SPE strategies, which includes a discussion of how the insights of the previous chapter reappear in the policy bargaining model with state variables.

The ability to destroy the entire pool of fish means that time paths with zero catches in one or more rounds for one of the countries have to be considered and this causes the technical problem that the logarithm of zero catches are not defined. In order to overcome these technical problems we rather ad hoc switch to the Cobb-Douglas representation $\prod_{t=0}^{\infty} c_i^{\delta_i^t}(t)$. By doing so, evaluating streams with zero catches then simply yields zero utility. The characterization of the set of SPE utilities is split into two steps. First, country 1's worst SPE strategy is discussed in which this country gives up fishing and country 2 becomes the monopolist. Giving up fishing is each country's minmax utility level. Second, application of equilibrium switching yields the set of SPE utilities.

Table 10.3 represents country 1's worst SPE strategies.

The proposed SPE weights in every bargaining round are such that country 2 becomes the *monopolist* at every bargaining phase, i.e., the SPE weights are $\lambda_1 = 0$ and $\lambda_2 = 1$ in proposition 10.5. In case of disagreement at odd numbered rounds, these strategies prescribe that *both* countries threaten to destroy the entire pool of fish, which is a noninterior (but linear) MPE of the great fish war. The latter can be seen as the cold-war mutual retaliation threat. It is this threat that guarantees that giving up fishing is an equilibrium proposal and that making an unacceptable proposal is not better for country 1. Obviously, these disagreement actions minmax country 1 in case of disagreement. At every even round, the disagreement actions prescribe that country 1 refrains from fishing and that country 2 behaves as if it already is a monopolist in the current round, i.e., $a^d(x) = (0, (1 - \alpha\delta_2)x)$. Country 1 anticipates zero consumption from the next round onward and positive catches in the current round do not yield any utility. So, $a_1^d(x) = 0$ is an equilibrium action. Country 2's utility of rejecting country 1's proposal is this country's utility as if it were already a monopolist, meaning that its bargaining position has become so strong that this country can even afford to turn down every proposal in which this country does not become the monopolist. It is easy to verify that country 2's strategy is optimal against country 1's strategy. Thus, we have derived the following result.

Proposition 10.8. *For all $\delta_1, \delta_2 \in (0,1)$ the strategies of table 10.3 constitute country 1's worst SPE and country 1's utility is equal to the minmax utility level of* 0.

By reversing the roles of the players (and the behaviour at odd and even rounds) we obtain country 2's worst SPE.

The structure of country 1's worst SPE strategies in table 10.3 is similar to the structure table 9.2. Whenever country 1 is the responding party the SPE disagreement actions minmax this country and in rounds where this country proposes the disagreement actions are such that these maximize its opponent's utility in case of disagreement. The strategies of table 10.3 are rather peculiar, because the utility functions and disagreement actions are such that country 1 already plays a best response and, therefore, no compensation for country 1 is needed.

To conclude this section, similar as in section 9.2.3 and 9.2.4 equilibrium switching can be applied to support SPE utilities that are strictly individually rational with respect to the worst SPE utilities. Since the worst SPE utilities coincide with the players' minmax levels this implies that the set of SPE utilities is equal to the set of individually rational utilities. Thus, a similar result as in the prisoners' dilemma of example 9.19 holds. Note however, that the result in this section holds for all $\delta_1, \delta_2 \in (0,1)$ due to the peculiar functional forms. Similar as in the prisoners' dilemma perpetual disagreement is an SPE outcome and every 'disagreement stream' of positive catches in every round can be supported as the SPE outcome path. This informal discussion yields the following theorem.

Theorem 10.9. *For all* $\delta_1, \delta_2 \in (0,1)$ *the set of SPE utility pairs* $E_i(\delta) = S^{IR}$, $i = 1, 2$.

In Ponsati and Sákovics (1998) a similar result is obtained, namely every individual rational pair of utilities can be sustained by some pair of SPE utilities. In the latter model, where both players can simultaneously leave the negotiations for ever in between bargaining rounds of the standard alternating offers procedure, the 'quit' game in between rounds is a normal-form game with two pure Nash equilibria: Both stay or both leave. In case the decisions would be sequential instead of simultaneous, then only the 'both stay' Nash equilibrium involves credible threats and the alternating offers procedure with quit options admits a unique SPE strategy, namely the MPE strategies where none of the players leave. With respect to the timing of the countries' catches in each season the same conclusion can be drawn: The strategies of proposition 10.8 crucially depend upon the assumption that the decisions on each season's catches are taken simultaneously. Of course, future research should be directed to analyzing a more realistic fish war model where none of the countries is able to empty the pool of fish.

10.4. Multiple state variables

In this section the policy bargaining model with a finite time horizon and the LQ game (with the same time horizon) underlying it is analyzed. Recall that for ordinary LQ games the unique SPE is a unique MPE with quadratic value functions and linear strategies. This particular structure breaks down in case of multiple state variables when negotiations are present, meaning that there does not exist a MPE in linear strategies. Moreover, the nonlinear MPE is no longer analytically tractable.[6] Nevertheless, we can salvage the quadratic structure of the value functions for the class with a single state variable. In order to illustrate the breakdown we start with discussing a numerical example and then proceed with the general analysis. The motivation to go through the full analysis is to show that the results of the numerical examples belong to a generic class of bargaining problems. Note that the emphasis is on MPE strategies for the simple reason that the unique MPE for policy bargaining problems with a single state variable is also the unique SPE.

In order to derive an MPE in quadratic value functions (or the breakdown of it) these quadratic value functions are postulated for each phase at time t, $t \in T$, and these value functions represent the continuation utilities of the MPE. The quadratic value functions $V_i^b(t,x)$ and $V_i^d(t,x)$ are defined as

$$V_i^b(t,x) = -\tfrac{1}{2}x^\top K_i^b(t)x \quad \text{and} \quad V_i^d(t,x) = -\tfrac{1}{2}x^\top K_i^d(t)x, \quad i = 1,2.$$

where $K_i^b(t)$ and $K_i^d(t)$ are positive definite matrices to be derived below.

10.4.1 Motivating Example

In this section a numerical example is discussed that illustrates the complete breakdown of the quadratic structure even for relatively simple matrices A, B_i, R_i and Q_i, $i = 1,2$, and only two state variables.

Consider the LQ game with $T = \{0,1\}$, dimension $n = k_1 = k_2 = 2$ for all vectors, the players' utility function given for player 1 by

$$-\tfrac{1}{2}\sum_{t=0,1}\left[a_1^\top(t)\begin{bmatrix}\frac{1}{3} & 0\\ 0 & 1\end{bmatrix}a_1(t) + x^\top(t)Ix(t)\right] - \tfrac{1}{2}x^\top(2)Ix(2),$$

where I denotes the identity matrix, and for player 2 by

$$-\tfrac{1}{2}\sum_{t=0,1}\left[a_2^\top(t)\begin{bmatrix}1 & 0\\ 0 & \frac{1}{2}\end{bmatrix}a_2(t) + x^\top(t)Ix(t)\right] - \tfrac{1}{2}x^\top(2)Ix(2).$$

Furthermore, the state transition is given by

$$x(t+1) = Ix(t) + Ia_1(t) + Ia_2(t), \quad t \in T, \quad x(0) = [1,1]^\top$$

and no restrictions on the set of actions meaning

$$a_i(t) \in \mathbb{R}^2, \quad i = 1,2, \quad t \in T.$$

Similar as before, it is assumed that player 1 is the proposing player in round 0, so that player 2 is the proposing player in round 1. It will be shown that the value functions $V_i^b(1,x)$, $i = 1,2$, are not quadratic in x, because λ depends on state x in round 1.

The Markov perfect equilibrium is found by applying dynamic programming and the first step consists of computing the MPE disagreement actions in the last round, i.e., $t = 1$. Given $x(1) = x$, each player i, $i = 1,2$, solves

$$\max_{a_i \in \mathbb{R}^2} -\tfrac{1}{2}a_i^\top R_i a_i - \tfrac{1}{2}\left[x + a_1 + a_2\right]^\top\left[x + a_1 + a_2\right]$$

over $a_i(1) \in \mathbb{R}^2$. The MPE disagreement actions are found by first deriving each player's first-order conditions and then solving both systems of first-order conditions simultaneously for $a(1)$. For $t = t_f$ this yields the MPE disagreement actions and associated value functions of proposition 10.3. For the negotiation process at round 1 only the responding player's MPE utility matters, which is player 1's utility and it is fully determined by the matrix $K_1^d(1)$. Proposition 10.3 implies

$$K_1^d(1) = I + E^{-1}(1)E^{-1}(1) + E^{-1}(1)R_1^{-1}E^{-1}(1),$$

where $E(1) = I + R_1^{-1} + R_2^{-1}$. It follows that,

$$E(1) = \begin{bmatrix} 5 & 0 \\ 0 & 4 \end{bmatrix} \quad \text{and} \quad K_1^d(1) = \tfrac{1}{400}\begin{bmatrix} 464 & 0 \\ 0 & 450 \end{bmatrix}.$$

Similar as before, player 2's optimal proposal at round 1 maximizes this player's utility over the set of joint policies that yield player 1 a utility of at least his threshold utility and the latter is equal to $x^\top K_1^d(1)x$ for $x(1) = x$. Thus, player 2's best joint policy $c^1(x) = (a(1)) \in \mathbb{R}^4$ solves

$$\max_{a \in \mathbb{R}^4} -\tfrac{1}{2}a_2^\top R_2 a_2 - \tfrac{1}{2}[x + a_1 + a_2]^\top [x + a_1 + a_2],$$

subject to

$$-x^\top K_1^d(1)x \leq -x^\top x - a_1^\top R_1 a_1 - [x + a_1 + a_2]^\top [x + a_1 + a_2].$$

The constraint is binding and for solving player 1's utility $-x^\top K_1^b(1,\lambda)x$, $x(1) = x$, as a function of the Lagrange multiplier λ yields (after tedious calculations)

$$K_1^b(1,\lambda) = \begin{bmatrix} \frac{3+(\lambda^2+5\lambda+3)^2+(2+3\lambda)^2}{(\lambda^2+5\lambda+3)^2} & 0 \\ 0 & \frac{1+(2\lambda^2+4\lambda+1)^2+4(1+2\lambda)^2}{(2\lambda^2+4\lambda+1)^2} \end{bmatrix}.$$

The constraint becomes $x^\top[K_1^b(1,\lambda) - K_1^d(1)]x = 0$ and the optimal λ is the only positive solution of this equation. Rewriting yields

$$0 = \left(2\lambda^2 + 4\lambda + 1\right)^2 \left[1200 + 400\,(2 + 3\lambda)^2 - 64\left(\lambda^2 + 5\lambda + 3\right)^2\right] x_1^2$$

$$+ \left(\lambda^2 + 5\lambda + 3\right)^2 \left[400 + 1600\,(1 + 2\lambda)^2 - 50\left(2\lambda^2 + 4\lambda + 1\right)^2\right] x_2^2.$$

Since the polynomial in λ is of degree eight no analytical expression for $\lambda > 0$ as a function of x exists.

In order to show that λ depends upon the state x it suffices to calculate λ twice for two different state vectors. For $x = (1,0)^\top$ the equation in λ becomes

$$4\lambda^4 + 40\lambda^3 - 101\lambda^2 - 180\lambda - 139 = 0,$$

which has one positive solution, namely $\lambda = 3.2204$. Similarly, for $x = (0, 1)^\top$ the equation in λ becomes

$$4\lambda^4 + 4\lambda^3 - 123\lambda^2 - 126\lambda - 39 = 0,$$

which has one positive solution, namely $\lambda = 5.5794$. Hence, λ is a function of x and we may write $\lambda(x)$. Since $K_i^b(1\lambda)$, $i = 1, 2$, is a function of λ, we obtain that player i's utility of the optimal contract is equal to $x^\top K_i^b(1, \lambda(x))\, x$ and this function is no longer quadratic in the state x. In the following sections it is shown that the breakdown of the quadratic value functions is generic for the class of LQ games with multiple state variables. However, for the case of a single state variable we are able to prove that $\lambda(x)$ is constant in the state x meaning that the quadratic structure is salvaged for this class of policy bargaining problems.

10.4.2 Optimal disagreement actions

In the derivation of the MPE with quadratic value functions we start at the last stage of bargaining round t and this means that we start the analysis by deriving the disagreement actions, which is done in this short section.

Dynamic programming requires that, given the state $x(t) = x$ at round t and the value functions $V_i^b(t + 1, f(x, a, t))$ one round later, the MPE disagreement actions $a^d(t, x) \in A(t)$ and the corresponding value functions $V_i^d(t, x)$ are derived. The derivation is essentially the same as in the derivation of the MPE in ordinary difference games without bargaining as derived in proposition 10.3 and the following result is therefore given without proof.

Proposition 10.10. *Let* $V_i^b(t, x(t+1)) = -\frac{1}{2}x^\top(t+1)\, K_i^b(t+1)\, x(t+1)$. *Then the MPE disagreement actions at time t,* $t \in T$, *given state* $x(t) = x$, *are given by*

$$a_i^d(t, x) = -R_i^{-1}B_i^\top K_i^b(t+1)E(t)^{-1}Ax, \quad i = 1, 2,$$

and the evolution of the state is given by

$$x(t+1) = E(t)^{-1}Ax$$

where

$$E(t) = I + B_1R_1^{-1}B_1^\top K_1^b(t+1) + B_2R_2^{-1}B_2^\top K_2^b(t+1)$$

and

$$K_i^d(t) = Q_i + A^\top E(t)^{-1}[K_i^b(t+1)B_iR_i^{-1}B_i^\top K_i^b(t+1) + K_i^b(t+1)]E(t)^{-1}A.$$

Moreover, $K_i^d(t)$ is positive definite.

We conclude this section by comparing the disagreement actions of proposition 10.10 with those of proposition 10.3. It is easy to observe that for round $t = t_f$ the disagreement actions coincide with the those derived in proposition 10.3, because the value functions at $t_f + 1$ are in both cases equal to $-\frac{1}{2}x^\top(t_f+1)\,Q_i x(t_f+1)$, $i = 1,2$. However, for all other bargaining rounds this no longer holds. The reason is that in the policy bargaining both players anticipate next round's agreement (and this agreement is Pareto efficient), while in the ordinary difference game the inefficient equilibrium continuation is anticipated. Hence, the matrices specifying the players' value functions differ in both models for all the rounds $t < t_f$ and different equilibrium anticipations simply imply different behaviour one round earlier. We state the following proposition without proof and refer to section 10.4.5 for a numerical example.

Proposition 10.11. *The MPE disagreement actions in proposition 10.10 for round $t < t_f$, given state $x(t) = x$ at round t, differ from the MPE actions in proposition 10.3 for the ordinary LQ game without bargaining.*

10.4.3 Optimal proposals

The second step in deriving MPE strategies consists of characterizing the responding player's optimal response and the optimal proposal for the proposing player at each bargaining round. Since the arguments involved duplicate those given in section 10.3.2 we will briefly go through the derivation in this section.

Consider round t, $t \in \mathbb{N}$, and the state $x(t) = x$. The value function . The aim is to derive the MPE proposal $c^t(x) = (a(t), \ldots, a(t_f+1))$ at time t and the corresponding value functions $V_i^b(t,x)$.

Consider round t, $t \in T$, and the state $x(t) = x$. The proposing player is player P, $P = 1,2$, and player R, $R = 1,2$ and $R \neq P$, is the responding player. Then subgame perfectness requires that player i's optimal joint policy $c^t(x)$ solves

$$\max_{c^t} u_P\left(x,c^t\right), \quad \text{s.t.} \quad u_R\left(x,c^t\right) \geq V_R^d(t,x), \tag{10.8}$$

where $V_i^d(t,x)$, $i = 1,2$, refers to proposition 10.10. Assumption 10.2 is a sufficient condition to guarantee that the constraint is binding in the optimum for every state $x \neq 0$.[7] This implies that the resulting solution is Pareto efficient.

In the next proposition a quadratic value function $V_i^d(t,x)$, $i = P,R$, is postulated and, then, the expressions for player P's best joint policy $c^t(x)$ as a function of the Lagrange multiplier $\lambda \geq 0$ and the state x are derived.

Proposition 10.12. *Let $V_R^d(t,x(t)) = -\frac{1}{2}x^\top(t)\,K_R^d(t)\,x(t)$, K_R^d is positive definite. Then the proposing player P's best joint policy at time t, $t \in T$,* *for state $x(t) = x$ is given by $c^t(x) = (a(t,x), \ldots, a(t_f, x))$, where $a(s,x)$, $s = t, \ldots, t_f$, is given by*

$$a_i(s,x) = -G_i(s)\Pi_{\tau=t}^{s-1}[A - BG(\tau)]x, \quad i = P, R,$$

with

$$B = [B_P, B_R] \quad \text{and} \quad G(s) = \begin{bmatrix} G_P(s) \\ G_R(s) \end{bmatrix}.$$

The matrices $G_i(s)$, $i = P, R$, $s = t, \ldots, t_f$, of $G(s)$ are found by solving backward iteratively the system of equations ($i = P, R$)

$$\begin{aligned} G(s) =& [R^\lambda + B^\top (K_P(s+1) + \lambda K_R(s+1))B]^{-1} B^\top (K_P(s+1) \\ & + \lambda K_R(s+1))A, \\ K_i(s) =& Q_i + G_i^\top(s) R_i G_i(s) + [A - BG(s)]^\top K_i(s+1)[A - BG(s)], \\ K_i(t_f+1) =& Q_i, \end{aligned}$$

where $R^\lambda = \begin{bmatrix} R_P & 0 \\ 0 & \lambda R_R \end{bmatrix}$ and the Lagrange multiplier $\lambda > 0$ solves

$$x^\top \left[K_R(t) - K_R^d(t)\right] x = 0.$$

The resulting total utilities are $-\frac{1}{2}x^\top K_i(t)x$, $i = P, R$, which will be denoted as $-\frac{1}{2}x^\top K_i^b(t,\lambda)x$. Moreover, there is only one positive solution for λ.

10.4.4 Breakdown quadratic value functions

In this section the breakdown of the quadratic structure is shown for the case of multiple state variables. The analysis also shows that in case of a single state variable the MPE admits the quadratic structure.

Recall from proposition 10.12 that the optimal proposal depends upon the Lagrange multiplier $\lambda > 0$ and that $\lambda > 0$ is the only positive solution of

$$x^\top \left[K_R^b(t,\lambda) - K_R^d(t)\right] x = 0.$$

Thus, in principle λ is some function of the state variable x. For the quadratic structure it matters whether or not this function is constant. If the Lagrange multiplier λ is independent of the state x, then the matrices $K_i(t,\lambda)$, $i = 1, 2$, can be seen as 'constants' and the value functions $V_i^b(t,x) = x^\top K_i(t,\lambda)x$ are quadratic. However, if the Lagrange multiplier λ depends on state x, which we write as $\lambda(x)$, then $K_i(t, \lambda(x))$ depends on state x and the value functions $V_i^b(t,x) = x^\top K_i(t,\lambda(x))x$ are certainly not quadratic, which destroys the quadratic structure.

Without going into details, in Houba and De Zeeuw (1995) it is shown that the elements of the matrix $\left[K_R(t,\lambda) - K_R^d(t)\right]$ can be written as a fraction of two polynomials in the Lagrange multiplier λ and that all elements have the same denominator, namely the common, positive denominator of each element of $K_R(t)$. Thus, rewriting of the matrix $K_R(t) - K_R^d(t)$ yields the symmetric matrix

$$x^\top \begin{bmatrix} a_{11}(\lambda) & \cdots & a_{1n}(\lambda) \\ \vdots & \ddots & \vdots \\ a_{n1}(\lambda) & \cdots & a_{nn}(\lambda) \end{bmatrix} x,$$

where each $a_{ij}(\lambda)$ is a polynomial in the Lagrange multiplier λ. In order to explain why the Lagrange multiplier $\lambda > 0$ depends on state x consider the states $x = e_k$ and $x = e_\ell$, $k, \ell = 1, \ldots, n$ and $k \neq \ell$, where e_k and e_ℓ denote the k-th and ℓ-th unit vector.[8] Then, $\lambda > 0$ has to solve $a_{kk}(\lambda) = 0$ in case $x = e_k$ and $\lambda > 0$ has to solve $a_{\ell\ell}(\lambda) = 0$ in case $x = e_\ell$. In general $a_{\ell\ell}(\lambda) \neq a_{kk}(\lambda)$ and, thus, the solution for $\lambda > 0$ depends on whether $x = e_\ell$ or $x = e_k$.[9] See the numerical example of section 10.4.1 to illustrates this. Hence, the quadratic structure breaks down.

From this exposition it is also immediately clear that the above can not occur if the state vector has dimension $n = 1$. Then $\lambda > 0$ solves a polynomial that is independent of the state x and, hence, it can simply not depend upon the state. This means that the unique MPE for $n = 1$ admits quadratic value functions as postulated. The next theorem summarizes the discussion thus far.

Theorem 10.13. *For $n = 1$ the unique MPE admits quadratic value functions $V_i^b(t, x(t))$ and $V_i^d(t, x(t))$. For $n \geq 2$ there does not exists a MPE with quadratic value functions.*

10.4.5 To negotiate or not?

As we have seen, for the class of bargaining problems with a single state variable it is possible to maintain the quadratic structure. In this section the MPE of a numerical example with two bargaining rounds is computed in order to illustrate the results obtained thus far. We conclude this section with a comparison of the MPE in the policy bargaining model with the MPE of proposition 10.3 for the ordinary LQ game (without negotiations). Rather surprisingly we conclude that player 2 prefers the outcome of proposition 10.3.

Consider the policy bargaining problem for the LQ game of example 10.4 once more. Recall that player 1 is the proposing player in round 0, so that player 1 is the proposing player in round 1. Proposition 10.11 implies that the MPE disagreement actions in the final round are equal to the disagreement actions of proposition 10.3. Then making use of example 10.4 yields

$$a^d(1, x) = (-0.25x, -0.5x), \quad x(2) = 0.25x$$

and the value functions

$$V_1^d(1,x) = -0.5625x^2, \quad V_2^d(1,x) = -1.1875x^2.$$

Player 1's value function is this player's threshold utility level meaning that he rejects every joint policy $c^1 = (a(1)) \in \mathbb{R}^2$ that falls short of it, i.e., $u_1(x, c^1) < -0.5625x^2$, and accepts otherwise. Application of proposition 10.12 for $t = 1$ and state $x(1) = x$ implies that player 2's equilibrium proposal, which is accepted by player 1, is equal to

$$c^1(x) = (-0.3221x, -0.5322x), \quad x(2) = 0.1457x$$

and the value functions are given by

$$V_1^b(1,x) = -0.5625x^2, \quad V_2^b(1,x) = -1.1628x^2$$

This solves the final bargaining round and we can now go one round back in time to the first round. Proceeding similar as in the final round, the equilibrium disagreement actions for $t = 0$ and state $x(0) = 1$, we obtain

$$a^d(0,1) = (-0.2528, -0.5225), \quad x(1) = 0.2247$$

and the value functions

$$V_1^d(0,1) = -0.5603, \quad V_2^d(0,1) = -1.1952.$$

The value function $V_2^d\,(0,1)$ specifies player 2's threshold level, which is equal to -1.1952. Player 1's equilibrium proposal for $t = 0$ and state $x(0) = 1$, is given by

$$c^0(0,1) = (\,(-0.2800, -0.5937),\ (-0.0347, -0.0736)\,).$$

The corresponding value functions are given by

$$V_1^b(0,1) = -0.5479, \quad V_2^b(0,1) = -1.1952$$

and the evolution of the state by

$$x(0) = 1, \quad x(1) = 0.1262, \quad x(2) = 0.0179.$$

In the MPE there is immediate agreement upon player 1's proposed joint policy. Note that the MPE disagreement actions $a^d(0,1)$ differ from the MPE actions $a\,(0,1)$ derived in example 10.4, which is in accordance with proposition 10.11.

Comparing player 2's utility $V_2^b(0,1)$ with this player's utility $V_2(0,1)$ in example 10.4 reveals that

$$V_2^b(0,1) = -1.1952 < -1.1911 = V_2(0,1).$$

So, the MPE outcome in the policy bargaining model may be worse for one of the two players than the MPE of the ordinary difference game. Although the bargaining outcome in any subgame is individually rational with respect to the disagreement outcome in this subgame, this result is due to the fact that the disagreement actions differ from the MPE actions of the ordinary difference game, as established in proposition 10.11. Player 2 cannot secure the utility $V_2(0,1)$ in the policy bargaining model for the following reason. Suppose player 2 announces prior to the negotiations that it will follow the strategy: Reject any joint policy in round 0, propose a joint strategy in round 1 and use player 2's actions of proposition 10.3 as the disagreement actions in both rounds. Since this strategy is not an equilibrium strategy it cannot be credible. Indeed, for every state $x(t) = x$ at round 1 it is advantageous for player 2 to reconsider the announced strategy and propose his best joint policy $c^1(x)$ in round 1. Therefore, rational expectations imply that both players will not use the actions of proposition 10.3 as their disagreement actions at round 0, but instead both players will use $a_i^d(0,1)$, $i = 1, 2$.

Hence, the fact that player 2 lacks a credible commitment to refrain from bargaining in round 1 makes this player worse off than in the MPE of proposition 10.3. Only if player 2 would be able to disconnect forever all communication channels before the negotiations start, then this player could secure the MPE utilities of proposition 10.3. Alternatively, if negotiations require some prior costly investment in setting up communication devices or assembling a negotiation delegation, then player 2 is better off by not making these investments and by doing so committing to refrain from negotiations. This example shows that the introduction of binding agreements and communication in the form of the policy bargaining process may not be beneficial for both players. Note that these results can never be obtained in the context of policy bargaining within repeated games, as analyzed in chapter 9. The reason is that the MPE proposals are always individually rational with respect to the disagreement utilities specified by the Nash equilibrium actions of the disagreement game, because in policy bargaining in a repeated game context current actions do not affect the future surpluses.

10.5. Concluding remarks

The analysis in this chapter was conducted for two classes of difference games that are known to be analytically tractable and this already reveals some remarkable results. This raises the question whether it is really necessary to consider difference games that are analytically tractable? And, how general are the results? Also, how to generalize the worst SPE strategies of chapter 9?

The fish war model in Levhari and Mirman (1980) and the class of trilinear games in Clemhout and Wan (1974) all belong to the class of difference games that are called state separable. In Dockner, Feightinger, and Jørgensen (1985) it is shown that Markov equilibria in this class of differences games are analytically tractable. For this class of games results similar to propositions 10.5 and 10.6 are expected to hold and, then, even uniqueness as established in theorem 10.7 seems within reach.

State-separable difference games are still quite restricted, but throughout this book we have seen that some results with respect to MPE strategies in bargaining models are persistent. In all MPE strategies characterized thus far the general features are that i) the responding player applies a threshold level and rejects every proposal made that is short of it ii) the proposing player maximizes his own utility over the set of contracts that are acceptable for his opponent given the opponent's threshold and iii) MPE proposals are Pareto efficient and result in immediate agreement in every subgame. These three types of results are also confirmed in this chapter, even though the vectors of weights λ^i, $i = 1, 2$, become state dependent if multiple state variables are present as in section 10.4 and that this can be expected to be the general case. Dependency on the state of the weights is only from a computational point of view more demanding but not conceptually different, as for example the proof of proposition 10.6 illustrates. So, the results obtained hint at that similar results will hold for MPE strategies in the policy bargaining model if the class of difference games with strictly concave utility functions and concave state transitions are considered and all functions being autonomous.

The general problem for MPE strategies can be formulated as follows. Suppose there exist sets $X \subset \mathbb{R}^k$ and $A \subset \mathbb{R}^{n_1+n_2}$ such that for all $x \in X$ the set $A(x) \subseteq A$. Similar as in proposition 10.6 we write the vector of MPE weights as $(\lambda_1(x), 1 - \lambda_1(x))$ and $(1 - \lambda_2(x), \lambda_2(x))$, which expresses that these are functions of the state. Furthermore, for $i, j = 1, 2$ we write $V_j^b(x, \lambda_i)$ for player j's Pareto efficient value functions corresponding to player i's weight $\lambda_i \in [0, 1]$ proposed by this player i. The general problem is to find suitable functions $\lambda_i : X \to [0, 1]$, $i = 1, 2$, and $\widehat{a}^d, \widetilde{a}^d : X \to A$, representing MPE proposals and MPE disagreement actions, and, simultaneously, to find suitable value functions $V_i^b : X \times [0, 1] \to \mathbb{R}$ such that:

1 for t odd, for all $x \in X$ and both $i = 1, 2$ the function $\widehat{a}^d$ solves

$$\max_{a \in A} d_i(a, x) + \delta_i V_i^b(f(x, a), \lambda_2(f(x, a))),$$

2 for t even, for all $x \in X$ and both $i = 1, 2$ the function $\widetilde{a}^d$ solves

$$\max_{a \in A} d_i(a, x) + \delta_i V_i^b(f(x, a), \lambda_1(f(x, a))),$$

3 the functions λ_1, λ_2, V_1^b and V_2^b simultaneously solve for t is odd, for all $x \in X$:

$$V_2^b(x, \lambda_1(x)) = d_2(\widehat{a}(x), x) + \delta_2 V_2^b(f(x, \widehat{a}(x)), \lambda_2(f(x, \widehat{a}(x)))),$$

and for t is even, for all $x \in X$:

$$V_1^b(x, \lambda_2(x)) = d_1(\widetilde{a}(x), x) + \delta_1 V_1^b(f(x, \widetilde{a}(x)), \lambda_1(f(x, \widetilde{a}(x)))).$$

This is the smallest system of equations that formulates the characterization of the MPE strategies and its resemblance with fixed point problem (3.1) and the formulation in proposition 3.14 is striking.

Solving this mathematical problem is an open problem.[10] Nevertheless, we conjecture that such functions exists and that these are even smooth provided the sets X and A are nonempty, compact and convex sets combined with sufficiently smooth functions d_i, $i = 1, 2$, and f that remain bounded upon their domain plus the standard second-order conditions upon these functions. The analysis of section 10.4 immediately implies that the value functions are quadratic for infinite LQ games with a single state variable and that this quadratic structure breaks down in case of multiple state variables. A computational impossibility for an analytical solution in the single state case and infinite-horizon LQ games is that the proposing player's weight has to solve a polynomial of degree higher than three, meaning that one has to resort to numerical methods. This is one reason why our analysis resorts to the finite horizon case only.

More interestingly then the existence of the above MPE strategies are its economic properties. For example, how robust is the 'patience is weakness' phenomenon? A key factor is the dependence of the value function V_i^b, $i = 1, 2$, and V_i^d upon the discount factors δ_i. This holds for every difference game replacing the fish war model, whether state separable or not. Necessary conditions for 'patience is weakness' are disagreement utilities V_i^d, $i = 1, 2$, that increase when the opponent becomes more patient. Future research could be directed toward deriving sufficient and necessary conditions for a 'patience is weakness' effect or its opposite. For difference games that are 'close' to infinitely repeated games, i.e., with 'dampened dynamics' of the actions upon the state, the standard 'patience is strength' result known from 6.3.1 is likely to result.

To conclude, the real challenge is in the characterization of player 1's worst SPE strategies. Although the example in section 10.2.4 can be heavily criticized, it does show that multiplicity of SPE utility pairs can be expected and that the underlying strategies can be interpreted in terms of player 1's worst SPE strategy discussed in chapter 9. We now briefly sketch how we conjecture that the generalization of player 1's worst SPE strategies looks like and what new phenomena not encountered before will show up.

The function $a^1 : X \to A$ denotes the disagreement actions in even rounds and, similarly, for the function $a^d : X \to A$ at odd rounds. Denote λ_i, $i = 1, 2$, as the agreement that yields player 1's worst SPE agreement at bargaining rounds where player i is the proposing player and let λ^* be the agreement that includes compensation proposed by player 2 if player 1 did not deviate from

$a^d(x(t))$ during disagreement in the last odd round. For explanatory reasons, suppose first that the functions a^1 and a^d are known. Then the functions λ_i, $i = 1, 2$, λ^*, and V_i^b solve for all $x \in X$:

$$V_2^b(x, \lambda_1(x)) = d_2(a^d(x), x) + \delta V_2^b(f(x, a^d(x)), \lambda_2(f(x, a^d(x)))),$$
$$V_1^b(x, \lambda_2(x)) = d_1(a^1(x), x) + \delta V_1^b(f(x, a^1(x)), \lambda_1 f(x, a^1(x))))$$

and

$$d_1\left(a^d(x), x\right) + \delta V_1^b(x, \lambda^*(x)) \geq$$
$$\max_{a_1}\left[d_1((a_1, a_2^d(x)), x) + \delta V_1^b(f(x, (a_1, a_2^d(x))), \lambda^*(f(x, (a_1, a_2^d(x)))))\right].$$

The latter inequality deters player 1 from deviating from the disagreement action $a_2(x)$ and in player 1's worst SPE the equal sign suffices. Note that the intertemporal linkage of disagreement actions to next round's state variables makes the generalization of the net gain function $g_1 : X \times A \to \mathbb{R}$ obsolete.

The above equations should hold given the functions a^1 and a^d and the question is what are the additional conditions that the functions a^1 and a^d should satisfy in player 1's worst SPE? The first issue is to correctly generalize the minmax actions a^1 needed in the characterization. The difficulty in difference games is that current actions influence next round's state and in player 1's worst SPE the players anticipate the agreement $\lambda_1(x(t+1))$ in round t, t is even. So, the minmax actions become the function $a^1 : X \to A$ implicitly given by

$$a^1(x) = \arg\min_{a_2}\max_{a_1}\left[d_1(a, x) + \delta V_1^b(f(x, a), \lambda_1(f(x, a)))\right].$$

This reveals that the minmax actions do not refer to the minmax actions in the difference game at all, but that these depend upon the equilibrium agreement in next round. The remaining issue is that the function a^d should minimize player 1's utility $V_i^b(x, \lambda_1(x))$ for every state $x \in X$ subject to the equations derived thus far.[11] It is an open problem under what conditions suitable functions λ_i and V_i^b, $i = 1, 2$, λ^*, a^1 and a^d do exist. Analytical tractability of the difference game eases the technical problems encountered, but at the expense of tedious calculations.

10.6. Related literature

The analysis in this chapter combines the analyses in Houba and De Zeeuw (1995) and Houba, Sneek, and Várdy (2000). The first reference considers the finite horizon LQ games underlying the policy bargaining and the second reference studies the negotiations for fish quota. For details on incorporating normalized discounting we refer to the latter reference. The exact formulation of the great fish war model follows Dutta and Sundaram (1993), where the notation also takes into account the case $a_1(t) + a_2(t) > x(t)$, which is not needed in the analysis of the interior and linear MPE in Levhari and Mirman (1980).

The other literature on destructive threats always assumes that the threats do destroy part of the future surplus but that the parties do not attach utility from using these destructive threats itself, i.e., the set of attainable utilities S depends upon $a \in A$ and $d_i(a(t)) = 0$, $i = 1, 2$, for all $a \in A$. Destructive threats are also referred to as models with 'money burning'. In Avery and Zemsky (1994) one party can 'burn money', whereas our model features two parties that can destroy the future surplus. The two-sided ability to destroy the entire surplus can also be found in Ponsati and Sákovics (1998) where similar results are obtained as in section 10.3.4. There is not much literature on the presence of state variables in strategic bargaining models, except for section 10.2 Muthoo (1999) who analyzes negotiations in the context of dynamic capital investment. The latter analysis is conducted in the steady state, whereas the analysis in this chapter is also valid outside the steady state.

For a recent textbook on difference and differential games we refer to Dockner, Jørgensen, Long, and Sorger (2001). LQ games can be considered as approximations to general difference games and applications often resort to this class of games, e.g. Fershtman and Kamien (1987), Van der Ploeg and De Zeeuw (1991) and Tsutsui and Mino (1990). The economic problem of the Tragedy of the Commons is well known in the literature, e.g., Dutta and Sundaram (1993), Fisher and Mirman (1992), Fisher and Mirman (1996), Levhari and Mirman (1980), Van der Ploeg and De Zeeuw (1992) and the references therein. To be precise, Dutta and Sundaram (1993) show that interior continuously differentiable MPE strategies always fail to be Pareto efficient, but that Pareto efficiency of MPE can be established in Markov strategies that are discontinuous in the state variable. Pareto efficient SPE strategies in trigger strategies are derived in Benhabib and Radner (1992), Haurie and Pohjola (1987) and Kaitala and Pohjola (1990).

Notes

1 For completeness, assumption 10.2 ensures that the matrix $E\,(t)$ in proposition 10.3 and 10.10 is regular and that its inverse $E^{-1}\,(t)$ exists. Furthermore, this assumption also guarantees that the constraint in optimization problem (10.8) is binding. As argued in Houba and De Zeeuw (1995) it is not possible to formulate the necessary and sufficient conditions for which $E\,(t)$ is regular. However, it is easy to see that $E\,(t)$ remains regular if assumption 10.2 is relaxed to allow for positive diagonal elements and off-diagonal elements that are relatively small in absolute values. Thus, the class of problems under consideration is generic.

2 In many economic applications the players are assumed to be symmetric and the multiplier is fixed at $\lambda = 1$, i.e., the symmetric Pareto efficient contract. This is typically the case in studies where the symmetric equilibrium of symmetric difference games is shown to be inefficient.

3 The reader can check for himself that application of system (10.3) yields Pareto efficient consumption paths with time-dependent fractions of $x\,(t)$ in which the impatient country's consumption decreases over time and eventually converges to zero over time. We rule out such paths as not credible and politically infeasible.

4 With respect to utilities the absence of normalized discounted utilities is somewhat troublesome. Normalized discount factors can easily be incorporated without redoing the calculations and, doing so, yields the adjusted weights $\mu = (\mu_1, \mu_2)$ with

$$\mu_i = \frac{\lambda_i(1-\delta_i)}{\lambda_1(1-\delta_1)+\lambda_2(1-\delta_2)}, \qquad i = 1, 2,$$

and $\frac{\partial \mu_i}{\partial \delta_i} < 0$ if $\frac{\partial \lambda_i}{\partial \delta_i} < 0$. Therefore, μ_1 also decreases in δ_1. Furthermore, normalized discounted utilities exhibit the patience is weakness result for sufficiently large state $x < 1$. For details we refer to Houba, Sneek, and Várdy (2000).

5 Note that the effect of $\frac{\partial (1-\alpha\delta_1)^{-1}\ln x}{\partial \delta_1} < 0$ is negligible in case $x \approx 1$.

6 Things might be even worse, because no results are known yet about the existence of nonlinear MPE strategies in case several state variables are present in the policy bargaining model.

7 We refer to Houba and De Zeeuw (1995) for necessary and sufficient conditions for a binding constraint in the proposing player's optimization problem.

8 A unit vector is a vector filled with zeros, except for one component that is equal to 1. The k-th unit vector has its k-th component equal to 1.

9 We refer to Houba and De Zeeuw (1995) for the necessary and sufficient conditions for the Lagrange multiplier $\lambda > 0$ to be independent of state x at time t. These conditions are nongeneric for $n \geq 2$.

10 For those interested in solving this problem, recall that the value functions V_i^b, $i = 1, 2$, are the value functions corresponding to the Pareto frontier of the difference game and this implies that the properties of these value functions can be derived directly.

11 We think it is wise to maximize over the functions a^d and a^1 simultaneously under the additional restriction that player 1 should not have an incentive to deviate from a_1^1 for all $x \in X$. By doing so, one can show that either the generalized minmax formulation does corresponds to player 1's worst SPE or that it does not.

Appendix A
Proofs of Selected Theorems

Proof of lemma 3.2
The convex, compact and nonempty set $S^{IR} \times S^{IR}$ and the continuous function $p \times q : S^{IR} \times S^{IR} \to S^{IR} \times S^{IR}$ satisfy the conditions of Brouwer's fixed point theorem, see e.g., the mathematical appendix M.I. in Mas-Colell, Whinston, and Green (1995). Therefore, the function $p \times q$ has at least one fixed point $(x^*, y^*) = (p(y^*), q(x^*))$. This establishes existence and the next step consists of proving uniqueness. Let (x^*, y^*) be a fixed point. Then it is easy to show that the points x^* and y^* have the same Nash product with disagreement point d, because $x_2^* = p_2(y^*)$ and $y_1^* = q_1(x^*)$ imply that

$$(x_1^* - d_1)(x_2^* - d_2) = \delta(x_1^* - d_1)(y_2^* - d_2)$$

and

$$(y_1^* - d_1)(y_2^* - d_2) = \delta(x_1^* - d_1)(y_2^* - d_2).$$

Furthermore, y^* lies to the left of x^*, because $d_1 < s_1^{\max}$ and $y_1^* = q_1(x^*)$. Then the convexity of the set S^{IR} and the increasing and strict-quasiconcave 'Cobb-Douglas' function $(s_1 - d_1)(s_2 - d_2)$ implies that each 'indifference' curve of Nash products intersects the Pareto frontier of S^{IR} at most twice. Suppose (x^*, y^*) and $(\hat{x}^*, \hat{y}^*)$ are both fixed points of $p \times q$, then x^* (y^*) and $\hat{x}^*$ $(\hat{y}^*)$ must lie on distinct curves of Nash products (if not, either the curve would intersect the Pareto frontier four times, which is a contradiction, or $x^* = \hat{x}^*$ and $y^* = \hat{y}^*$). Therefore, their Nash products differ. Without loss of generality assume that the Nash product of x^* (y^*) is strictly larger than the Nash product of $\hat{x}^*$ $(\hat{y}^*)$. This implies that $\hat{x}^*$ lies to the right of x^* and $\hat{y}^*$ lies to the left of y^* (all four points lie on the Pareto frontier of S^{IR}). Thus, $\hat{x}_1^* > x_1^*$ and $\hat{y}_2^* > y_2^*$. However, $\hat{x}_1^* > x_1^*$ implies that $\hat{y}_1^* = q_1(\hat{x}^*) > q_1(x^*) = y_1^*$ and, hence, $\hat{y}_2^* = f_2(\hat{y}_1^*) < f_2(y_1^*) = y_2^*$, which contradicts $\hat{y}_2^* > y_2^*$ above. Hence, the fixed point (x^*, y^*) is unique. This completes the proof. □

Proof of lemma 3.3
1. Both x^* and y^* are Pareto efficient and belong to S^{IR} by construction of the functions p and q. The vector x^* lies north-west of y^*, because $y_1^* = (1 - \delta)d_1 + \delta x_1^* \leq x_1^*$ and, consequently, $f_2(y_1^*) \geq f_2(x_1^*)$. The weak inequalities are both strict if $\delta > 0$, because then $x^* \neq (s_1^{\max}, d_2)$ and $y^* \neq (d_1, s_2^{\max})$. (Hint: Note that the latter proposals have a Nash product of 0.)

2. We show that x_1^* is increasing in d_1 and decreasing in d_2. The fixed point problem $(x, y) = (p(y), q(x))$ can be rewritten as the fixed point problem in the variable x_1 given by

$$x_1 = f_1((1-\delta)d_2 + \delta y_2) = f_1((1-\delta)d_2 + \delta f_2((1-\delta)d_1 + \delta x_1)).$$

The right hand side is decreasing in d_2 and increasing in x_1 and d_1, because both f_1 and f_2 are decreasing functions. Thus, the right hand side is smaller, respectively, larger than x_1 for all $x_1 < x_1^*$ and $x_1 > x_1^*$. So, an increase in d_1 to d_1' implies that

$$\begin{aligned} x_1^* &= f_1((1-\delta)d_2 + \delta f_2((1-\delta)d_1 + \delta x_1^*)) \\ &< f_1((1-\delta)d_2 + \delta f_2((1-\delta)d_1' + \delta x_1^*)). \end{aligned}$$

Hence, x_1^* corresponding to d_1 lies below the function on the second line and x_1 has to increase in order to become a fixed point. Thus, the fixed point for d_1' exceeds x_1^*. The opposite effect follows if d_2 increases to d_2'. Finally, the reverse effect holds for player 2's components in (x^*, y^*), because f_2 is decreasing, $x_2^* =_2 (x_1^*)$ and $y_2^* = f_2(y_1^*)$. □

Proof of lemma 3.6
Suppose to the contrary that there exists an optimal $y \notin T_1(V_1^1, V_2^1)$. Then player 1 will reject y and the resulting vector of utilities will be equal to $(1-\delta)d + \delta(V_1^1, V_2^1) \in S^{IR}$. The latter vector is not Pareto efficient, because in section 3.2 it is additionally assumed that d is not Pareto efficient. But then there exists a $y' \in S^{IR}$ such that $y' > (1-\delta)d + \delta(V_1^1, V_2^1)$. Moreover, $y' \in T_1(V_1^1, V_2^1)$, because $y_1' > (1-\delta)d_1 + \delta V_1^1 = q_1(V_1^1, V_2^1)$. So, player 1 would accept y'. This implies that the one-stage deviation in which player 2 proposes y' instead of y and continues to follow his Markov strategy from the next round onward is a profitable deviation from her Markov strategy that prescribes $y \notin T_1(V_1^1, V_2^1)$. Clearly, a contradiction. □

Proof of lemma 3.7
From (3.13) it follows that y solves $\max s_2^t$ subject to $T_1(V_1^1, V_2^1)$. Since this maximization problem has an objective function that is continuous and the domain of the maximization is a compact and nonempty set it follows that this problem admits a maximum. Similar as in the proof of lemma 3.6 there may not exist a $y' \in T_1(V_1^1, V_2^1)$ such that $y_2' > y_2$. So, $y \in T_1(V_1^1, V_2^1)$ must be Pareto efficient, i.e., $y_2 = f_2(y_1)$. Next, each Pareto efficient $(y_1, f_2(y_1)) \in T_1(V_1^1, V_2^1)$ such that $y_1 > q_1(V_1^1, V_2^1)$ cannot be an optimal proposal either, because (again) similar arguments as in the proof of lemma 3.6 imply that the one-stage deviation with $y' = (y_1 - \varepsilon, f_2(y_1 - \varepsilon)) \in T_1(V_1^1, V_2^1)$, where $0 < \varepsilon \leq y_1 - q_1(V_1^1, V_2^1)$, is profitable for player 2. The latter conclusion follows from f_2 is strictly decreasing and, therefore, $f_2(y_1 - \varepsilon) > f_2(y_1)$. Obviously, the only proposal y for which no profitable one-stage deviation exists is $y = (q_1(V_1^1, V_2^1), f_2(q_1(V_1^1, V_2^1))) = q(V_1^1, V_2^1)$. Then each proposal $y'' \in S$ would make player 2 worse of. □

Proof of lemma 3.17
The MPE strategies imply that $x^* \in E(h_1(t), t)$ for each t is even in T and $h_1(t) \in H_1(t)$. Similarly, $y^* \in E(h_1(t), t)$ for each t is odd in T and $h_1(t) \in H_1(t)$. So, $E(h_1(t), t) \neq \emptyset$ for all $t \in T$ and $h_t \in H_t$.
Next, it is shown that $E(h_1(0), 0) \subseteq E(h_1(t), t)$ for all t even and $h_1(t) \in H_1(t)$. This is accomplished by considering a pair of SPE strategies σ^* and reformulate these strategies such that the behaviour prescribed by σ^* at the information set h, $h \in H$, of the bargaining game is postponed to the information set $(h_1(t), h)$ of the bargaining game and then show that these reformulated SPE strategies are SPE strategies in the subgame $h_1(t)$ (but not necessarily SPE strategies outside this subgame). Note that $(h_1(t), h)$ corresponds to the information set of the

subgame starting at $h_1(t)$ followed by the history h in this particular subgame. The pair of reformulated strategies for the subgame $h_1(t)$ is denoted as σ' and it is related to σ^* as follows

$$\sigma'|_{(h_1(t),h)} = \sigma^*|_h, \quad h \in H,$$

where behaviour outside the subgame $h_1(t)$ is neglected. Discounting means that for all σ it holds that

$$U_i^h\left(p(\sigma);\delta\right) = U_i^{(h_1(t),h)}\left((h_1(t), p(\sigma));\delta\right), \quad i = 1, 2, \quad h \in H.$$

Then for all $h \in H$ and for all σ_i, $i = 1, 2$, it follows that

$$U_i^h\left(\sigma_i^*|_h, \sigma_{-i}^*|_h\right) \geq U_i^h\left(\sigma_i|_h, \sigma_{-i}^*|_h\right)$$

is equivalent to

$$U_i^{(h_1(t),h)}(\sigma_i'|_{(h_1(t),h)}, \sigma_{-i}'|_{(h_1(t),h)}) \geq U_i^{(h_1(t),h)}(\sigma_i|_{(h_1(t),h)}, \sigma_{-i}'|_{(h_1(t),h)}).$$

Since σ^* is SPE this implies that the strategies σ' are SPE strategies in the subgame $h_1(t)$. Thus,

$$U_i^{h_1(0)}\left(\sigma^*|_{h_1(0)}\right) = U_i^{(h_1(t),h_1(0))}\left(\sigma'|_{(h_1(t),h_1(0))}\right) \in E\left(h_1(t), t\right),$$

because $h_1(0) = \emptyset$ means that $(h_1(t), h_1(0)) = h_1(t)$. So, $E\left(h_1(0), 0\right) \subseteq E\left(h_1(t), t\right)$. The other way around, i.e., $E\left(h_1(t), t\right) \subseteq E\left(h_1(0), 0\right)$, goes similar. Consider a pair of strategies σ' that are SPE in the subgame $h_1(t)$ but that are not necessarily SPE outside this subgame. In general strategies are contingent upon the entire history $(h_1(t), h)$, but for the part of the strategies σ' that is restricted to the subgame $h_1(t)$ all histories share the first part, i.e., $h_1(t)$, as a common part that can be discarded. Then proceeding similar as before, given the pair of strategies σ' a pair of strategies σ^* for the subgame $h_1(0)$, i.e., the bargaining game itself, is constructed as follows

$$\sigma^*|_h = \sigma'|_{(h_1(t),h)}, \quad h \in H,$$

Then similar arguments as before yield that if σ' is a pair of SPE strategies restricted to the subgame $h_1(t)$ then σ^* is a pair of strategies for the entire bargaining game. Thus, $E\left(h_1(t), t\right) \subseteq E\left(h_1(0), 0\right)$. Combining the two inclusions yields $E\left(h_1(t), t\right) = E\left(h_1(0), 0\right)$ for each even $t \in T$ and $h_1(t) \in H_1(t)$. Take $E_1 = E\left(h_1(0), 0\right) \supseteq \{x^*\}$ and we have shown that there exists a nonempty set E_1 such that $E\left(h_1(t), t\right) = E_1$ for each even $t \in T$ and $h_1(t) \in H_1(t)$. Finally, consider t is odd. Similar arguments as before show that for fixed $h_1(1) = (s^0, no) \in H_1(1)$ and for all $h_1(t) \in H_1(t)$ such that $h_1(1)$ prescribes the history of $h_1(t)$ at round 0 it holds that

$$E\left(h_1(t), t\right) = E\left(h_1(1), 1\right).$$

Application of the arguments in Binmore (1987c), which we omit, to relate the sets $E\left(h_1(1), 1\right)$, $h_1(1) \in H_1(1)$, to the history independent sets $E\left(h_2(2), 1\right) = E_1$ yields

$$E\left(h_1(1), 1\right) = E\left(h_1'(1), 1\right), \quad h_1, h_1' \in H_1(1).$$

Thus, there exists a nonempty set E_2 such that each set $E\left(h_1(t), t\right) = E_2$ for each odd $t \in T$ and $h_1(t) \in H_1(t)$. □

Proof of proposition 3.18

The following four relations can be derived for $\underline{s}_2^1$, $\underline{s}_2^2$, $\overline{s}_1^1$ and $\overline{s}_1^2$.

1 $\underline{s}_2^1 \geq (1-\delta)d_2 + \delta\underline{s}_2^2$.
At t even, given any SPE strategy for player 1, the strategy in which player 2 rejects all proposals made by player 1 in the current round yields player 2 at least $(1-\delta)d_2 + \delta\underline{s}_2^2$. Thus, any best response to player 1's SPE strategy must at least be equal to this payoff.

2 $\overline{s}_1^2 \leq (1-\delta)d_1 + \delta\overline{s}_1^1$.
At t odd player 1 will always accept any proposal y such that $y_1 > (1-\delta)d_1 + \delta\overline{s}_1^1$, because accepting this proposal yields strictly more than any continuation SPE payoff after rejection. But then there cannot exist a SPE strategy in which player 2 makes a proposal y such that $y_1 > (1-\delta)d_1 + \delta\overline{s}_1^1$ and player 1 accepts, because for sufficiently small $\epsilon > 0$ the one-stage deviation that specifies the proposal $(y_1 - \epsilon, f_2(y_1 - \epsilon))$, that automatically satisfies $f_2(y_1 - \epsilon) > y_2$, will also be accepted by player 1 and yields a higher payoff for player 2.

3 $\underline{s}_2^2 \geq \max\left\{(1-\delta)d_2 + \delta\underline{s}_2^1, f_2(\overline{s}_1^2)\right\}$.
At t even player 2 can secure at least $(1-\delta)d_2 + \delta\underline{s}_2^1$ by making the proposal $y \in S$ such that $y_1 < d_1$ that will be rejected by player 1. Next, $\underline{s}_2^2$ cannot be smaller than $f_2(\overline{s}_1^2)$, because then any proposal $(\overline{s}_1^2 + \epsilon, f_2(\overline{s}_1^2 + \epsilon))$ for each sufficiently small $\epsilon > 0$, which will be accepted by player 1, is a profitable one-stage deviation for player 2.

4 $\overline{s}_1^1 \leq \max\left\{(1-\delta)d_1 + \delta\overline{s}_1^2, f_1(\underline{s}_2^1)\right\}$.
At t even player 1 can obtain at most $(1-\delta)d_1 + \delta\overline{s}_1^2$ by making the proposal $x \in S$ such that $x_2 < d_2$ that will be rejected by player 2. Furthermore, any SPE that prescribes agreement upon $s \in S^{IR}$ at t even satisfies $s_1 \leq f_1(s_2)$. Since player 2 will never accept $s_2 < \underline{s}_2^1$ it follows that player 1's payoff of any such agreement is bounded from above by $f_1(\underline{s}_2^1)$. Hence, $\overline{s}_1^1 \leq f_1(\underline{s}_2^1)$.

The proof proceeds with reducing the expressions on the right hand side of these four inequalities. First, it is shown that $f_1(\underline{s}_2^1) \geq (1-\delta)d_1 + \delta\overline{s}_1^2$ in the fourth inequality. Suppose not, then $\overline{s}_1^1 \leq (1-\delta)d_1 + \delta\overline{s}_1^2$ and substitution of the second inequality yields

$$\overline{s}_1^1 \leq (1-\delta)d_1 + \delta\left[(1-\delta)d_1 + \delta\overline{s}_1^1\right]$$

and thus $\overline{s}_1^1 \leq d_1$ which contradicts individual rationality $\overline{s}_1^1 \geq x_1^* > d_1$.
Second, $(1-\delta)d_2 + \delta\underline{s}_2^1 \leq f_2(\overline{s}_1^2)$ in the third inequality. Substitution of the second inequality, the previous result and concavity of f_2 imply

$$\begin{aligned} f_2(\overline{s}_1^2) &\geq (1-\delta)f_2(d_1) + \delta f_2(\overline{s}_1^1) \geq (1-\delta)d_2 + \delta f_2(f_1(\underline{s}_2^1)) \\ &= (1-\delta)d_2 + \delta\underline{s}_2^1. \end{aligned}$$

Substitution of all these results yields

$$\overline{s}_1^1 \leq f_1\left((1-\delta)d_2 + \delta f_2((1-\delta)d_1 + \delta\overline{s}_1^1)\right).$$

From lemma 3.2 it immediately follows that there is a unique value x_1^* such that $\overline{s}_1^1 \leq x_1^*$, because the function value on the right hand side for $\overline{s}_1^1$ is at least $\overline{s}_1^1$ and given the unique value x_1^* this implies that $\overline{s}_1^1$ has to lie to the left of this fixed point. Furthermore, the definition of $\overline{s}_1^1$ immediately implies $\overline{s}_1^1 \geq x_1^*$ and, hence, $\overline{s}_1^1 = x_1^*$. Similarly, $\underline{s}_2^2 = y_2^*$ and, hence, also $\overline{s}_1^2 = y_1^*$. Finally, these results and the first relation imply that $\underline{s}_2^1 \geq (1-\delta)d_2 + \delta y_2^* = x_2^*$. To conclude, $(\overline{s}_1^1, \underline{s}_2^1) = x^*$ and $(\overline{s}_1^2, \underline{s}_2^2) = y^*$. Similarly, $(\underline{s}_1^1, \overline{s}_2^1) = x^*$ and $(\underline{s}_1^2, \overline{s}_2^2) = y^*$. Hence, $E_1 = \{x^*\}$ and $E_2 = \{y^*\}$. □

Proof of proposition 4.9
Trivially, if $d_1' > N_1(S,d)$, then $N_1(S,d') \geq d_1' > N_1(S,d)$. Therefore, it is without loss of

generality to assume $d'_1 < N_1(S,d)$. From $d' \in S$ located to the right of the line through d and $N(S,d)$ it follows that

$$\frac{N_2(S,d) - d'_2}{N_1(S,d) - d'_1} > \frac{N_2(S,d) - d_2}{N_1(S,d) - d_1} \geq -f_2'^{-}(N_1(S,d)).$$

So, the line through d' and $N(S,d)$ is steeper than the line through d and $N(S,d)$. There are two different cases to be distinguished. First, $\frac{N_2(S,d)-d'_2}{N_1(S,d)-d'_1} \leq -f_2'^{+}(N_1(S,d))$. This implies that f_2 is not differentiable in $N_1(S,d)$ and that the line with slope $-\frac{N_2(S,d)-d'_2}{N_1(S,d)-d'_1}$ through $N(S,d)$ is also tangent to the Pareto frontier in $N(S,d)$. Hence, $N(S,d)$ also satisfies the geographical interpretation for d' and $N(S,d') = N(S,d)$. Second, $\frac{N_2(S,d)-d'_2}{N_1(S,d)-d'_1} > -f_2'^{+}(N_1(S,d))$. Then the argument for the first case cannot be valid and, therefore, $N(S,d') \neq N(S,d)$. Concavity of f_2 implies that $N_1(S,d') > N_1(S,d)$ is necessary in order to obtain

$$\frac{N_2(S,d') - d'_2}{N_1(S,d') - d'_1} \geq -f_2'^{-}(N_1(S,d')).$$

The second case includes f_2 is differentiable in $N_1(S,d)$. □

Proof of theorem 4.20
From theorem 4.17 we know that the generically unique $N(C, \{u_i\}_{i=1,2}, c^d)$ satisfies axiom 4.12-4.16. Utility functions that are unique up to an affine transformation trivially fulfil axiom 4.19. So, $N(C, \{u_i\}_{i=1,2}, c^d)$ is the unique solution for these utility functions. Consider the case that at least one of the utility functions fails to be unique up to a affine transformation but is unique up to a monotone transformation including all affine transformations. For such preferences theorem 4.17 already reduces the number of feasible solutions to $N(C, \{u_i\}_{i=1,2}, c^d)$ and replacing axiom 4.15 by the more restrictive axiom 4.19, for which $N(C, \{u_i\}_{i=1,2}, c^d)$ fails, also discards the NBS as the only candidate bargaining solution leaving an empty set of solutions. □

Proof of theorem 4.27
If $\hat{u}_i$ and $\hat{u}'_i$ represent the same preference relation on $\hat{C}$, then

$$\hat{u}_i(c,p) \geq \hat{u}_i(c',p') \quad \Leftrightarrow \quad \hat{u}'_i(c,p) \geq \hat{u}'_i(c',p')$$

and the same holds on both sides if the weak inequalities are equalities. Thus, if $\hat{u}_i$ and $\hat{u}'_i$ represent the same preference relation then the conditions $i)$ and $ii)$ hold in axiom 4.26. □

Proof of theorem 4.22
Consider $c \in C$ such that $u_1(c) > u_1(c^*)$. Then for all $p \in [0,1]$ player 2 has no objection and, trivially, player 1 can counter object each objection raised by player 2. Next, it holds that

$$pu_1(c) + (1-p)u_1(c^d) > u_1(c^*) \Leftrightarrow p > \frac{u_1(c^*) - u_1(c^d)}{u_1(c) - u_1(c^d)} \tag{A.1}$$

and

$$pu_2(c^*) + (1-p)u_2(c^d) \geq u_2(c) \Leftrightarrow p \geq \frac{u_2(c) - u_2(c^d)}{u_2(c^*) - u_2(c^d)}. \tag{A.2}$$

The definition requires that if player 1 is willing to face the risk of the objection c then player 2 should be willing to face the risk to counter object. This requirement can only be satisfied if the lower bound on p in (A.2) is larger than or equal to the lower bound (A.1), i.e.

$$\frac{u_1(c^*) - u_1(c^d)}{u_1(c) - u_1(c^d)} \geq \frac{u_2(c) - u_2(c^d)}{u_2(c^*) - u_2(c^d)}.$$

But then

$$(u_2(c^*) - u_2(c^d))(u_1(c^*) - u_1(c^d)) \geq (u_2(c) - u_2(c^d))(u_1(c) - u_1(c^d))$$

Similar arguments yield the same inequality for $c \in C$ such that $u_2(c) > u_2(c^*)$. □

Proof of theorem 4.31
Before proving the theorem, we first mention several properties of $\gamma(x)$ that are needed in this proof. Furthermore, we assume that the function f_2 is differentiable. In Van Damme (1991) it is shown that the function γ is convex, differentiable and that

$$\gamma(x) = (x_1 - d_1)\frac{\partial\gamma(x)}{\partial x_1} + (x_2 - d_2)\frac{\partial\gamma(x)}{\partial x_2}. \tag{A.3}$$

Furthermore, the line through x and d intersects S^{PE} and if $s^* \in S^{PE}$ is this point then $x - d = \gamma(x)(s^* - d)$ and $s_i^* = f_i(s_j^*)$. Or, $s^* = d + (\gamma(x))^{-1}(x - d)$ and $s_i^* = f_i(s_j^*)$. Therefore,

$$f_2(d_1 + (\gamma(x))^{-1}(x_1 - d_1)) = d_2 + (\gamma(x))^{-1}(x_2 - d_2)$$

and taking the partial derivative on both sides with respect to x_1 yields

$$f_2'(d_1 + (\gamma(x))^{-1}(x_1 - d_1))\frac{\gamma(x) - \frac{\partial\gamma(x)}{\partial x_1}(x_1 - d_1)}{(\gamma(x))^2} = -\frac{\frac{\partial\gamma(x)}{\partial x_1}(x_2 - d_2)}{(\gamma(x))^2}. \tag{A.4}$$

Although the function $p_\varepsilon(x)$ is not partially differentiable in x_i, $i = 1, 2$, at $x \in S^{PE} \cap S^{IR}$ player i's utility function $p_\varepsilon(x)(x_i - d_i)$ is partially differentiable in x_i at $x_i = f_i(x_j)$, because $\frac{\partial}{\partial x_i}p_\varepsilon(x)(x_i - d_i) = 1$ for $x_i < f_i(x_j)$ and $\lim_{x_i \downarrow f_i(x_j)} \frac{\partial}{\partial x_i}p_\varepsilon(x)(x_i - d_i) = 1$, where the latter follows from

$$\frac{\partial}{\partial x_i}p_\varepsilon(x)(x_i - d_i) = \varepsilon^{-1}p_\varepsilon(x)\left[\varepsilon - 2(\gamma(x) - 1)(x_i - d_i)\frac{\partial\gamma(x)}{\partial x_i}\right], \tag{A.5}$$

for $x_i > f_i(x_j)$, $\lim_{x_i \downarrow f_i(x_j)}\gamma(x) = 1$ and $\lim_{x_i \downarrow f_i(x_j)} p_\varepsilon(x_1) = 1$.

We are now ready to prove the theorem.
For $x_i \leq f_i(x_j)$ the partial derivative $\frac{\partial}{\partial x_i}p_\varepsilon(x)(x_i - d_i) = 1 > 0$ meaning that player i's best response against x_j satisfies $x_i > f_i(x_j)$. Then subtracting $i = 2$ from $i = 1$ in (A.5) yields

$$2\varepsilon^{-1}p_\varepsilon(x)(\gamma(x) - 1)\left[(x_2 - d_2)\frac{\partial\gamma(x)}{\partial x_2} - (x_1 - d_1)\frac{\partial\gamma(x)}{\partial x_1}\right] = 0$$

and combined with (A.3) this yields

$$(x_2 - d_2)\frac{\partial\gamma(x)}{\partial x_2} = (x_1 - d_1)\frac{\partial\gamma(x)}{\partial x_1} = \tfrac{1}{2}\gamma(x). \tag{A.6}$$

Then

$$(x_i - d_i)\frac{\partial\gamma(x)}{\partial x_j} = (x_j - d_j)\frac{\partial\gamma(x)}{\partial x_j}\frac{x_i - d_i}{x_j - d_j)} = \tfrac{1}{2}\gamma(x)\frac{x_i - d_i}{x_j - d_j}$$

implies that

$$(x_1 - d_1)\,\tfrac{\partial\gamma(x)}{\partial x_2} = \tfrac{1}{2}\gamma(x)\,\tfrac{x_1-d_1}{x_2-d_2}, \quad (x_2 - d_2)\,\tfrac{\partial\gamma(x)}{\partial x_1} = \tfrac{1}{2}\gamma(x)\,\tfrac{x_2-d_2}{x_1-d_1}. \tag{A.7}$$

Substitution of (A.6) and (A.7) into (A.4) yields

$$f_2'(d_1 + (\gamma(x))^{-1}(x_1 - d_1)) = -\frac{x_2 - d_2}{x_1 - d_1} = -\frac{\gamma(x))^{-1}(x_2 - d_2)}{\gamma(x))^{-1}(x_1 - d_1)}$$

and, hence,

$$f_2'(s_1^*) = -\frac{s_2^* - d_2}{s_1^* - d_1}.$$

Therefore, $s^* = d + \gamma(x)^{-1}(x - d)$ satisfies the first geometrical interpretation of the Nash product and every Nash equilibrium x is characterized by $N(S,d) = d + \gamma(x)^{-1}(x - d)$. Finally, substitution of ((A.6)) into (A.5) and solving for $\gamma(x)$ yields

$$\gamma(x) = \tfrac{1}{2} + \tfrac{1}{2}\sqrt{1 + 4\varepsilon} > 1$$

and a second root $\gamma(x) < 1$ that contradicts $x_i > f_i(x_j)$. Since $\gamma(x)$ is a constant it follows that the Nash equilibrium x is unique. □

Proof of theorem 4.34

We first show that $(x^*(\delta), y^*(\delta))$ is a continuous (vector) function in $\delta \in (0,1)$. Lemma 3.2 implies that $(x^*(\delta), y^*(\delta))$ is a function in $\delta \in (0,1)$. Consider a sequence $\langle\delta_k\rangle$ with $\lim_{k\to\infty}\delta_k = \delta \in (0,1)$. Let $\langle(x^*(\delta_k), y^*(\delta_k))\rangle$ in $\left(S^{PE} \cap S^{IR}\right) \times \left(S^{PE} \cap S^{IR}\right)$ be the corresponding sequence. Since the latter space is compact, there exists a convergent subsequence $\langle(x^*(\delta_\ell), y^*(\delta_\ell))\rangle$ with $\lim_{\ell\to\infty}(x^*(\delta_\ell), y^*(\delta_\ell)) = (x^*, y^*)$ for some (x^*, y^*). But then the continuity of the function $p \times q$ implies that

$$\begin{aligned}(x^*, y^*) &= \lim_{\ell\to\infty}(x^*(\delta_\ell), y^*(\delta_\ell)) = \lim_{\ell\to\infty}(p(y^*(\delta_\ell)), q(x^*(\delta_\ell))) \\ &= (p(\lim_{\ell\to\infty} y^*(\delta_\ell)), q(\lim_{\ell\to\infty} x^*(\delta_\ell))) = (p(y^*), q(x^*)),\end{aligned}$$

which implies that $\lim_{\ell\to\infty}(x^*(\delta_\ell), y^*(\delta_\ell))$ is also a fixed point of $p \times q$. Since this holds for each convergent subsequence it follows that each convergent subsequence has the same limit. Therefore, the sequence $\langle(x^*(\delta_k), y^*(\delta_k))\rangle$ is also convergent and $\lim_{k\to\infty}(x^*(\delta_k), y^*(\delta_k))$ is the fixed point $(x^*(\delta), y^*(\delta))$. So, $(x^*(\delta), y^*(\delta))$ is a continuous function in $\delta \in [0,1)$ and $\lim_{\delta\uparrow 1}(x^*(\delta), y^*(\delta))$ exists. Note that at $\delta = 1$ each (s,s), $s \in S^{PE} \cap S^{IR}$, is a fixed point of $p \times q$.

Next, we characterize $\lim_{\delta\uparrow 1}(x^*(\delta), y^*(\delta))$. First, the points $\lim_{\delta\uparrow 1} x^*(\delta) = \lim_{\delta\uparrow 1} y^*(\delta)$, because

$$\begin{aligned}0 &\le \lim_{\delta\uparrow 1}[x_1^*(\delta) - y_1^*(\delta)] = \lim_{\delta\uparrow 1}(1-\delta)\,[f_1((1-\delta)d_2 + \delta y_2^*(\delta)) - d_1] \\ &\le \lim_{\delta\uparrow 1}(1-\delta)[s_1^{\max} - d_1] = 0.\end{aligned}$$

Furthermore, the points $x^*(\delta)$ and $y^*(\delta)$ are two distinct points that lie on the same Nash product curve for each $\delta \in [0,1)$ and these points have a common limit $s^* \in S^{PE} \cap S^{IR}$, which must be the unique Pareto efficient point where the Nash product curve $(s_1^* - d_1)(s_2^* - d_2)$ is tangent to the Pareto frontier. Hence, $\lim_{\delta\uparrow 1} x^*(\delta) = \lim_{\delta\uparrow 1} y^*(\delta) = N(S,d)$. □

Proof of proposition 5.4

($\Rightarrow$) Basically, this part of the proof would duplicate each of the steps of sections 3.3.2-3.3.5. The arguments in section 3.3.3 are valid for every nonempty set S and, thus, also for S^{IR} is a strongly comprehensive set. Similar, the arguments and lemmas in section 3.3.4 also hold in case S^{IR} is a strongly comprehensive set. Lemma 3.6 is valid for all nonempty sets S^{IR}, whereas the uniqueness of $y = q\left(V_1^1, V_2^1\right)$ in lemma 3.7 requires that f_2 is strictly decreasing, i.e., S^{IR} is strongly comprehensive. In the proof of the latter lemma the concavity of f_2 is not used. Furthermore, the arguments in section 3.3.4 are valid if S^{IR} is nonempty and a connected set, i.e., for every $s, s' \in S^{IR}$ there exists a finite number n of points $s^i \in S^{IR}, i = 1, \ldots, n$, such that $s^1 = s$ and $s^n = s'$ and for every $\lambda \in [0,1]$ it holds that $\lambda s^{i-1} + (1-\lambda)s^i \in S^{IR}$, $i = 1, \ldots, n$ (each section of the line between its points s^{i-1} and s^i belongs to S^{IR}). Finally, the uniqueness result of lemma 3.2 is derived under the assumption of S^{IR} is a convex set and this property is explicitly used in its proof. So, the latter lemma cannot be applied here. Wrapping up the results then yields that: If a pair of Markov strategies is MPE, then (x, y) are a fixed point of $p \times q$ and the sets of acceptable proposals are given by $T_1(x)$ and $T_2(y)$ as stated in the proposition.

($\Leftarrow$) It is sufficient to consider one-stage deviations only, because the one-stage-deviation principle applies. First, $(x, y) = (p(y), q(x))$ implies that $x \in T_2(y)$ and $y \in T_1(x)$, i.e., player 2 always accepts x and player 1 always accepts y. Thus, not deviating at round t means immediate agreement in this round and a one-stage deviation resulting in disagreement at round t means that the players follow their Markov strategies from the next round onward resulting in an agreement at round $t+1$.

Consider the subgame $(h_1(t), s^t)$ at round t, t even. There are two cases.

1 $s_2^t < x_2$. Then the Markov strategies prescribe *no* (reject) in this subgame and agreement upon y in the next round which gives player 2 a utility of $(1-\delta)d_2 + \delta y_2 = p_2(y) = x_2$. The only one-stage deviation prescribes *yes* (accept) in this subgame which yields $s_2^t < x_2$. So, every one-stage deviation does worse.

2 $s_2^t \geq x_2$. Then the Markov strategies prescribe *yes* (accept) in this subgame which gives player 2 a utility of s_2^t. The only one-stage deviation prescribes *no* (reject) in this subgame and following the Markov strategy from round $t+1$ onwards, which results in agreement upon y at $t+1$. The latter yields player 2 a utility of $(1-\delta)d_2 + \delta y_2 = p_2(y) = x_2 \leq s_2^t$. So, this one-stage deviation is not profitable.

So, for each subgame $(h_1(t), s^t)$, t even, player 2 does not have a one-stage deviation that improves upon her Markov strategy against player 1's Markov strategy.

Consider the subgame $h_1(t)$ at round t, t even, i.e., player 1 proposes. The Markov strategies prescribe agreement upon $x \in S^{IR}$. Player 1 has two types of one-stage deviations to consider. Either propose $s^t \in T_2(y)\backslash\{x\}$ or $s^t \in S\backslash T_2(y)$ at round t and follow the Markov strategy from round $t+1$ onward.

1 $s^t \in T_2(y)\backslash\{x\}$. Then player 2 accepts s^t. However, $s^t \in T_2(y)\backslash\{x\}$ implies $s_2^t > x_2$ and, therefore, $s_1^t \leq f_1(s_2^t) < f_2(x_2) = x_1$, because f_2 is strictly decreasing. So, player 1 is worse off.

2 $s^t \in S\backslash T_2(y)$. Then player 2's Markov strategy prescribes *no* (reject) and both players Markov strategies from $t+1$ onward result in agreement upon y at $t+1$. So, player 1's utility is equal to

$$(1-\delta)d_1 + \delta y_1 = (1-\delta)d_1 + \delta q_1(x) = (1-\delta^2)d_1 + \delta^2 x_1 < x_1$$

and this one-stage deviation is worse for player 1.

Thus, player 1 does not have a profitable one-stage deviation at the subgame $h_1(t)$, t even. Finally, similar arguments establish that there do not exist one-stage deviations that yield a higher utility at the subgames $h_1(t)$ and $(h_1(t), s^t)$, for all t is odd. So, the one-stage-deviation principle implies that the Markov strategies are SPE, i.e., MPE. □

Proof of proposition 5.6

Since $x^* \in E_1$ for each fixed point (x^*, y^*) it immediately follows that

$$\overline{s}_1^1 \geq \max\{x_1^* \,|(x^*, y^*) = (p(y^*), q(x^*))\}, \tag{A.8}$$

Following the arguments in the proof of proposition 3.18 yields, as before,

$$\overline{s}_1^1 \leq f_1\left((1-\delta)d_2 + \delta f_2((1-\delta)d_1 + \delta \overline{s}_1^1)\right).$$

Furthermore, $f_1\left((1-\delta)d_2 + \delta f_2((1-\delta)d_1 + \delta s_1^{\max})\right) < s_1^{\max}$ and combined with (A.8) yields that the weak inequality in (A.8) has to be an equality. The other equalities stated in the proposition follow similarly. □

Proof of proposition 5.10

The strategies in state 1 and 2 are MPE strategies by construction. Next, consider the subgame $h_1(t(\sigma))$ for $t(\sigma)$ is even. Then proposition 5.9 implies that $s^* \geq (\underline{s}_1^1, \underline{s}_2^1)$. For $t(\sigma)$ is odd at the subgame $h_1(t(\sigma))$ similar arguments as in section 5.4.3 imply that $s^* \geq (\underline{s}_1^2, \underline{s}_2^2)$, i.e., $E_2 = \{s \in S^{IR} \,|s_1^* \geq \underline{s}_1^2, s_2^* \geq \underline{s}_2^2\}$. Since $\underline{s}_i^j > d_i$, $i, j = 1, 2$, E_j is a strict subset of S^{IR}. Consider the subgame $h_1(t)$, $t \leq t(\sigma)$ and t is even. Then similar arguments as in section 5.4.3 imply that proposing $(s_1^{\max}, d_2)$ yields $\left(1 - \delta^{t(\sigma)-t}\right) d_1 + \delta^{t(\sigma)-t} s_1^*$, whereas deviating by proposing $s^t \in S$ yields at most $\underline{s}_1^1$ if player 2 accepts s^t and $(1-\delta)d_1 + \delta \underline{s}_1^2 < \underline{s}_1^1$ if player 2 rejects s^t. So, player 1 does not deviate if and only if $\left(1 - \delta^{t(\sigma)-t}\right) d_1 + \delta^{t(\sigma)-t} s_1^* \geq \underline{s}_1^1$ for even $t \leq t(\sigma)$. Since either $s^* \in E_1$ or $s^* \in E_2$ (and thus $\underline{s}_1^j > d_1$) player 1's discounted utility is increasing in t even and it attains its minimum at $t = 0$. So, $\left(1 - \delta^{t(\sigma)-t}\right) d_1 + \delta^{t(\sigma)-t} s_1^* \geq \underline{s}_1^1$ for all t is even if and only if $\left(1 - \delta^{t(\sigma)}\right) d_1 + \delta^{t(\sigma)} s_1^* \geq \underline{s}_1^1$.

Finally, at the subgame $h_1(t)$, $t \leq t(\sigma)$ and t is odd similar arguments as for t is even imply that $\left(1 - \delta^{t(\sigma)-t}\right) d_2 + \delta^{t(\sigma)-t} s_2^* \geq \underline{s}_2^2$ is satisfied for all odd t if this inequality holds at $t = 1$, i.e., $\left(1 - \delta^{t(\sigma)-1}\right) d_2 + \delta^{t(\sigma)-1} s_2^* \geq \underline{s}_2^2$. Then

$$\begin{aligned}(1-\delta^{t(\sigma)})d_2 + \delta^{t(\sigma)} s_2^* &= (1-\delta)d_2 + \delta\left[(1-\delta^{t(\sigma)-1})d_2 + \delta^{t(\sigma)-1} s_2^*\right] \\ &\geq (1-\delta)d_2 + \delta \underline{s}_2^2 = \underline{s}_2^1.\end{aligned}$$

This completes the proof. □

Proof of theorem 5.20

Theorem 5.17 implies that it suffices to show that $T(S, d)$ consists of a single element. Choose the utility representation u_i, $i = 1, 2$, of player i's preferences such that $u_i\left(c^d\right) = 0$. Next, a monotone transformation does not affect the maximizer of a function. In particular the monotone transformation $f(z) = \ln(z)$ implies that

$$\left\{\arg\max_{c \in C^{IR}} u_1(c)\, u_2(c)\right\} = \left\{\arg\max_{c \in C^{IR}} \ln u_1(c) + \ln u_2(c)\right\}.$$

The latter maximization problem admits a unique solution, because its objective function is strictly concave and C^{IR} is a convex set. □

Proof of proposition 6.1
If $S^+ = S^{IR}$ the lemma is trivial. Therefore, assume $S^+ \subset S^{IR}$. In every even bargaining round independent of $h_1(t)$ player 1 can secure s_1^+ by proposing $(s_1^+, s_2^{\max})$, because rejecting yields player 2 at most $(1-\delta) d_2 + \delta s_2^{\max} < s_2^{\max}$. Obviously, one round earlier at an odd round player 1 can secure $(1-\delta) d_1 + \delta s_1^+$ simply by rejecting all proposals made by his opponent and proposes $(s_1^+, s_2^{\max})$ in the next bargaining round. This security level converges to s_1^+ as δ goes to 1. So, $s \in \lim_{\delta\to 1} E(h_1(t), t)$ implies that $s_1 \geq s_1^+$. Similar arguments show that $s_2 \geq s_2^+$ for each $s \in \lim_{\delta\to 1} E(h_1(t), t)$. □

Proof of proposition 6.11
Since $d^{even} \in S$ we have that $d_2^{even} \geq f_2(d_1^{even})$ and, therefore,

$$x_2^* = (1-\delta) d_2^{even} + \delta y_2^* \leq (1-\delta) f_2(d_1^{even}) + \delta y_2^*.$$

Combining this with $x_1^* = f_1(x_2^*)$ and f_1 is decreasing and concave yields

$$\begin{aligned} x_1^* &= f_1(x_2^*) \geq f_1((1-\delta) f_2(d_1^{even}) + \delta y_2^*) \\ &\geq (1-\delta) f_1(f_2(d_1^{even})) + \delta f_1(y_2^*) = (1-\delta) d_1^{even} + \delta y_1^*, \end{aligned}$$

where the last equality follows from f_1 and f_2 are each others inverse functions and y^* is Pareto efficient and, therefore, $y_1^* = f_1(y_2^*)$. So, player 1 is better of by proposing $x^* \in T_2(y^*)$, than to propose some unacceptable proposal $s^t \in S$. Furthermore, combining this result with $y_1^* = p_1(x^*)$ yields

$$x_1^* \geq (1-\delta) d_1^{even} + \delta(1-\delta) d_1^{odd} + \delta^2 x_1^* \Leftrightarrow x_1^* \geq m_1^{even}.$$

Moreover, $y_1^* = p_1(x^*) \geq m_1^{odd}$. Similar arguments hold for player 2. □

Proof of proposition 6.13
($\Rightarrow$) If (x, y) is a fixed point of $p \times q$ and $d \in S$, then lemma 3.3 implies that 1 prefers x to y. So, if 1 prefers y over x, then it can not be the case that $d \in S$. Thus, $d \notin S$.
($\Leftarrow$) Since f_1 is concave and $d_1 > f_1(d_2)$, i.e., d above the Pareto frontier, it follows that

$$\begin{aligned} x_1^* &= f_1((1-\delta)d_2 + \delta y_2^*) \geq (1-\delta)f_1(d_2) + \delta y_1^* \\ &= (1-\delta)[f_1(d_2) + \delta d_1] + \delta^2 x_1^* > (1-\delta^2)f_1(d_2) + \delta^2 x_1^*. \end{aligned}$$

Hence, $x_1^* > f_1(d_2)$ and $y_1^* = q_1(x^*) > f_1(d_2)$. Pareto efficiency of y^* and f_1 decreasing imply $y_2^* < d_2$. But then $x_2^* = p_2(y^*) > y_2^*$ or, equivalently, $x_1^* < y_1^*$. □

Proof of proposition 6.17
Application of theorem 6.7 implies that, as Δ goes to 0, both $x^*(\Delta)$ and $y^*(\Delta)$ converge to $N_\alpha(S, d)$ with $\hat{\alpha} = \lim_{\Delta\to 0} \ln\delta_1 (\ln\delta_1 + \ln\delta_2)^{-1}$. Note that $\lim_{\Delta\to 0} \frac{\ln\delta_1}{(\ln\delta_1 + \ln\delta_2)} = \frac{0}{0}$. However, for $i, j = 1, 2$ and $i \neq j$:

$$\begin{aligned} \lim_{\Delta\to 0} \tfrac{\partial}{\partial\Delta} \ln\delta_i &= \lim_{\Delta\to 0} \tfrac{\partial}{\partial\Delta} \ln \frac{e^{-r\Delta} p_{ij}(\Delta)}{1-e^{-r\Delta}+e^{-r\Delta} p_{ij}(\Delta)} \\ &= \lim_{\Delta\to 0} \frac{-r+(1-e^{-r\Delta}) p'_{ij}(\Delta)[p_{ij}(\Delta)]^{-1}}{1-e^{-r\Delta}+e^{-r\Delta} p_{ij}(\Delta)} = -\frac{r}{\rho_{ij}} \neq 0, \end{aligned}$$

because $p_{ij}(\Delta)$ continuously differentiable implies that $\lim_{\Delta\to 0} p'_{ij}(\Delta)$ is bounded. Then application of L'Hôpital's rule yields

$$\alpha = \lim_{\Delta\to 0} \frac{\ln \delta_1}{\ln \delta_1 + \ln \delta_2} = \lim_{\Delta\to 0} \frac{\frac{\partial}{\partial\Delta}\ln \delta_1}{\frac{\partial}{\partial\Delta}\ln \delta_1 + \frac{\partial}{\partial\Delta}\ln \delta_2} = \frac{\rho_{12}}{\rho_{12}+\rho_{12}}.$$

This completes the proof. □

Proof of proposition 6.19

($\Leftarrow$) First, note that (s^1, s^2) is a fixed point of $p \times q$ for $\delta = e^{-r(1-\rho)\Delta}$ means that

$$s_i^j = \left(1 - e^{-r(1-\rho)\Delta}\right) d_i + e^{-r(1-\rho)\Delta} s_i^i, \quad i, j = 1, 2,\ j \neq i,$$

and, therefore,

$$s^j \in T_i(t + \rho\Delta, t) = \left\{ s \in S \mid s_i \geq (1 - e^{-r(1-\rho)\Delta}) d_i + e^{-r(1-\rho)\Delta} s_i^i \right\}$$

meaning that player i matches $s^j(t) = s^j$ at time $t + \rho$. There are two types of histories to consider.

Consider the history h_t such that $(s^1(t_1), s^2(t_2)) = (\emptyset, \emptyset)$. If none of the players deviates, the vector of expected utilities is given by

$$\left(1 - e^{-r\rho\Delta}\right) d + e^{-r\rho\Delta} \left(\tfrac{1}{2} s^i + \tfrac{1}{2} s^j\right).$$

There are two types of deviations. The first type of deviation is that player i, $i = 1, 2$, waits at time t. Since player j proposes $s^j(t) = s^j$, player i has to wait on $(t, t + \rho\Delta)$ also while his opponent is committed to $s^j(t) = s^j$ on $[t, t + \Delta)$. Then the strategies prescribe player i to match $s^j(t) = s^j$ at $t + \rho\Delta$, which yields player i a present value at t of

$$\left(1 - e^{-r\rho\Delta}\right) d_i + e^{-r\rho\Delta} s_i^j < \left(1 - e^{-r\rho\Delta}\right) d_i + e^{-r\rho\Delta} \left(\tfrac{1}{2} s_i^i + \tfrac{1}{2} s_i^j\right).$$

Thus, waiting at t is not a profitable deviation. The second type of deviation has $s^i(t) \neq s^i$ such that $s_i^i(t) > s_i^i$. How does player j respond to $s^i(t)$ if this proposal is validated? Since $s_j^i(t) < s_j^i$ we have that

$$s^i(t) \notin T_j(t + \rho\Delta, t) = \left\{ s \in S \mid s_j \geq (1 - e^{-r(1-\rho)\Delta}) d_j + e^{-r(1-\rho)\Delta} s_j^j \right\}$$

and, therefore, for all $t' \in [t + \rho\Delta, t + \Delta - \rho\Delta)$ ($\rho \leq \frac{1}{2}$ implies nonemptiness)

$$s^i(t) \notin T_j(t', t) = \left\{ s \in S \mid s_j \geq (1 - e^{-r(t+\Delta-t')}) d_j + e^{-r(t+\Delta-t')} s_j^j \right\}.$$

This means that player j prefers to wait until $t' = t + (1-\rho)\Delta$ and then offer $s^j(t + (1-\rho)\Delta) = s^j$, which will be matched by player i at $t + \Delta$. So, $s^i(t) \neq s^i$ yields agreement upon s^j at time $t + \Delta$. The present value at time t for player i corresponding to this deviation is

$$\tfrac{1}{2}\left(1 - e^{-r\rho\Delta}\right) d_i + \tfrac{1}{2} e^{-r\rho\Delta} s_i^j + \tfrac{1}{2}\left(1 - e^{-r\rho\Delta}\right) d_i + \tfrac{1}{2} e^{-r\Delta} s_i^j$$

and this is smaller than

$$\left(1 - e^{-r\rho\Delta}\right) d_i + e^{-r\rho\Delta} \left(\tfrac{1}{2} s_i^i + \tfrac{1}{2} s_i^j\right).$$

Thus, the deviation $s^i(t) \neq s^i$ is not profitable.
Next, consider the history h_t such that $s^i(t_i) = \emptyset$ and $s^j(t_j) \neq \emptyset$. There are three cases to be distinguished.
First, $s^j(t_j) \notin T_i(t, t_j)$ and $t \in [t_j + (1-\rho)\Delta, t_j + \Delta)$. If player i does not deviate, then the strategies prescribe that this player proposes $s^i(t) = s^i$ and player j will match this proposal at $t + \rho\Delta > t_j + \Delta$. This yields player i a present value at t equal to $\left(1 - e^{-r\rho\Delta}\right) d_i + e^{-r\rho\Delta} s_i^i$. Consider the deviation in which player i waits until $t_j + \Delta$, then both players are free at that particular time and they simultaneously propose s^i and s^j. This yields player i a present value at t of

$$\left(1 - e^{-r\left(t_j+\Delta-t+\rho\Delta\right)}\right) d_i + e^{-r\left(t_j+\Delta-t+\rho\Delta\right)} \left(\tfrac{1}{2}s_i^1 + \tfrac{1}{2}s_i^2\right) < \left(1 - e^{-r\rho\Delta}\right) d_i + e^{-r\rho\Delta} s_i^i.$$

So, waiting is not a profitable deviation. The second type of deviation is $s^i(t) \neq s^i$ such that $s_i^i(t) > s_i^i$. Similar as before, player j respond to $s^i(t)$ by waiting until $t + (1-\rho)\Delta$ and then make the counter proposal $s^j(t + (1-\rho)\Delta) = s^j$, which will be matched by player i at $t + \Delta$. The deviation $s^i(t) \neq s^i$ is not profitable.
Second, $s^j(t_j) \notin T_i(t, t_j)$ and $t \in [t_j + \rho\Delta, t_j + (1-\rho)\Delta)$. The strategies prescribe player i to wait until $t_j + (1-\rho)\Delta$ in order to make his counter proposal s^i, which is optimal at $t_j + (1-\rho)\Delta$ as the previous case showed. So, player i's present value at t is

$$\left(1 - e^{-r\left(t_j+\Delta-t\right)}\right) d_i + e^{-r\left(t_j+\Delta-t\right)} s_i^i.$$

If player i deviates at t by proposing $s^i(t)$ then $s^j(t_j) \notin T_i(t, t_j)$ excludes that player i matches player j's standing proposal, i.e., $s^i(t) = s^j(t_j)$. So, consider $s^i(t) \neq s^j(t_j)$ such that $s_i^i(t) > s_i^i$. (The reason being that player j has to wait until $t_j + \Delta > t + \rho\Delta$ and player i's present value of $s_i^i(t)$ has to be larger than that of s^i proposed at $t_j + (1-\rho\Delta) > t$ in order to be profitable at all.) At $t_j + \Delta$ player j accepts all proposals $s^i(t)$ that yield at least

$$(1 - e^{-r\left(t+\Delta-t_j-\Delta\right)})d_j + e^{-r\left(t+\Delta-t_j-\Delta\right)} s_j^j > s_j^i = f_j\left(s_i^i\right) > f_j\left(s_i^i(t)\right).$$

Clearly, there does not exist a $s^i(t)$ that is matched by player j at $t_j + \Delta$ and that yields player i a higher present value than her Markov strategies prescribe.
Third, $s^j(t_j) \in T_i(t, t_j)$. This case is trivial, because player i prefers to match $s^j(t_j)$ rather than to make his counter proposal s^i at $\max\{t, t_j + (1-\rho)\Delta\}$.
Finally, note that $\rho\Delta - t + \max\{t, t_j + (1-\rho)\Delta\} = \max\{\rho\Delta, t_j + \Delta - t\}$.
($\Leftarrow$) This part is omitted. □

Proof of proposition 7.15
Recall that subgame perfectness is checked by applying the one-stage-deviation principle.
First, in case of a breakdown, none of the players can credibly deviate because a Nash equilibrium is played. Second, if player 2 rejects a proposal $x \in S$ at t even the strategies prescribe Nash equilibrium actions $\bar{a}^2$ in case of a breakdown and, if bargaining continues, agreement on $y^*(\delta, (n_1, N_2))$ at t odd. Thus, player 2's (expected) continuation payoff after rejection is equal to $(1-\delta)N_2 + \delta y_2^*(\delta, (n_1, N_2)) = x_2^*(\delta, (n_1, N_2))$ and hence player 2's behaviour to accept any $x \in S$ if and only if $x_2 \geq x_2^*(\delta, (n_1, N_2))$ is optimal.
Third, player 1 will not propose $x \in S$ such that $x_2 > x_2^*(\delta, (n_1, N_2))$ because acceptance by player 2 implies $x_1 < x_1^*(\delta, (n_1, N_2))$. Similarly, if player 1 proposes $x \in S$ such that $x_2 < x_2^*(\delta, (n_1, N_2))$, then the strategies prescribe rejection of x, Nash equilibrium actions

$\bar{a}^2$ in case of a breakdown and, if bargaining continues, agreement on $y^*(\delta,(n_1,N_2))$ at the next bargaining round. Therefore, player 1's expected payoff is equal to $(1-\delta)d_1(\bar{a}^2)+\delta y_1^*(\delta,(n_1,N_2)) \leq x_1^*(\delta,(n_1,N_2))$, because of the concavity of f_1 and $d(\bar{a}^2) \in S^{IR}$. Hence, proposing $x^*(\delta,(n_1,N_2))$ is optimal for player 1. Analogously, player 2's strategy to propose $y^*(\delta,(n_1,N_2))$ at t odd and player 1's strategy to accept any $y \in S$ if and only if $y_1 \geq y_1^*(\delta,(n_1,N_2))$ at t odd is also optimal. □

Proof of theorem 7.17

The proof proceeds along the lines of the method of Shaked and Sutton, as described in section 3.4.1. Let V_i, $i=1,2$, denote the supremum of all SPE payoffs to player i in a subgame in which this player proposes and v_i denote the infimum of all SPE payoffs to player i in this subgame. Also, let W_i, $i=1,2$, denote the supremum of all SPE payoffs to player i in a subgame in which this player responds to an offer and w_i denote the infimum of all SPE payoffs to player i in this subgame. Recall that some $a^N \in A^N$ is played in case of any breakdown.
In any SPE player 2 does not offer player 1 more than $(1-\delta)N_1+\delta V_1$, because this would certainly be accepted by player 1. Furthermore, in any SPE player 2 can secure $(1-\delta)n_2+\delta w_2$, simply by proposing $y=(0,1)$ which will be rejected by player 1 and playing its worst Nash equilibrium, i.e. playing $\underline{a}^2$, in case of breakdown. Thus, $W_1 \leq (1-\delta)N_1+\delta V_1$ and

$$v_2 \geq \max\{f_2\left((1-\delta)\,N_1+\delta V_1\right),(1-\delta)n_2+\delta w_2\}.$$

Also, in any SPE player 2 rejects every proposal x if $x_2 < (1-\delta)n_2+\delta v_2$. Thus, $w_2 \geq (1-\delta)n_2+\delta v_2$ and

$$V_1 \leq \max\{f_1\left((1-\delta)\,n_2+\delta v_2\right),\ (1-\delta)N_1+\delta W_1\}.$$

Substituting $W_1 \leq (1-\delta)N_1+\delta V_1$and $v_2 \geq f_2\left((1-\delta)\,N_1+\delta V_1\right)$ into the last expression first and then making use of the concavity twice yields

$$\begin{aligned} V_1 &\leq \max\{f_1\left((1-\delta)\,n_2+\delta f_2\left((1-\delta)\,N_1+\delta V_1\right)\right),\ (1-\delta^2)N_1+\delta^2 V_1\} \\ &\leq \max\{f_1((1-\delta)n_2+\delta(1-\delta)f_2(N_1)+\delta^2 f_2(V_1)),\ (1-\delta^2)N_1+\delta^2 V_1\} \end{aligned}$$

The first term is the largest term meaning that V_1 is at most equal to the fixed point of $p \times q$, i.e., $V_1 \leq x_1^*(\delta,(N_1,n_2))$. Then immediately $v_2 \geq y_2^*(\delta,(N_1,n_2))$. But then also $W_1 \leq (1-\delta)N_1+\delta x_1^*(\delta,(N_1,n_2)=y_1^*(\delta,(N_1,n_2))$ and $w_2 \geq (1-\delta)n_2+\delta y_2^*(\delta,(N_1,n_2))=x_2^*(\delta,(N_1,n_2))$. Similar arguments yield the other lower and upper bounds on the set of SPE payoffs. □

Proof of theorem 8.6

First, consider $w^{\min}(t)$. Since the union chooses the least costly option, i.e. holds out, the union has no incentive to deviate. Then $w^{\min}(t)$ is identical to player 1's unique SPE proposal in round t of the standard alternating offer model in which one dollar is disputed, utility functions are $\delta^t s_i$, $i=1,2$, and disagreement point $(w_0,\alpha-w_0)$.
Second, the maximum equilibrium wage under the threat of strike is given by $w_0+\frac{1-w_0}{1+\delta}$ at t even and $w_0+\delta\frac{1-w_0}{1+\delta}$ if t is odd. The only relevant equilibrium condition requires that strike is credible in case of disagreement at t even, i.e.,

$$\delta\left(w_0+\delta\frac{1-w_0}{1+\delta}\right) \geq (1-\delta)\,w_0+\delta\left(w_0+\delta\frac{1-\alpha}{1+\delta}\right), \tag{A.9}$$

where $w_0+\delta\frac{1-\alpha}{1+\delta}$ is $w^{\min}(t)$ at t odd. This condition reduces to $\delta^2 \geq w_0/\alpha$.
Third, if strike is not credible, then in terms of Haller (1991) we have that $a=\beta-w_0$,

$b = (1-\gamma)\,w_0$, $1 - r = w_0$ and the union demands $1-\alpha^* = 1 - \frac{1}{1+\delta}[r+\delta a]$ and the firm offers $1-\beta^* = 1-\frac{1}{1+\delta}[a+\delta r]$. The only relevant equilibrium condition requires that work-to-rule is credible in case of disagreement at t is even, i.e.,

$$(1-\delta)(w_0 - \gamma w_0) + \delta\left(w_0 + \delta\frac{1-\beta}{1+\delta}\right) \geq (1-\delta)\,w_0 + \delta\left(w_0 + \delta\frac{1-\alpha}{1+\delta}\right),$$

which yields $\delta^2 \geq \gamma w_0/(\alpha - \beta + \gamma w_0)$. Finally, the interval $[\gamma w_0/(\alpha-\beta+\gamma w_0)\,, w_0/\alpha)$ is not empty if and only if $\gamma < \frac{\alpha-\beta}{\alpha - w_0}$. □

Proof of proposition 8.5

See the arguments in section 8.3.2 for the first part of the proposition. Let us consider the second part. First, in order for the union not to deviate from its strike decision in any round as long as no agreement is reached, we must have

$$w_0 \;\leq\; \delta_u(1-\delta_u)w_0 \;+\; \delta_u^2\left(\frac{1+\delta_f w_0}{1+\delta_f}\right).$$

For δ_u sufficiently large this condition is satisfied. Second, from an odd round onward the firm's no-concession payoff is $(1-w_0)(1+\delta_f)^{-1}$, given the strategy of the union. Hence, the firm will not offer more to the union than $1-(1-w_0)(1+\delta_f)^{-1} = (\delta_f + w_0)(1+\delta_f)^{-1}$, so $z_1(t) \leq (\delta_f + w_0)(1+\delta_f)^{-1}$.

The union will reject any offer of the firm followed by a holdout if $w^t \leq (\delta_f + w_0)(1+\delta_f)^{-1}$, in particular $w^t = 0$, t odd, because

$$\frac{\delta_f + w_0}{1+\delta_f} < (1-\delta_u)w_0 + \delta_u\left(\frac{1+\delta_f w_0}{1+\delta_f}\right) \text{ iff } \delta_f < \delta_u.$$

Third, in even rounds the union prefers to make the offer $w^t = \delta_f(1-w_0)(1+\delta_f)^{-1}$ because by not doing so it will obtain the (normalized) payoff w_0 which is less than $(1+\delta_f w_0)(1+\delta_f)^{-1}$. The firm is indifferent between accepting and rejecting; in equilibrium the firm accepts. □

Proof of theorem 8.7

Consider T is even. The relevant equilibrium conditions are $s_1^*(T-t, w, \delta) \geq w^{\min}(t)$ and $s_2^*(T-t, w, \delta) \geq 1 - w^{\max}(t)$ for all $t = 0, \ldots, T$. First, for $t = T$ we obtain $w \in [w^{\min}(T), w^{\max}(T)] = [w^{\min}(0), w^{\max}(0)]$, because T is even. Second, $w \geq w^{\min}(0) \geq w_0$ implies that the union's utility $s_1^*(T-t, w, \delta)$ increases in t and, therefore, the most profitable deviation for the union is at $t = 0$. Rewriting $s_1^*(T, w, \delta) \geq w^{\min}(0)$ yields

$$w \geq w_0 + \delta^{-T}\left(\frac{1-\alpha}{1+\delta} + \left(1-\delta^T\right)\gamma w_0\right) \geq w^{\min}(0).$$

Third, $s_2^*(T-t, w, \delta)$ strictly decreases in t if and only if $w > w_0 + 1 - \beta$. The presence of either decreasing or increasing payoffs makes it necessary to distinguish two cases.

Case 1 $w \leq w_0 + 1 - \beta$. Then $s_2^*(T-t, w, \delta)$ increases in t and the most profitable deviation for the firm is at $t = 1$. Rewriting $s_2^*(T-1, w, \delta) \geq 1 - w^{\max}(1)$ yields

$$w \leq w_0 + 1 - \beta + \delta^{-T}\left(\beta - \frac{w_0 + \delta}{1+\delta}\right) \tag{A.10}$$

and $\beta \geq \frac{1+\delta w_0}{1+\delta} > \frac{w_0+\delta}{1+\delta}$ implies that the right hand side is larger than $w_0 + 1 - \beta$. Therefore, (A.10) is not binding.

Case 2 $w > w_0 + 1 - \beta$. Then $s_2^*(T-t, w, \delta)$ strictly decreases in t and, therefore, the most profitable deviation for the firm is at $t = T-1$. Rewriting $s_2^*(1, w, \delta) \geq 1 - w^{\max}(T-1)$ yields

$$w \leq w_0 + \frac{1-w_0}{1+\delta} - \frac{(1-\beta)(1-\delta)}{\delta} \leq w^{\max}(0).$$

Then the interval $\left(w_0 + 1 - \beta, w_0 + \frac{1-w_0}{1+\delta} - \frac{(1-\beta)(1-\delta)}{\delta}\right]$ is not empty if and only if $\beta > \frac{1+\delta w_0}{1+\delta}$. The latter is assumed. □

Proof of theorem 8.8
Fix $s \in S^*$. Then for any $\Delta > 0$ there exists a unique real number of periods $T(s,\Delta)$ with work-to-rule and wage $w(s,\Delta)$ such that $s^*\left(T(s,\Delta), w(s,\Delta), e^{-r\Delta}\right) = s$, where $s^*(T, w, \delta)$ is defined in (8.8). Solving for $w(s,\Delta)$ and $\delta^{T(s,\Delta)}$ and making use of $s \in S^*$ yields $w(s,\Delta) = w(s)$, where $w(s)$ is given in (8.12) and $\delta^{T(s,\Delta)} = \frac{s_2+s_1-\beta+\gamma w_0}{1-\beta+\gamma w_0} \leq 1$. Making use of $\delta = e^{-r\Delta}$ and $T(s,\Delta) = \ell(s)/\Delta$ yields the expression for $\ell(s)$ given in (8.11). Next, given $T(s,\Delta)$ and $w(s)$ we have to show that the equilibrium conditions in the proof of Theorem 8.7 hold for sufficiently small Δ's. Given S^* and $s^*\left(T(s,\Delta), w(s), e^{-r\Delta}\right) = s$ we have that every $s \in S^*$ is a convex combination of $((1-\gamma)w_0, \beta - w_0)$ and $(w(s), 1-w(s))$, where both points also belong to S^*. Therefore, $(w(s), 1-w(s))$ lies on the Pareto frontier in between $(\underline{s}_1, 1-\underline{s}_1)$ and $(1-\underline{s}_2, \underline{s}_2)$. Hence, $w(s) > \underline{s}_1$ and $1 - w(s) > \underline{s}_2$. Consider Case 2 in the proof of Theorem 8.7. The two relevant equilibrium conditions for Case 2 are:

$$s_1 = s_1^*\left(T(s,\Delta), w(s), e^{-r\Delta}\right) \geq w^{\min}(0) \text{ and } s_2^*\left(1, w(s), e^{-r\Delta}\right) \geq 1 - w^{\max}(1).$$

The first condition holds for sufficiently small $\Delta > 0$, because $s_1 > \underline{s}_1$ and $w^{\min}(0)$ converges to $\underline{s}_1$ as Δ goes to 0. The second condition also holds for sufficiently small $\Delta > 0$, because $s_2^*\left(1, w(s), e^{-r\Delta}\right) \to 1 - w(s) > \underline{s}_2$ and $1 - w^{\max}(1) \to \underline{s}_2$ as Δ goes to 0. For Case 1 in the proof of Theorem 8.7 similar arguments apply. □

Proof of theorem 8.9
It is without loss of generality to assume $\delta^2 \geq w_0/\alpha$ and consider $s_1^{\max}(h_t, t)$ only. The union's problem at t even is given by

$$s_1^{\max}(h_t, t) = \max_{w_U} w_0 + (1 + (1-\delta)h_t)(w_U - w_0),$$

s.t. $1 - w_0 - (1 + (1-\delta)h_t)(w_U - w_0) \geq \delta s_2^{\min}(h_t + 1, t+1)$, because $h_T = T$ implies that $h_{t+1} = t + 1 = h_t + 1$. Solving yields the boundary solution

$$w_U - w_0 = \frac{1 - w_0 - \delta s_2^{\min}(h_t+1, t+1)}{1 + (1-\delta)h_t}.$$

Substitution into the union's objective function and rewriting yields

$$s_1^{\max}(h_t, t) = w_0 + \left(1 - w_0 - \delta s_2^{\min}(h_t + 1, t+1)\right).$$

Similar, at $t+1$ odd under $h_{t+2} = h_{t+1} + 1$ the firm's problem given by

$$s_2^{\min}(h_{t+1}, t+1) = \max_{w_F} 1 - w_0 - (1 + (1-\delta)h_{t+1})(w_F - w_0),$$

s.t. $w_0 + (1 + (1-\delta) h_{t+1}) (w_F - w_0) \geq (1-\delta) w_0 + \delta s_1^{\max} (h_{t+1} + 1, t+2)$, yields

$$s_2^{\min} (h_{t+1}, t+1) = 1 - w_0 - \delta \left(s_1^{\max} (h_{t+1} + 1, t+2) - w_0 \right).$$

Substitution of $s_2^{\min} (h_{t+1}, t+1)$ into $s_1^{\max} (h_t, t)$ and rewriting yields

$$s_1^{\max} (h_t, t) - w_0 = (1-\delta)(1-w_0) + \delta^2 \left(s_1^{\max} (h_t + 2, t+2) - w_0 \right),$$

which admits $w^{\max}(t)$, t even, as its solution. Substitution into $s_1^{\max} (h_{t+1}, t+1) = (1-\delta) w_0$ $+\delta \left(s_1^{\max} (h_{t+1} + 1, t+2) - w_0 \right)$, t even, yields the expression stated for $t+1$ odd. Finally, $w^{\max}(h_t, t)$ follows from

$$s_1^{\max} (h_t, t) - w_0 = (1 + (1-\delta) h_t) \left(w^{\max} (h_t, t) - w_0 \right).$$

This completes the proof. □

Proof of theorem 8.10
Minor modification is the arguments of the proof of Theorem 8.7 show that every $s \in S^*$ is a vector of equilibrium utilities. Furthermore, for every $s \in S^*$ and $\Delta > 0$ the backdated wage $w^B (s, \Delta, T(s, \Delta))$ satisfies

$$w(s) = w_0 + \left[1 + \left(1 - \delta^{-r\Delta} \right) T(s, \Delta) \right] \left(w^B (s, \Delta, T(s, \Delta)) - w_0 \right),$$

where $T(s, \Delta) = \ell(s) / \Delta$. Thus, $w^B (s, \Delta, T(s, \Delta)) = w_0 + \frac{w(s) - w_0}{1 + \ell(s)\left(1 - e^{-r\Delta}\right)/\Delta}$. Finally, application of L'Hopitâl's rule yields $\lim_{\Delta \to 0} \left(1 - e^{-r\Delta} \right) / \Delta = r$. □

Proof of proposition 9.4
The correspondence $F_{x_1}(\delta, a)$ is upper semi-continuous in a and δ, and it remains so on the domain A if we fix $\delta < 1$. Every upper semi-continuous correspondence on a bounded domain attains its minimum (and maximum), e.g., mathematical appendix M.H in Mas-Colell, Whinston, and Green (1995). Hence, for every $\delta < 1$ there exists an $a^*(\delta) \in A$ such that $F_{x_1}(\delta, a)$ attains its minimum. □

Proof of lemma 9.6
First, note that for every $a^N \in A^N$ we have that

$$\lim_{\delta \to 1} x_1^m (\delta, a^*(\delta)) \leq \lim_{\delta \to 1} x_1^m (\delta, a^N) \leq \lim_{\delta \to 1} x_1^M (\delta, a^N) \leq \lim_{\delta \to 1} x_1^M (\delta, a^{**}(\delta)).$$

Furthermore, for every $a^N \in A^N$: $\left(m_1, d_2 \left(a^N \right) \right) \in S^{IR}$ and $\left(d_1 \left(a^N \right), m_2 \right) \in S^{IR}$. Thus, the standard results of chapter 3 and 4 imply that $x^m(\delta, a^N)$ converges to $N\left(S, \left(m_1, d_2 \left(a^N \right) \right) \right)$ as δ goes to 1 and $x^M(\delta, a^N)$ converge to $N\left(S, \left(d_1 \left(a^N \right), m_2 \right) \right)$. Finally, application of proposition 4.9 implies that $N(S, m)$ lies in between these two NBS's. Combining all inequalities yields the stated result. □

Proof of lemma 9.7
($\Rightarrow$) First, if $d(a^N) \neq m$ for some $a^N \in A^N$, then the arguments used to prove lemma 9.6 would yield

$$\lim_{\delta \to 1} x_1^m (\delta, a^*(\delta)) \leq \lim_{\delta \to 1} x_1^m (\delta, a^N) < \lim_{\delta \to 1} x_1^M (\delta, a^N) \leq \lim_{\delta \to 1} x_1^M (\delta, a^{**}(\delta)),$$

i.e., a contradiction. Second, if there exists a $a^N \in A^N$ such that $a^N \notin A^*$, then

$$\lim_{\delta \to 1} x_1^m (\delta, a^*(\delta)) < \lim_{\delta \to 1} x_1^m (\delta, a^N)$$

which contradicts uniqueness. Hence, $A^N \subseteq A^*$. Similarly, $A^N \subseteq A^{**}$.
($\Leftarrow$) Trivial, because $(m_1 + g_1(a^N), d_2(a^N)) = m$ and

$$\lim_{\delta \to 1} x^m(\delta, a^*(\delta)) = \lim_{\delta \to 1} x^m(\delta, a^N) = N(S, m).$$

This completes the proof. □

Proof of proposition 9.9
Without loss of generality we assume that deviations are punished with the deviator's worst SPE. For convenience we take t is even, i.e., $t+1$ is odd.
Consider the beginning of bargaining round $t+1$ (odd). Obviously, player 1 can secure the payoff $(1-\delta)m_1 + \delta \underline{s}_1^1$ by rejecting all proposals y and using a_1^1 as his disagreement action at $t+1$. Thus, $\underline{s}_1^2 \geq (1-\delta)m_1 + \delta \underline{s}_1^1$. Furthermore, there are two possible outcomes associated with player 1's worst SPE, namely either agreement at $t+1$ or delay at $t+1$ followed by a^1. These two outcomes induce the following upper bounds upon $\overline{s}_2^2$, namely

$$\overline{s}_2^2 \leq \max\left\{ f_2((1-\delta)m_1 + \delta \underline{s}_1^1), (1-\delta)d_2(a^1) + \delta f_2(\underline{s}_1^1) \right\}.$$

Since f_2 is concave, the first term on the right-hand side is the largest. Hence,

$$\overline{s}_2^2 \leq f_2((1-\delta)m_1 + \delta \underline{s}_1^1)$$

. Next, consider the disagreement game at t. Denote $\hat{a} \in A$ as the SPE disagreement actions at t in player 1's worst SPE payoff. Obviously, by using a best response to $\hat{a}_2$ player 1 can secure $(1-\delta)[d_1(\hat{a}) + g_1(\hat{a})] + \delta^2 \underline{s}_1^1$. In order to prevent player 1 from using a best response, SPE requires either $g_1(\hat{a}) = 0$ or applying the arguments that support the strategies of table 9.2. Since $g_1(\hat{a}) \geq 0$ includes the case $g_1(\hat{a}) = 0$ there is no loss in stating that player 2's SPE payoff is bounded from above by $(1-\delta)d_2(\hat{a}) + \delta f_2((1-\delta)\delta^{-1}g_1(\hat{a}) + \delta \underline{s}_1^1)$.
Finally, consider the beginning of the bargaining round at t. Player 2 can secure his supremum SPE payoff by rejecting every proposal $s^t \in S$. As above it can be argued that agreement at t is preferred by player 1 to delay and, hence,

$$\underline{s}_1^1 \geq f_1\left((1-\delta)d_2(\hat{a}) + \delta f_2\left((1-\delta)\delta^{-1}g_1(\hat{a}) + \delta \underline{s}_1^1\right)\right).$$

Thus, $\underline{s}_1^1$ is equal to or larger than the smallest fixed point of the function $p_1 \circ p_2 \circ q_2 \circ q_1$ with $d = \left(m_1 + (1-\delta)\delta^{-1}g_1(\hat{a}), d_2(\hat{a})\right)$. Note that $x_1^m(\delta, a)$ is also defined as the fixed point of the function $p_1 \circ p_2 \circ q_2 \circ q_1$ for the same disagreement point, after implicitly solving system 9.2 for $x_1^m(\delta, a)$. Hence, $\underline{s}_1^1 \geq \min_{a \in A}\{\min_{x_1} F_{x_1}(\delta, a)\} = x_1^m(\delta, a^*(\delta))$. But then also $\overline{s}_2^2 \leq y_2^m(\delta, a^*(\delta))$. Similar arguments yield the other bounds. □

Proof of proposition 10.3
Dynamic programming for player i, $i = 1, 2$, implies

$$V_i(t,x) = \max_{a_i \in \mathbf{R}^{n_i}} \left[-\tfrac{1}{2} a_i^\top R_i a_i - x^\top Q_i x + V_i(t+1, x(t+1)) \right], \qquad \text{(A.11)}$$

subject to

$$x(t+1) = Ax + B_1 a_1 + B_2 a_2.$$

Substitution of $V_i(t+1, x(t+1)) = -\frac{1}{2} x^\top (t+1) K_i(t) x(t+1)$ and the state transition into the right-hand side yields the first-order condition

$$R_i a_i + B_i^\top K_i(t+1)(Ax + B_1 a_1 + B_2 a_2) = 0. \qquad \text{(A.12)}$$

To obtain the following equation, first premultiply both sides by $B_iR_i^{-1}$, then sum up over $i = 1, 2$ and add Ax to both sides, and finally rewrite the equation. This yields

$$Ax + B_1a_1 + B_2a_2 = E(t)^{-1}Ax,$$

where

$$E(t) := I + B_1R_1^{-1}B_1^\top K_1(t+1) + B_2R_2^{-1}B_2^\top K_2(t+1).$$

The term on the left hand side is equal to $x(t+1)$ and, therefore, $x(t+1) = E(t)^{-1}Ax(t)$. Substitution of these results into (A.12) and premultiplication by R_i^{-1} yields the expressions for $a_i(t,x)$. Finally, substitution of $a_i(t,x)$ and $x(t+1)$ into the right-hand side of (A.11) yields $V_i(t,x)$ from which the expression for $K_i(t)$ can be obtained. Given $K_i(t+1)$ is positive definite, it is easy to show that $K_i(t)$ possesses the same property. □

Proof of proposition 10.5

Postulate value functions of the form $a_i \ln x + A_i$, $i = 1, 2$. Applying Bellman's dynamic programming principle implies

$$\sum_{i=1,2} \lambda_i (a_i \ln x + A_i) = \max_{c_1,c_2} \sum_{i=1,2} \lambda_i \left(\ln c_i + \delta_i a_i \ln (x - c_1 - c_2)^\alpha + \delta_i A_i\right).$$

The first-order conditions can be rewritten as the *regular* linear system

$$\begin{bmatrix} \lambda_1 + \sum_{j=1}^2 \lambda_j \alpha \delta_j a_j & \lambda_1 \\ \lambda_2 & \lambda_2 + \sum_{j=1}^2 \lambda_j \alpha \delta_j a_j \end{bmatrix} \begin{bmatrix} c_1 \\ c_2 \end{bmatrix} = \begin{bmatrix} \lambda_1 x \\ \lambda_2 x \end{bmatrix}.$$

It is easy to verify that the unique solution is given by

$$c_i^b(x) = \frac{\lambda_i}{1 + \sum_{j=1}^2 \lambda_j \alpha \delta_j a_j} x, \quad i = 1, 2. \tag{A.13}$$

Substitution into the state transition yields that the state evolves as

$$x(t+1) = \left(\frac{\sum_{j=1}^2 \lambda_j \alpha \delta_j a_j}{1 + \sum_{j=1}^2 \lambda_j \alpha \delta_j a_j}\right)^\alpha x^\alpha. \tag{A.14}$$

Substitution into $\ln c_i^b(x) + \delta_i (a_i \ln x(t+1) + A_i)$ and rewriting yields

$$(1 + \alpha\delta_i a_i) \ln x + \ln \frac{\lambda_i \left(\sum_{j=1}^2 \lambda_j \alpha \delta_j a_j\right)^{\alpha\delta_i a_i}}{\left(1 + \sum_{j=1}^2 \lambda_j \alpha \delta_j a_j\right)^{1+\alpha\delta_i a_i}} + \delta_i A_i.$$

This latter expression is equal to $a_i \ln x + A_i$ and, therefore, it follows that

$$a_i = \frac{1}{1 - \alpha\delta_i}, \quad \text{and} \quad A_i = \ln \frac{\lambda_i (\sum_{j=1}^2 \lambda_j \alpha \delta_j a_j)^{\alpha\delta_i a_i}}{(1 + \sum_{j=1}^2 \lambda_j \alpha \delta_j a_j)^{1+\alpha\delta_i a_i}} + \delta_i A_i.$$

Solving for A_i yields $A_i = \ln A_i^b$. Then $c_i^b(x)$ and $x(t+1)$ are obtained by one final substitution of a_i into (A.13) and (A.14). □

Proof of proposition 10.6
It is without loss of generality to consider the disagreement phase at t, t odd. This means that country 2 proposes $\lambda_2(x(t+1)) = \lambda_2$ at $t+1$. In order to obtain insight into the problems in calculating nonlinear MPE's in differentiable strategies we treat the nonlinear MPE weights $\lambda_2(x(t+1))$ proposed by country 2 as a C^1-function and impose $\lambda_2(x(t+1)) = \lambda_2$ for all $x(t+1)$ at the very end of this proof. Then the value functions of proposition 10.5 can be seen as the function $\ln A_i^b \lambda_2(x(t+1)) + \frac{1}{1-\alpha\delta_i} \ln x(t+1)$, $i = 1, 2$, after substitution of $\lambda_1(x(t+1)) = 1 - \lambda_2(x(t+1))$. The interior and nonlinear MPE disagreement catches of fish $c_i^d(x)$, $i = 1, 2$, are computed by applying backward induction in a straight forward manner, i.e., $c^d(x)$ are NE actions of the normal-form game in which country i, $i = 1, 2$, chooses $c_i \in [0, x]$ and its utility function given by

$$\ln c_i + \delta_i \left[\ln A_i^b \lambda_2\left((x - c_1 - c_2)^\alpha\right) + \frac{1}{1-\alpha\delta_i} \ln (x - c_1 - c_2)^\alpha \right].$$

The NE conditions applied to the logarithmic form of the utility functions in this normal form game reduce to

$$\frac{1}{c_i} - \frac{1}{x - c_1 - c_2} \cdot \frac{\alpha\delta_i}{1-\alpha\delta_i} + \frac{\partial \ln A_i^b \left(\lambda_2\left(x - c_1 - c_2\right)^\alpha\right)}{\partial c_i} = 0, \quad i = 1, 2.$$

Linearity of the MPE imposes that $\lambda_2(x(t+1))$ is independent of $x(t+1)$, i.e., $\frac{\partial \lambda_2(z)}{\partial z} = 0$, and, therefore, the third term equals 0. Then solving for c_1 and c_2 yields the expressions for the linear MPE disagreement actions as stated. The value functions $V_i^d(x)$, $i = 1, 2$, are country i's NE utility. □

Proof of theorem 10.7
First, equation (10.7) is derived. Combining proposition 10.5 and 10.6 and keeping in mind that the responding player's weight $\lambda_j = 1 - \lambda_i$ in each round (see the proof of proposition 10.6) yields

$$V_j^b(x, \lambda_i) = V_j^d(x, \lambda_j) \Leftrightarrow \ln A_j^b(\lambda_i) = \ln A_j^d + \delta_j \ln A_j^b(\lambda_j) \tag{A.15}$$

which can be rewritten as

$$A_j^b(\lambda_i) = A_j^d \cdot \left[A_j^b(\lambda_i)\right]^{\delta_j}, \quad i, j = 1, 2, \quad i \neq j. \tag{A.16}$$

Substitution of the expressions for $A_j^b(\lambda)$ and A_j^d, and then raising both sides to the power $(1 - \alpha\delta_j)(1 - \delta_j)$ yields (10.7). Second, we prove $\lambda_i > 1 - \lambda_j$. Since the MPE disagreement actions are inefficient we have that the proposing player does strictly better by proposing an acceptable offer, i.e., $V_i^b(x, \lambda_i) > V_i^d(x, \lambda_j)$. Combined with proposition 10.5 and 10.6 these inequalities yield $A_i^b(\lambda_i) > A_i^d A_i^b(\lambda_j)^{\delta_i}$. Therefore, making use of (A.16) yields

$$\begin{aligned} A_i^b(\lambda_i)^{1-\delta_i^2} &> A_i^d A_i^b(\lambda_j)^{\delta_i} \cdot A_i^b(\lambda_i)^{-\delta_i^2} \\ &= A_i^d \left(A_i^d A_i^b(\lambda_i)^{\delta_i}\right)^{\delta_i} \cdot A_i^b(\lambda_i)^{-\delta_i^2} = \left(A_i^d\right)^{1+\delta_i}. \end{aligned}$$

But then $A_i^b(\lambda_i)^{1-\delta_i} > A_i^d \Leftrightarrow A_i^b(\lambda_i) > A_i^d A_i^b(\lambda_i)^{\delta_i} = A_i^b(\lambda_j)$, because of (A.16). Then we obtain $\lambda_i > 1 - \lambda_j$ as claimed. Third, we prove existence. Equation (A.15) with $i = 1$ and $j = 2$ defines λ_1 as the implicit C^1-function $f_1 : [0, 1] \to [0, 1]$ of λ_2 with derivative

$$\tfrac{\partial}{\partial \lambda_2} f_1(\lambda_2) = \delta_2 \left(\tfrac{\partial}{\partial \lambda_2} \ln A_2^b(\lambda_2)\right) \cdot \left(\frac{\partial}{\partial \lambda_1} \ln A_2^b(\lambda_1)\right)^{-1} \neq 0.$$

Similarly, equation (A.15) with $i = 2$ and $j = 1$ defines λ_2 as the implicit C^1-function $f_2 : [0,1] \to [0,1]$ of λ_1 with derivative

$$\frac{\partial f_2(\lambda_1)}{\partial \lambda_1} = \delta_1 \frac{\partial \ln A_1^b(\lambda_1)}{\partial \lambda_1} \left(\frac{\partial \ln A_1^b(\lambda_2)}{\partial \lambda_2} \right)^{-1}.$$

Substitution of f_2 into f_1 yields $\lambda_1 = f_1(f_2(\lambda_1))$. Since $f_1 \circ f_2 : [0,1] \to [0,1]$ and $f_1 \circ f_2$ is continuous Brouwer's fixed point theorem applies and, hence, there exists a solution. Finally, in Houba, Sneek, and Várdy (2000) it is shown that $\frac{\partial f_1(f_2(\lambda_1))}{\partial \lambda_1} < 1$ and this suffices to prove uniqueness. □

Proof of proposition 10.12

The optimal proposal $c^t(x)$ maximizes the Lagrange function

$$u_P(x, c^t) - \lambda \left[u_R(x, c^t) - V_R^d(t, x) \right],$$

with $\lambda > 0$. Dynamic programming requires for all $s = t, \ldots, t_f$ solving

$$\max_{a(s) \in A(x(s))} -\tfrac{1}{2} \Big[a^\top(s) R^\lambda a(s) + x^{*\top}(s, x)(Q_P + \lambda Q_R) x^*(s, x) +$$

$$(A x^*(s, x) + B a(s))^\top (K_P(s+1) + \lambda K_R(s+1))(A x^*(s, x) + B a(s)) \Big],$$

where $K_P(s+1) + \lambda K_R(s+1)$ is the parameter of the quadratic value function of this problem and the vector a is the stacked vector $[a_P^\top, a_R^\top]^\top$. This yields

$$a(s) = -G(s) x^*(s, x) \Leftrightarrow \begin{bmatrix} a_P(s) \\ a_R(s) \end{bmatrix} = - \begin{bmatrix} G_P(s) \\ G_R(s) \end{bmatrix} x^*(s, x),$$

with state transition

$$x^*(s+1, x) = [A - BG(s)] x^*(s, x) \Rightarrow x^*(s, x) = \Pi_{\tau=t}^{s-1} [A - BG(\tau)] x.$$

Substitution of these results into the dynamic programming equation yields the backward recursive equations in K_i, $i = P, R$. The resulting total utilities are $-\frac{1}{2} x^\top K_i(t) x$, $i = P, R$. Finally, the value of the Lagrange parameter λ can be found because the constraint $u_R(x, a^t(x)) - V_R^d(t, x) = 0$ has to be satisfied implying that $x^\top [K_R(t) - K_R^d(t)] x = 0$ has to hold.

Finally, the players' utility functions are strictly concave and the contract space is convex (due to the linear state transition) and therefore the optimal joint proposal is unique. This implies that $x^\top [K_R(t) - K_R^d(t)] x = 0$ admits just one positive solution for λ. □

References

Abreu, D. (1988), 'On the theory of infinitely repeated games with discounting', *Econometrica*, **56**, 383–396.

Andrelini, L. and L. Felli (2001), 'Costly bargaining and renegotiation', *Econometrica*, **69**, 377–411.

Asheim, G. (1991), 'Extending renegotiation-proofness to infinite horizon games', *Games and Economic Behavior*, **3**, 278–294.

Avery, C. and P. Zemsky (1994), 'Money burning and multiple equilibria', *Games and Economic Behavior*, **7**, 154–168.

Benhabib, J. and R. Radner (1992), 'The joint exploitation of a productive asset: a game-theoretic approach', *Economic Theory*, **2**, 155–190.

Benoît, J.P. and V. Krishna (1993), 'Renegotiation in finitely repeated games', *Econometrica*, **61**, 303–324.

Bernheim, B., B. Peleg, and M. Whinston (1987), 'Coalition-proof Nash equilibria I', *Journal of Economic Theory*, **42**, 1–12.

Bernheim, B. and D. Ray (1989), 'Collective dynamic consistency in repeated games', *Games and Economic Behavior*, **1**, 295–326.

Billera, L. and R. Bixby (1973), 'A characterization of Pareto surfaces', *Proceedings American Mathematical Society*, **41**, 261–267.

Binmore, K. and P. Dasgupta (1987), *The economics of bargaining*, Oxford: Basil Blackwell.

Binmore, K., A. Rubinstein, and A. Wolinsky (1986), 'The Nash bargaining solution in economic modeling', *Rand Journal of Economics*, **17**, 176–188.

Binmore, K. (1987a), 'Nash bargaining theory I', in K. Binmore and P. Dasgupta (eds), *The economics of bargaining*, Oxford: Basil Blackwell, pp. 27–46.

Binmore, K. (1987b), 'Nash bargaining theory II', in K. Binmore and P. Dasgupta (eds), *The economics of bargaining*, Oxford: Basil Blackwell, pp. 61–76.

Binmore, K. (1987c), 'Perfect equilibria in bargaining models', in K. Binmore and P. Dasgupta (eds), *The economics of bargaining*, Oxford: Basil Blackwell, pp. 77–105.

Bolt, W. and H. Houba (1998), 'Strategic bargaining in the variable threat game', *Economic Theory*, **11**, 57–77.

Bolt, W. (1995), 'Striking for a bargain between two completely informed agents: Comment', *American Economic Review*, **85**, 1344–1347.

Busch, L.A. and I. Horstman (1997), 'A comment on issue-by-issue negotiations', *Games and Economic Behavior*, **19**, 144–148.

Busch, L.A. and Q. Wen (1995), 'Perfect equilibria in a negotiation model', *Econometrica*, **63**, 545–565.

Busch, L.A. and Q. Wen (2001), 'Negotiations games with unobservable mixed disagreement actions', *Journal of Mathematical Economics*, **35**, 563–579.

Carlsson, H. (1991), 'A bargaining model where parties make errors', *Econometrica*, **59**, 1487–1496.

Clemhout, S. and H. Wan (1974), 'A class of trilinear differential games', *Journal of Optimization Theory and Applications*, **14**, 419–424.

Corominas-Bosch, M. (2000), 'Bargaining with asymmetric threat points', *Economics Letters*, **69**, 333–339.

Cramton, P. and J. Tracy (1992), 'Strikes and holdouts in wage bargaining: theory and data', *American Economic Review*, **82**, 100–121.

Cramton, P. and J. Tracy (1994a), 'The determinants of us labour disputes', *Journal of Labor Economics*, **12**, 180–209.

Cramton, P. and J. Tracy (1994b), 'Wage bargaining with time-varying threats', *Journal of Labor Economics*, **12**, 594–617.

Crawford, V. (1982), 'A theory of disagreement in bargaining', *Econometrica*, **50**, 607–638.

Dixit, A. and B. Nalebuff (1991), *Thinking strategically: the competitive edge in business, politics and everyday life*, New York: Norton.

Dockner, E., G. Feightinger, and S. Jørgensen (1985), 'Tractable classes of nonzero-sum open-loop Nash differential games: Theory and examples', *Journal of Optimization Theory and Applications*, **14**, 179–197.

Dockner, E., S. Jørgensen, N. Long, and G. Sorger (2001), *Differential games in economics and management science*, Cambridge: Cambridge University Press.

Dutta, P. and R. Sundaram (1993), 'Tragedy of the commons?', *Economic Theory*, **3**, 413–426.

Farrel, J. and E. Maskin (1989), 'Renegotiation in repeated games', *Games and Economic Behavior*, **1**, 327–360.

Farrel, J. (1988), 'Communication, coordination and Nash equilibrium', *Economics Letters*, **27**, 209–214.

Fernandez, R. and J. Glazer (1991), 'Striking for a bargain between two completely informed agents', *American Economic Review*, **81**, 240–252.

Fershtman, C. and M. Kamien (1987), 'Dynamic duopolistic competition with sticky prices', *Econometrica*, **55**, 1151–1164.

Fershtman, C. and D. Seidmann (1993), 'Deadline effects and inefficient delay in bargaining with endogenous commitment', *Journal of Economic Theory*, **60**, 306–321.

Fershtman, C. (1990), 'The importance of the agenda in bargaining', *Games and Economic Behavior*, **2**, 224–238.

Fishburn, P. and A. Rubinstein (1982), 'Time preferences', *International Economic Review*, **23**, 677–694.

Fisher, R. and L. Mirman (1992), 'Strategic dynamic interactions: Fish wars', *Journal of Economic Dynamics and Control*, **16**, 267–287.

Fisher, R. and L. Mirman (1996), 'The complete fish wars: Biological and dynamic interactions', *Journal of Environmental Economics and Management*, **30**, 34–42.

Fudenberg, D. and E. Maskin (1986), 'The folk-theorem in repeated games with discounting or with incomplete information', *Econometrica*, **54**, 533–554.

Fudenberg, D. and J. Tirole (1991), *Game theory*, London: MIT Press.

Furusawa, T. and Q. Wen (2001), 'Flexibility of disagreement actions in negotiations', *International Journal of Game Theory*, **30**, 19–39.

Furusawa, T. and Q. Wen (2002), 'Disagreement points in trade negotiations', *Journal of International Economics*, **57**, 133–150.

Gale, D. (2000), *Strategic foundations of general equilibrium*, Cambridge: Cambridge University Press.

Games and Economic Behavior (1997), *Special issue on imperfect recall, vol. 20 No. 1*, New York: Academic Press.

Gintis, H. (2000), *Game theory evolving: A problem centered introduction to modelling strategic interaction*, Princeton: Princeton University Press.

Gul, F. and H. Sonnenschein (1988), 'On delay in bargaining with one-sided uncertainty', *Econometrica*, **56**, 601–611.

Haller, H. and S. Holden (1990), 'A letter to the editor on wage bargaining', *Journal of Economic Theory*, **52**, 232–236.

Haller, H. (1991), 'Wage bargaining as a strategic game', in R. Selten (ed.), *Game theoretic equilibrium models III: Strategic bargaining*, Berlin: Springer Verlag, pp. 230–241.

Hanany, E. and Z. Safra (2000), 'Existence and uniqueness of ordinal Nash outcomes', *Journal of Economic Theory*, **90**, 254–276.

Haurie, A. and M. Pohjola (1987), 'Efficient equilibria in a differential game of capitalism', *Journal of Economic Dynamics and Control*, **11**, 65–78.

Hendon, E., H.J. Jacobsen, and B. Sloth (1996), 'The one-shot-deviation principle for sequential rationality', *Games and Economic Behavior*, **12**, 274–284.

Herrero, M. (1989), 'The Nash program: Non-convex bargaining problems', *Journal of Economic Theory*, **49**, 266–277.

Hicks, J. (1963), *The theory of wages* (Second ed.), London: MacMillan.

Hoel, M. (1986), 'Perfect equilibria in sequential bargaining games with nonlinear utility functions', *Scandinavian Journal of Economics*, **88**, 383–400.

Hoel, M. (1987), 'Bargaining games with a random sequence of who makes the offers', *Economics Letters*, **24**, 5–9.

Holden, S. (1989), 'Wage drift and bargaining: Evidence from Norway', *economica*, **56**, 419–432.

Holden, S. (1994), 'Wage bargaining and nominal rigidities', *European Economic Review*, **38**, 1021–1039.

Holden, S. (1997), 'Wage bargaining, holdout and rigidities', *Oxford Economic Papers*, **49**, 235–255.

Houba, H. and E. Bennett (1997), 'Odd man out: the proposal-making model', *Journal of Mathematical Economics*, **28**, 375–396.

Houba, H. and W. Bolt (2000), 'Holdouts, backdating and wage negotiations', *European Economic Review*, **44**, 1783–1800.

Houba, H. and A. De Zeeuw (1995), 'Strategic bargaining for the control of a dynamic system in state-space form', *Group Decision and Negotiations*, **4**, 71–96.

Houba, H. and G.V. Lomwel (2001), 'Counter intuitive results in a simple model of wage bargaining', *Economic Theory*, **17**, 81–99.

Houba, H., K. Sneek, and F. Várdy (2000), 'Can negotiations prevent fish war?', *Journal of Economic Dynamics and Control*, **24**, 1265–1280.

Houba, H., A.F. Tieman, and R. Brinksma (1998), 'The Nash bargaining solution for decision weight utility functions', *Economics Letters*, **60**, 41–48.

Houba, H. (1993), 'An alternative proof of uniqueness in noncooperative bargaining', *Economics Letters*, **41**, 253–256.

Houba, H. (1994), 'Game theoretic models of bargaining', Ph.D. thesis, Tilburg University.

Houba, H. (1997), 'The policy bargaining model', *Journal of Mathematical Economics*, **28**, 1–27.

Jansen, M. and S. Tijs (1981), 'Arbitration games: A survey', in *DGOR Operations Research Proceedings 1980*, Berlin: Springer Verlag.

Kaitala, V. and M. Pohjola (1990), 'Economic development and agreeable redistribution in capitalism: Efficient game equilibria in a two-class neoclassical growth model', *International Economic Review*, **31**, 421–438.

Kalai, E. and M. Smorodinsky (1975), 'Other solutions to Nash's bargaining problem', *Econometrica*, **43**, 513–518.

Kalai, E. (1977), 'Nonsymmetric Nash solutions and replication of 2-person bargaining', *International Journal of Game Theory*, **6**, 129–133.

Kaneko, M. (1980), 'An extension of the Nash bargaining solution and the Nash social welfare function', *Theory and Decision*, **12**, 135–148.

Karlin, S. and H. Taylor (1975), *A first course in stochastic processes*, New York: Academic press.

Kennan, J. and R. Wilson (1989), 'Strategic bargaining models and interpretation of strike data', *Journal of Applied Econometrics*, **4**, 87–130.

Kennan, J. (1986), 'The economics of strikes', in O. Ashenfelter and R. Layard (eds), *Handbook of labour economics*, Amsterdam: North Holland.

Kihlstrom, R., A. Roth, and D. Schmeidler (1991), 'Risk aversion and solutions to Nash's bargaining problem', in O. Moeschlin and D. Pallaschke (eds), *Game theory and mathematical economics*, Amsterdam: North Holland, pp. 34–47.

Kreps, D. (1990), *A course in microeconomic theory*, New York: Harvester Wheatsheaf.

Levhari, D. and L. Mirman (1980), 'The great fish war: An example using a dynamic Cournot-Nash solution', *Bell Journal of Economics*, **11**, 322–334.

Mao, W. (1993), 'Commitments to threat strategies in Nash bargaining', *International Journal of Game Theory*, **22**, 141–151.

Mariotti, M. (1999), 'Fair bargainings: Distibutive justice and Nash bargaining theory', *Review of Economic Studies*, **66**, 733–741.

Mas-Colell, A., M. Whinston, and J. Green (1995), *Microeconomic theory*, Oxford: Oxford University Press.

Ma, C.T. and M. Manove (1993), 'Bargaining with deadline effects and imperfect player control', *Econometrica*, **61**, 1313–1340.

Merlow, A. and C. Wilson (1995), 'A stochastic model of sequential bargaining with complete information', *Econometrica*, **63**, 371–399.

Meyerson, R. (1991), *Game theory, analysis of conflict*, Cambridge: Harvard University Press.

Moene, K. (1988), 'Unions' threats and wage determination', *Economic Journal*, **98**, 471–483.

Moulin, H. (1984), 'Implementing the Kalai-Smorodinsky bargaining solution', *Journal of Economic Theory*, **33**, 32–45.

Muthoo, A. (1990), 'Bargaining without commitment', *Games and Economic Behavior*, **2**, 291–297.

Muthoo, A. (1991), 'A note on bargaining over a finite number of feasible agreements', *Economic Theory*, **1**, 290–292.

Muthoo, A. (1992), 'Revocable commitment and sequential bargaining', *Economic Journal*, **102**, 378–387.

Muthoo, A. (1996), 'A bargaining model based upon the commitment tactic', *Journal of Economic Theory*, **69**, 134–152.

Muthoo, A. (1999), *Bargaining theory with applications*, Cambridge: Cambridge University Press.

Nash, J. (1950), 'The bargaining problem', *Econometrica*, **18**, 155–162.

Nash, J. (1953), 'Two-person cooperative games', *Econometrica*, **21**, 128–140.

Okada, A. (1991a), 'A non-cooperative approach to the Nash bargaining problem', in R. Selten (ed.), *Game theoretic equilibrium models III: Strategic bargaining*, Berlin: Springer Verlag, pp. 7–33.

Okada, A. (1991b), 'A two-person repeated bargaining game with long term contracts', in R. Selten (ed.), *Game theoretic equilibrium models III: Strategic bargaining*, Berlin: Springer Verlag, pp. 34–47.

Ok, E. and L. Zhou (2000), 'The Choquet bargaining solutions', *Games and Economic Behavior*, **33**, 249–264.

Osborne, M. and A. Rubinstein (1990), *Bargaining and markets*, Boston: Academic Press.

Osborne, M. and A. Rubinstein (1994), *A course in game theory*, Cambridge: MIT Press.

Perry, M. and P. Reny (1993), 'A noncooperative bargaining model with strategically timed offers', *Journal of Economic Theory*, **59**, 50–77.

Peters, H. (1992), *Axiomatic bargaining game theory*, Theory and Decision Library. Series C: Game Theory Mathematical Programming and Operations Research, vol. 9., Norwell, Mass. and Dordrecht: Kluwer Academic.

Ponsati, C. and J. Sákovics (1998), 'Rubinstein bargaining with two-sided outside options', *Economic Theory*, **11**, 667–672.

Ray, D. (1994), 'Internally renegotiation-proof equilibrium sets: Limit behaviour with low discounting', *Games and Economic Behavior*, **6**, 162–177.

Roemer, J. (1988), 'Axiomatic bargaining theory on economic environments', *Journal of Economic Theory*, **45**, 1–31.

Roth, A. and U. Rothblum (1982), 'Risk aversion and Nash's solution for bargaining games with risky outcomes', *Econometrica*, **50**, 639–647.

Roth, A. (1979), *Axiomatix models of bargaining*, Berlin: Springer Verlag.

Roth, A. (1985), 'A note on risk aversion in a perfect equilibrium model of bargaining', *Econometrica*, **53**, 207–211.

Rubinstein, A., Z. Safra, and W. Thomson (1992), 'On the interpretation of the Nash bargaining solution and its extension to nonexpected utility preferences', *Econometrica*, **60**, 1171–1186.

Rubinstein, A. (1980), 'Strong perfect equilibrium in supergames', *International Journal of Game Theory*, **9**, 1–12.

Rubinstein, A. (1982), 'Perfect equilibrium in a bargaining model', *Econometrica*, **50**, 97–109.

Rubinstein, A. (1991), 'On the interpretation of game theory', *Econometrica*, **59**, 909–924.

Safra, Z., L. Zhou, and I. Zilcha (1990), 'Risk aversion in the Nash bargaining problem with risky outcomes and risky disagreement points', *Econometrica*, **58**, 961–965.

Sákovics, J. (1993), 'Delay in bargaining games with complete information', *Journal of Economic Theory*, **59**, 78–95.

Schelling, T. (1960), *The strategy of conflict*, Oxford: Oxford University Press.

Schultz, C. (1994), 'A note on strongly renegotiation-proof equilibria', *Games and Economic Behavior*, **6**, 469–473.

Selten, R. (1975), 'Reexamination of the prefectness concept for equilibrium points in extensive games', *International Journal of Game Theory*, **4**, 25–55.

Shaked, A. and J. Sutton (1984), 'Involuntary unemployment as a perfect equilibrium in a bargaining model', *Econometrica*, **62**, 1351–1364.

Shalev, J. (1997), 'Loss aversion and bargaining', CORE Discussion Paper 9706, University Catholique de Louvain.

Shalev, J. (2000), 'Loss aversion equilibrium', *International Journal of Game Theory*, **29**, 269–287.

Sorin, S. (1985), 'Repeated games with complete information', *Mathematics of Operation Research*, **11**, 147–161.

Ståhl, I. (1972), 'Bargaining theory', discussion paper, Economics Research Institute, Stockholm School of Economics.

Stefanski, J. and K. Cichocki (1986), 'Strategic bargaining in a dynamic economy', discussion paper, System Research Institute, Polish Academy of Sciences, Warshaw.

Tirole, J. (1988), *The theory of industrial organization*, London: MIT Press.
Tsutsui, S. and K. Mino (1990), 'Nonlinear strategies in dynamic duopolistic competition with sticky prices', *Journal of Economic Theory*, **52**, 136–161.
Van Damme, E., R. Selten, and E. Winter (1990), 'Alternating bid bargaining with a smallest money unit', *Games and Economic Behavior*, **2**, 188–201.
Van Damme, E. (1989), 'Renegotiation-proof equilibria in the repeated prisoners' dilemma', *Journal of Economic Theory*, **89**, 206–217.
Van Damme, E. (1991), *Stability and perfection of Nash equilibria* (Second ed.), Berlin: Springer Verlag.
Van der Ploeg, F. and A. De Zeeuw (1991), 'Difference games and policy evaluation: A conceptual framework', *Oxford Economic Papers*, **43**, 612–636.
Van der Ploeg, F. and A. De Zeeuw (1992), 'International aspects of pollution control', *Environmental and Resource Economics*, **2**, 117–139.
Van de Wijngaert, R. (1994), 'Trade unions and collective bargaining in The Netherlands', Ph.D. thesis, Vrije Universiteit, Amsterdam.
Van Ours, J. and R. Van de Wijngaert (1996), 'Holdouts and wage bargaining in the netherlands', *Economics Letters*, **53**, 83–88.
Varian, H. (1984), *Microeconomic analysis* (Second ed.), New York: W. W. Norton & Company, Inc.
von Neumann, J. (1928), 'Zur theorie de gesellschaftspiele', *Mathematische Annalen*, **100**, 295–320.
Wang, G. (1998), 'Bargaining over a menu of wage contracts', *Review of Economic Studies*, **65**, 295–305.
Watson, J. (1998), 'Alternating-offer bargaining with two-sided incomplete information', *Review of Economic Studies*, **65**, 573–594.
Weinberger, C. (2000), 'Selective acceptance and inefficiency in a two-issue complete information bargaining game', *Games and Economic Behavior*, **31**, 262–293.
Windmuller, J. (1987), 'Collective bargaining in industrialised market economies: a reappraisal', Technical Report, International Labour Office, Geneva.
Young, P. (1998), *Individual strategy and social structure*, Princeton: Princeton University Press.
Zhou, L. (1996), 'The Nash bargaining theory with non-convex problems', *Econometrica*, **65**, 681–685.

Index

Alternating offers, 9, 50
model, 28, 176
procedure, 9, 10, 49
robustness, 11
Alternating-strike strategy, 212, 215
Axiom, 82
alternative, 99
consistency across dimensions, 90
continuity, 132
efficiency, 83, 89
independence of irrelevant alternatives, 83
invariance of affine transformations, 83, 90
invariance of utility representation, 96
monotonicity, 101
symmetry, 83, 89
Axiomatic approach, 8, 81

Backdating, 204, 208, 223
Bargaining power, 2, 154
Bargaining problem, 15, 16
bargaining process, 3
content, 16
essential, 21
exogenous disagreement outcome, 50
fixed disagreement outcome, 19
stationary, 152
strongly comprehensive, 118
symmetric, 89
symmetry, 83
Bargaining procedure, 30, 188, 208
alternating offers, 9, 204
alternative, 160
continuous time, 164
Markov process, 161
Bargaining process, 3
axiomatic approach, 8
strategic approach, 8

Closure (of a set), 41
Coase conjecture, 225
Collective bargaining, 3, 203, 205
Commitment, 6, 11, 175
credible, 6, 176
endogenous threats, 175
Common interests, 21
Comprehensive, 118
Conflicting interests, 1, 21, 146
Consistency across dimensions, 90
Continuity
axiom, 132
Continuous time procedure, 165
Contract, 16
backdating of, 208
binding, 16
curve, 21
individual rational, 23
maximum-wage, 210
minimum-wage, 210
Pareto efficient, 21
space, 16, 64, 88, 138

Difference game, 257
exponential-logarithmic, 261
general, 259
linear-quadratic (LQ), 263
Disagreement actions, 15, 18, 207, 230
endogenous, 175
fixed, 49
optimal, 237, 278
threats, 3
Disagreement game, 19, 207, 209, 211, 230
Disagreement outcome, 19
Disagreement point, 27, 50, 234
alternating, 155
exogenous, 50
fixed, 27
nonstationary, 156
Discount factor
common, 204

individual, 152, 215
Divide a dollar, 73, 139
Dynamic programming, 56, 59

Equilibrium switching, 126, 193, 212
Exogenous disagreement outcome, 50
Expected utility, 20

Feasible payoffs
set of, 27
First-mover
disadvantage, 159
First-mover advantage, 70
Fixed point, 51, 190, 192, 210, 231
definition, 52
problem, 51, 120
uniqueness of, 52

Game theory, 2, 38
Genericity, 24
Great fish war, 261

History, 32, 33
independence, 210, 230
reconstructed, 32
Holdout, 203, 205
efficient, 209
inefficiency, 207

Individual rational payoffs
set of, 27
Individual rationality, 23
Industrial action, 203, 204
holdout, 203
strike, 203
work-to-rule, 204
Information set, 32, 33
Invariance of affine transformations
axiom, 90
Invariance of utility representation, 96
Irrelevant alternatives, 83

Joint policy, 244, 245
difference game, 260
Pareto efficient, 266

Linear-quadratic difference game, 263
Lotteries, 17
LQ difference game, 263

Markov process, 160
Markov strategies, 41
Maximum-wage contract, 210
Method of Shaked and Sutton, 67, 124, 190, 242
Minimum-wage contract, 210
Minmax, 274
actions, 23, 234
payoffs, 23
Monotonicity
axiom, 101
MPE, 50, 51, 62, 120, 210, 230, 262
interior, 262
linear, 262
Mutual interests, 1, 21, 146

Nash equilibrium, 38
Nash product, 52
asymmetric, 153
Nash program, 8, 82, 132
Nash's bargaining procedure, 10
Nash's bargaining solution, 82, 84
Nash's demand game, 10, 81, 103
Nash's solution
asymmetric, 154
Nash's variable-theat game, 254
Nash's variable-threat game, 176, 177, 185, 236
Nash-product curve, 52
No-concession payoff, 216
No-concession strategy, 216
Nongenericity, 24

One-stage deviation, 39, 56
principle, 40
Outcome, 35
Outcome path, 35, 130, 213
payoff relevant, 35

Pareto efficiency, 21
Pareto efficient
paths, 36
set of, 21
Pareto frontier, 28
Payoff relevant, 35
Perfect recall, 31
Policy bargaining, 229, 244
difference game, 260
model, 229, 245
problem, 12, 230, 244
Probability of breakdown, 152

Reconstruct history, 32
Regular, 114
Renegotiation, 246, 247
Risk of breakdown
nonstationary, 152

Second-mover disadvantage, 70
SPE, 38, 50, 67, 124, 230, 246, 273
worst, 210, 231, 274
State, 42
absorbing, 43, 234
initial, 43, 234
transition, 43, 259
variable, 259
Strategic approach, 8, 81, 102

Strategies, 34
- alternating-strike, 212
- history-independent, 41
- Markov, 41, 59, 231
- maxmin, 179
- minmax, 179
- no-concession, 216
- punishment, 127, 233
- stationary, 41
- subgame perfect, 39
- worst SPE, 192, 231, 233, 274

Strike, 203, 205

Subgame perfect equilibrium, 38
- characterization of, 62, 69, 124, 242, 246
- dynamic programming, 56
- equilibrium switching, 127
- method of Shaked and Sutton, 67
- uniqueness of, 62, 241
- with delay, 130, 196

Symmetry
- axiom, 83, 89
- function, 89

Threats, 3, 175, 205
- credible, 184, 218
- destructive, 257
- optimal, 177
- variable, 18, 188

Tradegy of the Commons, 257, 262

Utility function, 16, 19

Utility representation, 9, 26, 82, 122

Utopia payoff, 27

Variable threats, 15, 18

Variable-threat game, 177, 236

Wage bargaining, 3, 203, 244
- backdating, 204
- Dutch, 11, 217
- industrial action, 3, 203
- model, 205, 207

Work-to-rule, 204, 205

Zero-sum game, 179